Seismic Survei
for Reservoir D ... y

From a Practitioner's Point of View

O.I. Barkved

ISBN 978-90-73834-24-8

EAGE Publications bv
PO Box 59
3990 DB HOUTEN
The Netherlands

Acknowledgements

I would like to thank my employer, BP, for allowing me to undertake the 2011 EET VI Lectures Tour. Special thanks to Stijn Dejongh, Grete Vagle, Trevor Ricketts, John G. Fletcher, Nigel Robinson, Dave Whitcombe and Tim Summers for their support and valuable suggestions. Thanks to the Valhall Licensees BP Norge AS and Hess Norge AS for allowing me to use the Valhall material. Also thanks to NPD for helping out with figures, statistics and facts from the Norwegian Continental Shelf.

I'm truly in debt to a number of people who have influenced my current appreciation for use of seismic data in developing hydrocarbon reservoirs. Thanks to Lars Sønneland, who introduced me to the seismic industry and provided the initial excitements for use of 3D & 4 D seismic, to Phil Christie, who got me interested in 4C seismic, to Leon Thomsen for his simple explanation of complex technical issues and to Ian Jack for his energizing 4D seismic focus.

Mark Anderson introduced me to geomechanics when I joined Amoco in 1992, Tron Kristiansen has helped me continuing developing my awareness of the fact that the earth are continuously changing, at least across a compacting reservoir.

Special thanks to Martin Landrø, Colin MacBeth, Mike Kendal, Rune Holt and Erling Fjær for sharing illuminating and motivating insights.

My view on the potential of seismic surveillance is strongly influenced by hands on experiences from Valhall field, and especially from 10 years of working the Valhall Life of Field Seismic project. An active partnership made for many challenging but enlightening discussions: thanks to Kenneth Zwilgmeyer, Phil Rennison, Jerome Guilbot, Ruben Sanchez, Geir Drivenes, Frans Smith, Paul Hatchell, Dave Rampton, Peter Wills, Conny Gilbert, Andrew Davids, Stuart Fairhead and Ben Vaughn.

Thanks to colleagues within BP who have offered constructive advices and shared their experiences. Special thanks to all that helped implement and carried forward the LoFS project: Tim Jackson, Gary Nicol, Dave Howe, Arnfinn Bærheim, Dan Nordberg, Mike Sheen, Larry Walter, Nirina Haller, Christian Eligon, Sverre Brandsberg Dahl, OJ Askim, John Etgen, Joe Dellinger, Richard Clarke, Ian Stockden, Kevin D Best, H Scott Lane, Rune Martinsen, Morten Heir, Bob Smith, JP van Gestel, Einar Kjos, Pete Heavey, Kjell Inge Halten, Roger Bakke, Ross Milne, Kent Andorsen, Luke Rawsthorne, Mark Shaly, Torbjørn Darre, amongst others. And not at least thanks to Jan Kommedal, who has been my closest colleague in progressing the seismic agenda for Valhall.

Thanks to EAGE' Education Committee for asking to take on this task, and to the EAGE staff helping out, Karin Merkens and Laura van Kal. Thanks also to all the contributors, who have offered material and figures for use in this book.

Finally, special thanks to my wife, Eli Tove, and to Lene, Kari and Haavard for their patience during the last year's efforts to compile this material.

Olav Inge Barkved,
February, 2012.

General disclaimer

The purpose of this manual is to educate and entertain. The text is written to complement a training class given by the author, on behalf of EAGE. The author and EAGE shall have neither liability nor responsibility to any person or entity with respect to any loss or damage caused, or alleged to have been caused, directly or indirectly, by the information contained in this book or from written or oral information provided in the complementary training classes.

This book is designed to provide information on the surveillance of certain hydrocarbon reservoirs. It includes reviews of existing published material and contains references to a number of technical papers, expanded abstracts and text books. The intent has not been to produce a reprint of this material but instead, use this combined with the author's own experiences to illustrate the practical use of seismic surveillance information.

We recommend complementing the book by reading the referenced articles, to ensure you have the best possible context and to help tailor the information to your specific needs. You are urged to consult with experts on all aspects you are in doubt about, prior to starting to utilize any of the ideas and concepts referred to in the textbook and the associated lecture. Furthermore, you should keep in mind that all reservoirs have their genuine issues and are placed in a specific geological setting. The reservoirs presented in this book have been picked due to their ability to illustrate relevant seismic surveillance aspects, they are unique and might represent end members of their type, keep this in mind if using them as analogues.

At the time of writing this textbook, the author is employed by an oil company and is applying seismic surveillance to a certain type of reservoir as part of his current job. However, the book and the associated training class are prepared and arranged solely between the author and EAGE. The opinions and statements made are the author's own and the material coverage is biased to the type of experience the author has.

Every effort has been made to make this work as accurate as possible. However, there may be mistakes, both typographical and in content. Therefore, this text should be used only as a general guide and not used as the ultimate source for commercial work.

Contents

1 Introduction

"Cross country skiing is great if you live in a small country." - *Steven Wright.*

We will start by outlining the book. I will try to explain what seismic surveillance is by providing some context. We will discuss a few of the "historic" 4D seismic cases and share some impressive results. Then we will review the status of permanent seismic monitoring. Marine towed streamer is by far the most common seismic acquisition technique used today and permanent seismic installations are still limited to a few cases. Permanent seismic array favor frequent monitoring and is a relevant option in areas where accurate geometrical repeatability of the sensors and source is critical. The permanent seismic installation at Valhall, frequent seismic time-lapse acquisition and seismic response of production induced geomechanical effects will be key elements of the cases stories and examples presented.

1.1 Outline

The topic for EET 6 is seismic surveillance. I have elected to use this title although most of the time we will be discussing methods and techniques that may fall in the category traditionally called four-dimensional (4D) seismic.

Four-dimensional seismic are used to help us understand what happens to reservoir properties in space and time in our oil- and gas-producing reservoirs. The technology has also been used to monitor the storage of carbon dioxide from carbon sequestration projects. Four-dimensional seismic are intuitively linked to the acquisition of several 3D seismic data sets over time. The results are often presented by the changes, using differencing as the primary technique. This has been the most important seismic surveillance technique for the last two decades.

We do have other time-lapse seismic data, this could be data acquired in a well bore or at the surface and may not always be true 3D. Passive data recording and dedicated "forensic" type studies, where seismic methods are used to deduce production-induced changes, are emerging. As an example, the monitoring of hydraulic fracturing in "shale" has developed into a multi-million dollar industry just during the last 10–15 years.

It is also important to keep in mind that the production induced changes that influence the seismic responses, inside or outside our reservoirs will take place whether seismic are acquired or not. In some cases, these changes might be so strong that a single 3D data set will capture valuable insights.

In 1998 Ian Jack presented "Time-lapse Seismic in Reservoir Management", as part of the EAGE/SEG distinguished Instructor Short Course. In 2005, seven years later, Rodney Calvert presented "Insights and Methods for 4D Reservoir Monitoring and Characterization" as part of the same series. I have had the pleasure of working with both of them, Ian as a colleague in BP and Rodney through the Valhall license, where BP and Shell were partners.

For EET 6, EAGE wanted an updated one-day 4D seismic class. It will be impossible not to revisit some of the material that Ian and Rodney presented in their well-recognized Distinguished Instructor Short Courses. While Ian and Rodney had key roles in BP and Shell on a corporate level, my experience is primarily from working with a local business unit and predominately from specific fields.

When consulting with colleagues and friends in the 4D seismic world on how to approach this challenge (and whether I should be brave enough to take it on), the feedback has been pretty unanimous:

"Share the Valhall experience!" And in all modesty this is a story I know – and it covers the "new" elements that EAGE asked me to include: 4D seismic response of geomechanical changes, Permanent Seismic Monitoring, Continuous/Frequent Monitoring, Passive Recording, Examples of Capturing Value through Integrating Across disciplines and New Tools to support these efforts.

Chapter One starts with a brief summary of the development of seismology and key 4D seismic / seismic surveillance examples, and permanent seismic installations will also be presented. Additional figures outlining basic seismic principles, for those not too familiar with seismic methods, will be reviewed in the class based on needs. For others; what about google it?

Examples from the Valhall field appear throughout the book, so a short overview of this field will be provided in Chapter Two. There is no obvious reason for why the Valhall reservoir should be the ultimate candidate for seismic surveillance, due to the gas charges in the overburden it is very difficult to make a static image of part of the area. However, business needs and dedicated commitment to technology has resulted in some surprising results.

In Chapter Three we look at the changes taking place in the reservoir during production and how these effects may impact the surrounding rocks. We will very briefly discuss how to link these changes in fluid and stress state to a seismic signal and also introduce the key tools we have available to predict these changes.

Chapter Four addresses primarily the acquisition and processing of seismic time-lapse data - 4D imaging and examples of dedicated 4D analysis. We also look at how these issues are dealt with in case of using a permanent array for recording the seismic data.

Chapters Five, Six and Seven will address integration to capture value and I have elected to divide this into three activities: reservoir management, planning and drilling of a new wells and base management or production optimizations.

Chapter Eight includes examples of emerging technologies, yet to be used on a routine basis.

Chapter Nine sums up the rationale for doing 4D, how we can quantify value and provide some food for thoughts around where we are heading with seismic surveillance technology.

Finally, we will round up the story and try briefly to look into the future in Chapter 10.

I have deliberately elected not to dive into the details of geomechanics. This is the topic of EET 5 "Seismic Geomechanics", and in fact the underlying examples and parameterization used to illustrate some of the key concepts is based on a simplified Valhall analog.

The EAGE /SEG Distinguished Instruction Course by Colin M. Sayers, "Geophysics Under Stress" provides an excellent coverage of stress state changes and redistribution of stress which is a key production induced effect in weak and compacting reservoirs. Finally, for the rock physics details I recommend the recent book and training classes offered by Per Avseth et al. on quantitative seismic interpretations.

It is my intention that this book will be of inspiration to others using seismic methods for maximizing recovery from a reservoir in a safe and cost-effective manner.

1.2 Seismic surveillance

Seismic surveillance for reservoir delivery is about "watching" and listening to the effects of producing hydrocarbons, to help understand fluid movements and their impact on surrounding rocks. Does the fluid flow make enough noise for us to notice? Possibly, but the indirect effects might be easier for us to capture. This could be the effects of a steam-flood front in an advanced enhanced oil recovery process or simply the effects of changing the fluid type or the pressure in the reservoir.

Reading through the vast amount of recent papers on the topic, we find that many fields have multiple 4D seismic surveys, acquired with the aim of maximizing the value of the field. In some regions, like the North Sea, the view is that 4D is a necessary and integral part of prudent reservoir management. Identifying bypassed oil in a water-flooded field or un-drained compartments has made significant returns by investing in these technologies. The many reports of the successful use of seismic time-lapse data used for history matching, Walker et al. (2006), has resulted in more and more reservoir models being based on physics that are consistent with the acoustic response and not only fluid flow physics.

Seismic on demand is a phrase used by the industry. This was meant to illustrate seismic monitoring technologies that allowed for the cost-effective acquisition of seismic surveillance data at short notice, using a permanent recording system, flexible source solutions and quick turnaround processing schemes, Barkved (2003). But more importantly, the field's need for seismic surveillance data normally changes during its lifetime. For more information in planning a "Life-of-Field" seismic surveillance program and what information is needed to support this decision, see Watts et al. (2006).

Figure 1.1 links a "Life of Field" seismic strategy to the production stages of the field. 4D seismic is the most used seismic surveillance concept. This approach is based on a baseline survey, a 3D seismic survey that should ideally be acquired before the start of any production. Prior to production start up it is important to have a view on whether seismic might help to reduce the reservoir uncertainties and manage the risks. The information gathered during initial production stage could be essential in defining the future of the field. The first dynamic data is essential to calibrate the simulator, which will be an important tool in defining future depletion plans. At this stage, it is important to understand the overall connectivity of the system. Pressure diffusions out into the reservoir might be picked up by the seismic responses and could in that case provide very important information about the risk of compartmentalization and the possible need for more wells to effectively drain the field. Timing and frequency of acquisition will be dependent on the sensitivity of the seismic waves to the pressure induced changes.

During the early production stage, the impact of the aquifer may be picked up by the seismic signal. This is often one of the key dynamic uncertainties. If the development is phased, then early information on the aquifer's strength may significantly influence decisions in the later phases. If a gas cap is present or expected to develop during production, monitoring the gas oil contact could also be useful to understand compartmentalization, reservoir energy issues and ensuring the right capacity is available for future gas handling.

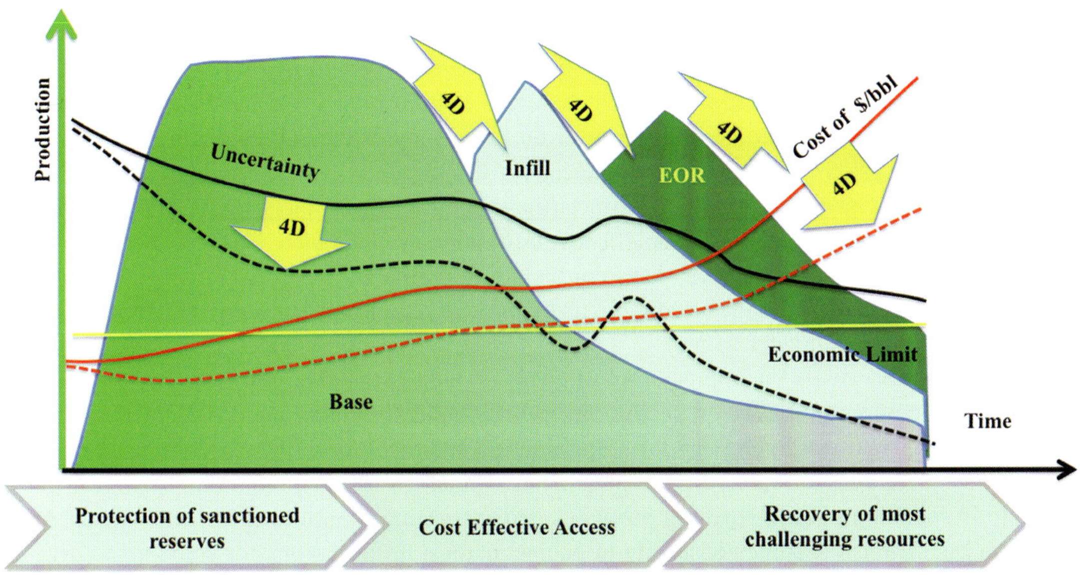

Figure 1.1: 4D Seismic has the potential to help reduce the uncertainties, reduce cost and improve the recovery of a field.

In the mid-production period, a number of general reservoir management issues related to production optimization and potential future upside options need to be considered. Most reservoirs are heterogeneous, due to changes in sedimentology; the presence of faults and even the field's fill history might have impacted this. If the fluid and pressure changes result in changes in the seismic properties, 4D seismic may be the ideal tool for localizing enhanced or reduced transmissibility effects, which may be vital in defining the optimal, enhanced recovery scheme for the field.

When the field has passed the production plateau, there might be capacity available for new infill wells or complementary injectors in case we have bypassed oil. The "easy and most valuable wells" have typically been picked up in earlier drilling campaigns and the right placement of the well may be critical for the well's economic success. This is an area were seismic surveillance, especially 4D seismic, has demonstrated its potential as a value-adding technology.

Unfortunately, too many examples show that 4D seismic came as an afterthought rather than being a fundamental element of any development scheme. In the past, it was almost impossible to justify a seismic time-lapse survey to an asset based on the value of the dynamic data alone, and uplift in image quality or resolutions was often the key driver for surveys that became successful 4D seismic projects. These successes caused technology leaders in companies like BP and Statoil to ask "Why are you not using 4D?", rather than: "Why are you using 4D?".

An important role of the seismic surveillance program is to provide context for the other surveillance data recorded at the surface or in the well. Some wells might even have multiple sensors that measure the temperature along the well (DTS) and which allow the changes in temperature readings to be used to interpret the relative contribution of particular intervals to the production along the well. The recent evolution in sensors and surveillance technology are very much driven by managing the field more effectively. This also includes ensuring that operations are safe and sustainable. The cost of intervening in a well, going in with a dedicated tool to acquire surveillance data, is costly and not always completely without risk. Permanently installed sensors not only reduce this risk, but they also deliver data on a regular basis, which may help in establishing critical trends in a well's behavior and identifying issues at an earlier stage. As illustrated in figure 1.2 below, current surveillance systems range from real time to ones invoked years apart.

Our focus is on seismic surveillance data. Today's established technology is fundamentally based on 4D or time-lapse seismic. 4D seismic based on a retrievable system is a powerful tool and is now established as a mature technology. Although there are still surprises, we have developed a reasonable understanding of where and how it works. If the frequency of the surveys is high, a permanently installed recording system might be the most economic. This could often provide an easier route to making 4D images of production-induced effects than using a retrievable system. However, it is still considered an emerging technology, as only a few field-wide systems have been installed. Beyond the permanently recording system, there is a system including permanent sources. This solution is already in place onshore; while offshore the alternative using passive recordings has shown interesting potential for shallow overburden characterization.

During the last ten years, the appreciation of changes outside the reservoir has grown considerably within the seismic communities. Compacting reservoirs, typically chalk, unconsolidated sandstone or deep high-pressure fields may compact during production and result in sudden stress release outside the reservoir. If we were able to record the micro-seismic events, we would have a rich source of information to help in reducing possible drilling problems and increasing the lifetime of these assets. A useful insight from the "strong" seismic sensitivity to geomechanical deformations for some reservoris, is the increased awareness of these effects. Changes in hydrostatic stress in the reservoir will simply result in induced effects outside the reservoir. These effects should not be ignored, unless strongly supported by the physical properties of the reservoir rocks.

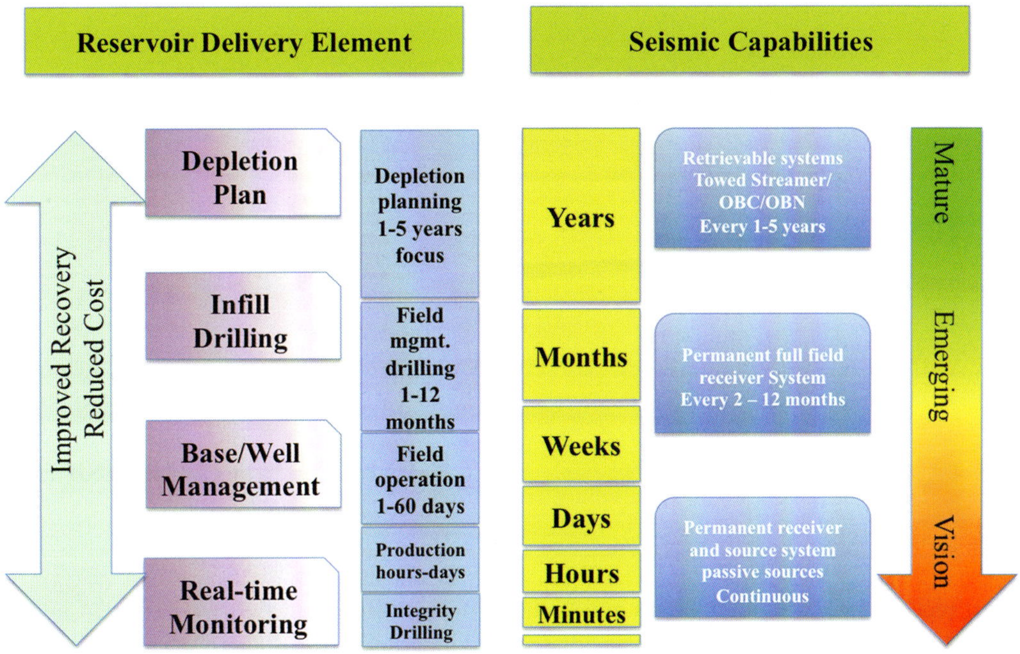

Figure 1.2: The time element for alternative surveillance technologies ranges from sampling the data in real time to doing checks several years apart.

1.3 Brief history of seismology

Modern seismology is a relative young science. Its birth is linked to the Lisbon earthquake in 1755, Ben-Menahem (1995). The event that killed maybe as many as 10, 000 people triggered some sobering thinking around the causes of earthquakes and suggestions to design apparatus for recording them. Since then, major earthquakes have continued to trigger new insights on what causes these events.

Ernst von Rebeur-Pasebwitz at Potsdam made the first recording of a teleseismic event in 1889. He was experimenting with a pendulum-based seismometer and happened to record an earthquake that took place in Japan, figure 1.3. Suddenly, the tracking of earthquakes became a global issue. A geologist and mining engineer, John Milne, built new and improved seismometers, which were simple and compact enough to be used in many places around the world and the amount of useful seismic data from earthquakes started to accumulate.

Seismographs became a useful tool for the geoscientist looking into inaccessible areas, in the same manner that the telescope is the tool of the astronomer. The second big discovery that impacted seismology was the introduction of the computer in the 1950s. In the meantime, a lot of important contributions were provided by other disciplines. Physicists, mathematicians and engineers provided most of the theory needed to interpret the seismic recordings before and after the invention of the seismograph.

The recordings from earthquake generated seismic waves also contained reflection and refractions of seismic waves. John Milne (1885) used dropped weights as energy sources to test his seismographs and before this Robert Mallet used gunpowder as an energy source and determined the velocity of surface waves. Turning into the early 1900s, further developments came through and the founder of applied seismic exploration Ludger Mintorp successfully applied learnings from earthquake seismology to make seismic a commercial technology for use in oil and gas exploration.

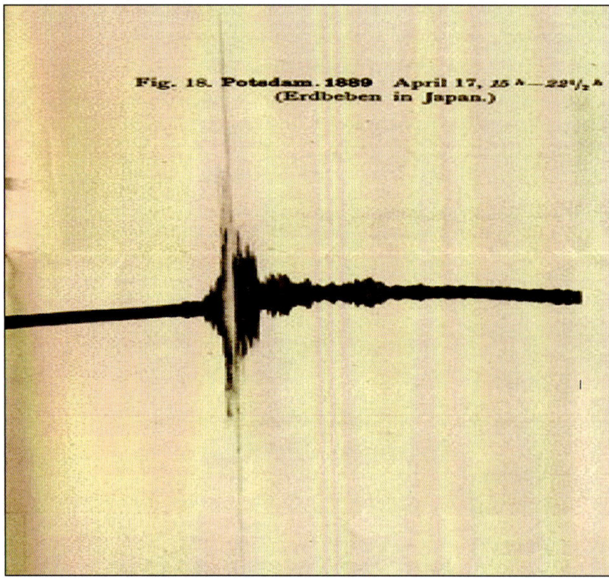

Figure 1.3: World's first seismograph. From Nature July 25, 1889.

The first recorded use of sound waves and "echoes" to identify an oil field was in 1924 in Brazoria County, Texas, just south of Houston. Oil was discovered in a field beneath the Nash salt dome based on single-fold seismic data. Before that, oil exploration was based primarily on surface geology and other indicators. It was crude and very much a guessing game and losses due to dry holes could be devastating.

What Mintrop and other contemporary pioneers utilized was a light tool that recorded the head waves with enough precision to be able to calculate the depth and dip of a refractive layer. Proof of the concept of using seismic reflection work in the search of oil and gas was not obtained before the 1930s (ref. Weatherby.) Of course, all the geophysical analysis was based on the assumption of a flat Earth.

During the 1950s new the digital revolutions made it possible to record the seismic recordings on tapes, which led to totally new ways of collecting and processing seismic data. At the same time, W. Harry (source SEG Virtual Geoscience Center) invented common mid-point data stacking and vibroseis was introduced as a replacement for dynamite.

In 1961 GSI introduced the first digital recording system and the use of computers for processing the data. Since then, the oil and gas exploration geophysics industry has continued to be one of the largest users of IT. Up until now, all seismic had been based on acquiring 2D lines. Exxon acquired the first onshore 3D data set over the Friendswood field, Texas, in 1967, JPT Online (1999). Following this, GSI initiated a Joint Industry Project acquiring 3D seismic data over the Bell Lake Field. This include six major oil companies and the results were remarkable. The dataset depicted the 3D structure that confirmed the nine producers and the three dry holes that had been drilled in the area. In addition, several new infill locations were identified.

The first marine 3D dataset was acquired in 1975, by Sun Oil and GSI. The introduction of the air gun source in the 1960s made a significant change to safety, repeatability and cost of marine seismic sources, figure 1.4

The strength of 3D surveys is the enhanced lateral detail. The data is presented in a cube, which can be sliced in any direction. There is a wealth of data examples where the structural changes between the early 2D interpretations and the later 3D are quite dramatic. Quite a few wells have been placed sub-optimally due to this. The first 3D seismic data set was acquired across a producing field. Today, 3D technology is used across the entire value change supporting exploration, appraisals and production drilling. The significant reduction in time used to cover a square kilometer has reduced the cost of 3D surveys dramatically and turned the marine tow seismic vessels into fascinating high-tech factories, Aylor (1995).

In the oil and gas industry, controlled seismic source seismic is by far the most dominant technology. However, during the last 10–15 years we have seen significant increases in seismic monitoring during the completion stage of tight or "shale" gas wells. In addition, we have some exotic examples of what type of information we could obtain from passive monitoring, using natural or industry-induced noise. In a way, this is going back to where it all started, listening to the seismic signals from natural or induced "rock deformations".

The first repeat 3D seismic survey was acquired at the beginning of the 1980s to monitor an enhanced oil recovery process. The first example of marine tow time-lapse was probably done a few years later. Seismic monitoring of gas-water contact rises and sweep efficiency at the Frigg Field on the Norwegian Continental Shelf, (NCS,) was done using repeated 2D seismic lines acquired in 1979, 1981 and 1983, Maritvold (1990). Apart from an unswept area near the platforms locations, the seismic ruled out the existence of larger unswept areas. Seismic time-lapse or 4D seismic became very popular in the North Sea very early on. No one knows who invented the wheel but the term 4D seismic is often attributed to Dr Amos Nur (Nur, 1984, 1987, 1989). The Norwegian Licensing scheme favored the introduction of new technologies and emerging concepts were very quickly embraced. Almost before the industry was able to effectively process and interpret 3D data, there were talks on 4D. The concept was clearly stimulated by the connections with Stanford.

By 1990 most marine tow vessels had been upgraded with modern-type navigation systems. The stage was set for launching 4D seismic in marine settings. However, very few of the early reported 4D surveys were "sold" on the basis of the value of mapping the changes in the reservoir. The argument for new re-acquisitions across major oil fields was driven by potential uplift in the static images due to the improved marine tow technology and the need for better structural imaging driven by the introduction of drilling long, horizontal wells.

The re-recording of a new 3D data set gave an opening for including 4D testing as part of the program. We recognized early that re-occupying shot and receiver position had an impact and at the very minimum the acquisition of a monitor survey needed to be the same orientation (azimuth) as the base-line survey. Most important though was the demonstration that the seismic data seemed to be very sensitive to relative changes and many cases clearly stated the commercial viability of the method.

Today, 4D towed streamers and monitoring of hydraulic-fracturing are part of our day-to-day business. Work is progressing in understanding the issues of using retrievable seabed systems like nodes and Ocean Bottom Cables and combining various systems like OBC and towed-streamer. One of the first Node 4D seismic datasets was acquired by BP in the Gulf of Mexico, Beaudoin (2010).

Figure 1.4: One of the early air guns. Courtesy Bolt Technologies.

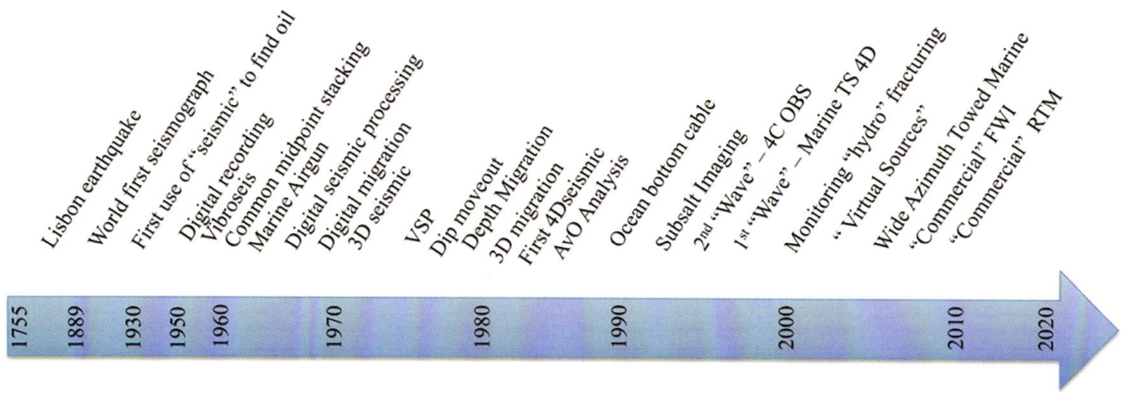

Figure 1.5: Milestones in the development of seismic methods for the oil and gas industry.

1.4 Some milestones in the development of 4D seismic

The Holt Reservoir in the northern part of Texas is recognized as being the very first 4D or seismic time-lapse survey and was acquired in 1982/83. Greaves and Fulp (1987) discuss the monitoring of an *in-situ* combustion process implemented to improve the recovery (EOR) process of producing heavy oil. Three surveys were acquired over a period of one year; the objective was to support effective management of an EOR process. Greaves and Fulp concluded that: "Reflection seismic surveying can be used to monitor the progress of some EOR processes. In this case study, a fireflood process was detected, its propagation direction and extent were determined and an estimate of net burn volume was made." "Temperature monitored wells and post-burn coring confirmed the shape, orientation and volume of the burn interpreted from the seismic data. It was concluded that the attenuation increase, due to high-temperature alterations of the reservoir rock and pore fluid changes, was the best seismic indicator of the combustion process. The subtraction of the baseline (pre-burn) data from the mid-burn and post-burn data for the interpretation of dynamic anomalies proved to be a very powerful technique. The subtraction technique has great potential for detecting anomalous seismic responses related to active reservoir processes." See figure 1.6.

The single geophone receiver group spacing was 6 m, a total of 182 geophones were buried 6 meters into the ground. One hundred and sixty-five shots were fired along crossed lines with 12 m between the shots. The source charges, made by 2.5 kg of dynamite, were fired at 23 m depth. The burial of the source and receiver eliminated the airwaves and reduced other source-generated noise. The data were recorded at one msec and band-limited to 50–320 Hz to further reduce noise. The setup resulted in uneven fold and azimuth contribution, but due to favorable geological settings, this did not generate too much of a problem for static and residual move-out corrections. In short, a fairly straightforward simple processing flow was applied but they still recovered high-quality time/lapse information.

The new data were used to describe the reservoir in more detail. Identical acquisition and processing was applied and seismic-derived attribute and difference volumes were used to support the interpretations. The Holt Sand is a 12 m thick reservoir with a this 2.5 m cap of limestone "encased in a unit in thick shale", with certain elements of heterogeneity, including a thin, fractured sand layer that most likely would influence the process. The reservoir was not very deep, around 500 m and the target area only around 800 square meters. Special focus was kept on repeating the acquisition as identically as possible, as "the amplitude change and spatial extent of the seismic anomaly would probably be

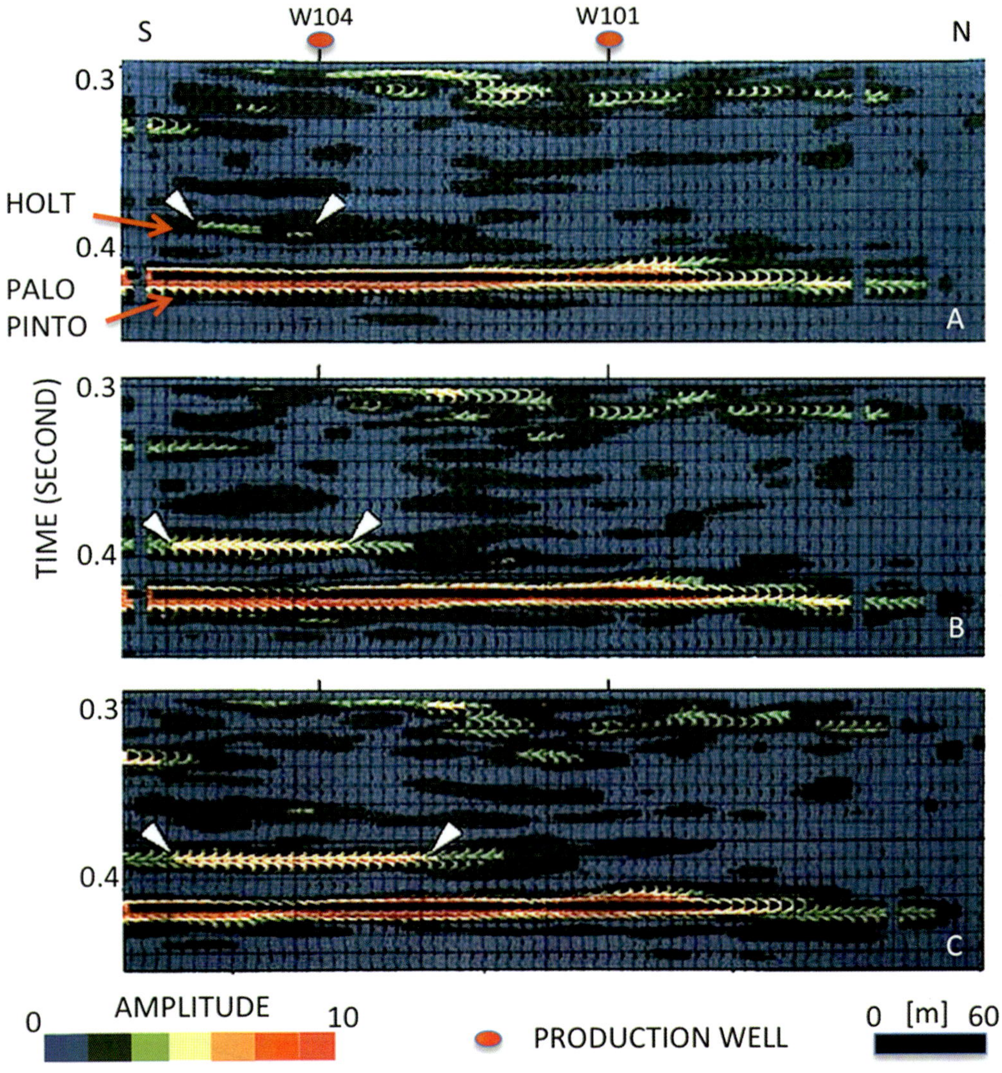

Figure 1.6: One seismic line showing from data from (a) pre-burn, (b) mid-burn, and (c) post-burn 3-D seismic data volumes. A color scale of the calculated envelope amplitude overlies the reflection wiggle traces. A bright spot was initiated at the top of the reservoir sand (Holt sandstone) and extended through the burn. A dim spot in the reflection from the Palo Pinto limestone formed just below the peak of the bright spot. From Greaves and Fulp (1987).

quite small." The pre-survey modeling work indicated that a central frequency of 100 Hz was needed to resolve the individual layers and to support quantitative solutions. Greaves and Pulp used a simple *in-situ* combustion model based on a tube experiment , Tadema (1959), as a basis for defining the process in time and space and used this to set clear expectations as to what they expected to observe. Today, we would probably have built a reservoir model to support these hypotheses. This early project had all the elements of a modern 4D project and the results were unambiguous, but still the technology was slow to progress.

1.4.1 Draugen

"Arguably one of the most successful 4D projects in the 1990s", is a phrase often used about the Draugen time-lapse seismic, Guderian (EAGE 2003).

The Draugen field is situated offshore mid-Norway, in 250 m of water. The reservoir is at a depth of 1600 m. The field is a north-south elongated low relief anticline, 21 km long and 6 km wide. Vertical closure is reported to be a bit more than 50 m and the main reservoir is excellent quality Jurassic sand. Production started in 1993 and by around 2000 the average production was about 225 000 b/day. The reservoir management strategy was to laterally displace the oil by water from north and south. There was some uncertainty about an aquifer below the reservoir, which may or may not be connected in parts. Various scenarios for connection to this aquifer through small throw faults, which also could act as barrier in the reservoir, led to a quite different production forecast.

To date 4 monitor surveys have been acquired.

In 1999 a new well was to be drilled and key decisions on the development of the southern and northern end of the field had to be made. Shell and their partners BP, Chevron and Petoro, decided to acquire a time-lapse towed streamer 3D dataset in 1998. Offset and azimuth ranges were kept the same as the 1990 baseline survey, but technology development meant that the number of streamers, streamer spacing and receiver group interval being had changed.

Before they acquired the time-lapse seismic results, all of the reservoir models had assumed that water movement could occur along both flanks of the field. The 4D results showed that the risks of water

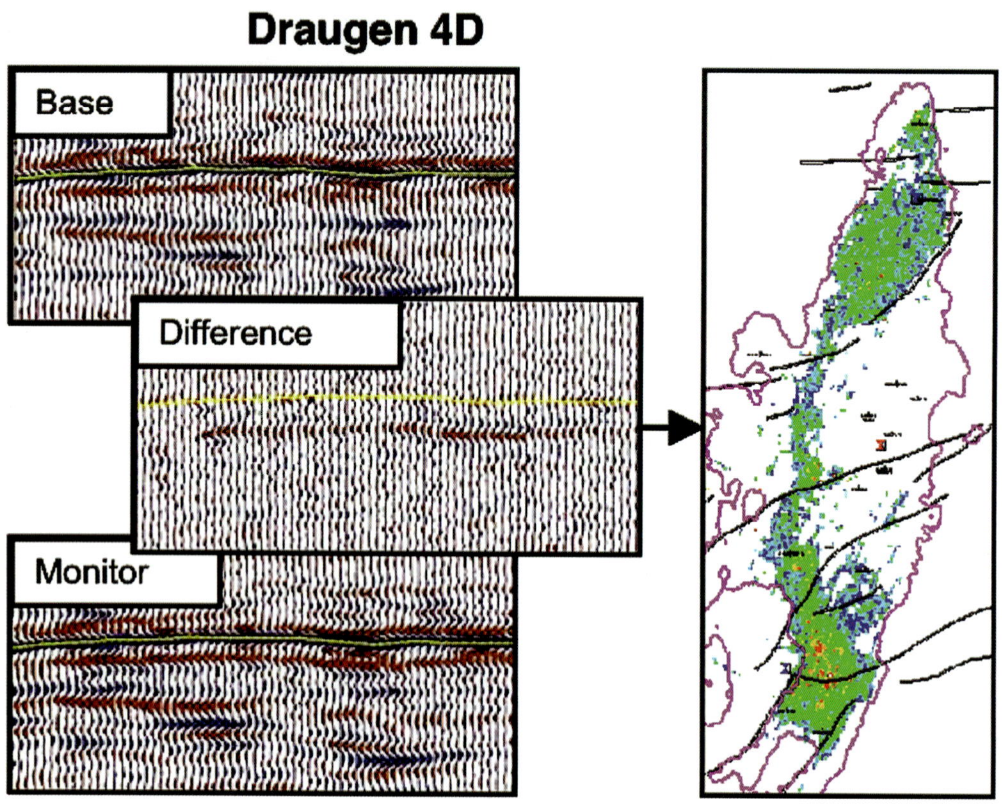

Figure 1.7: An example of the 4D seismic response observed at Draugen, Calvert (prospecting 2005).

encroachment were significantly higher on the western than the eastern flank. The 4D seismic quality was very good. The data was used for history matching of the reservoir model and resulted in a history-matched model that matched both the production and the seismic data.

The new reservoir model was used to test alternative location for the new well. The 1998 survey showed an unexpected E/W asymmetry in the waterfronts, invalidating the original position for the planned high capacity well. The optimal well location was found far away from the originally planned locations, which was west of the platform and close to water encroachment. The well was drilled in the new location and resulted in a record well that tested 76 775 b/day, Koster, (TLE 1990). The new model also predicted an extension of the plateau production for one additional year, estimated to represent a value of 84 MUSD. The cost of the time-lapse project was 4MUSD. In addition, the new time-lapse seismic resulted in an new optimized depletion plan, with a phased development schedule, reduced uncertainties in hydrocarbon volumes and more realistic water handling plans. The reduced uncertainties on water production also impacted the installation of water handling equipment.

The 1998 survey showed an unexpected E/W asymmetry in the waterfronts, invalidating the original position for the planned high capacity well.

The 2001 survey helped balance the water injection strategy. A further four satellite wells were drilled in the west and south, in areas not designed to be drained by the platform wells.

The 2004 High Resolution survey showed the main waterfront closing in further on the platform, with some attics beginning to form. Difference 4D signals were noted from all of the satellite wells, including those with only a few months of production. A further 2 satellite wells were drilled in the west and south.

The 2009 survey is extensively used for dynamic reservoir model history matching and is also being used to plan infill drilling, targeting some of the smaller attic volumes.

1.4.2 Gullfaks

A celebrated 4D example from the North Sea is the Gullfaks field, discovered in 1978 and still producing. The oilfield is located in the Tampen area in the northern part of the Norwegian North Sea. Statoil is the operator, with Petoro as co-venturer. The water depth in the area is 130 – 220 meters. The Gullfaks reservoirs consist of Middle Jurassic sandstones of the Brent Group, and Lower Jurassic and Upper Triassic sandstones of the Cook, Statfjord and Lunde Formations. The reservoirs lie 1 700 – 2 000 meters below sea level. The Gullfaks reservoirs are located in rotated fault blocks in the west and a structural horst in the east, with a highly faulted area in-between. The drive mechanisms are water injection, gas injection or water/alternating gas injection (WAG). The drive mechanism varies among the drainage areas in the field but water injection constitutes the main strategy. Ref (NPD fact-pages, http://factpages.npd.no).

The Gullfaks field has served as a full field geophysical reservoir monitoring lab and has been extensively used for piloting new 4D acquisition and processing techniques. On the Gullfaks field, time-lapse (4D) seismic data play a significant role in the management of the tail production, El Ouair et al. (2006), Williams (2008).

The operator, Statoil, reports: "4D seismic data have increased our understanding of both the static properties and dynamic behavior of the field, challenging the production strategy and driving the reservoir management decision process. The data have identified bypassed volumes in unswept reservoir compartments, improved the estimate of remaining reserves and the associated uncertainties and has helped to identify drilling hazards that have evolved over the lifetime of the field. Since the first repeat seismic survey in 1996, 4D seismic data have been integrated into a multi-disciplinary increased oil recovery (IOR) effort that aims to recover 70% of the in-place reserves from a structurally complex and heterogeneous reservoir sequence. 4D seismic has contributed directly to 19 successful infill wells and generated a net present value in excess of US$1 billion, See figure 1.8.

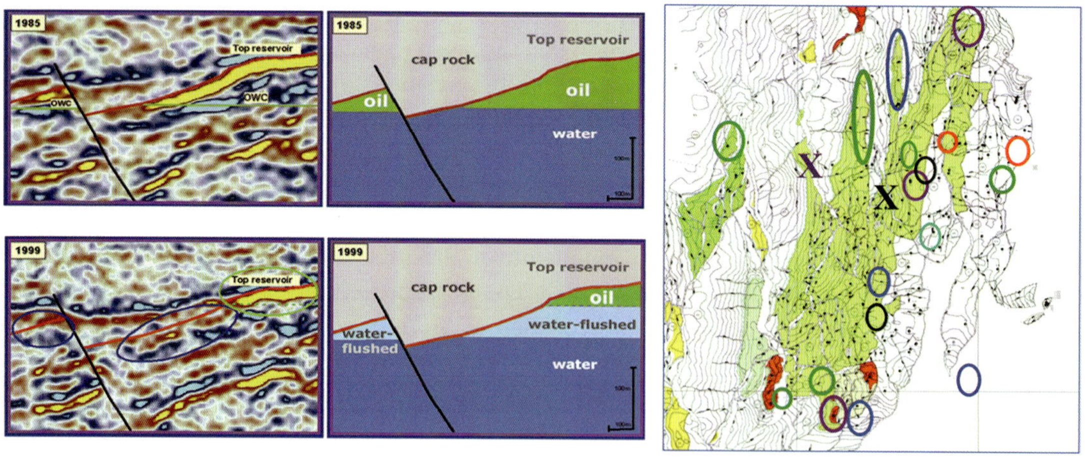

Figure 1.8: The upper left seismic section is from before production started in 1985, the lower left seismic section is from the 1999 seismic survey and reflects the water flushed zone, as illustrated by the two middle cartoons. To the right is a structural map of the top Tarbert, with the 4D driven successful drilling targets. The colors of the circles indicate the vintage of monitoring survey, blue is after 1996, purple after 1999, red after 2001, green after 2003 and black after 2005. The X's reflect cancelled wells. Figure courtesy of Statoil, Copyright Statoil.

Subsequent monitoring and flow control measures have increased the recovery targets from 50 to over 70%.

The changes in seismic reflection amplitudes between the two surveys in 1985 and 1999, as shown in figure 1.8, result from a significant depletion of the oil due to production. The difference in the reflection strength of the top of the reservoir is related not only to reduction in oil saturation, but also to the original oil column height. Where water replaces oil, the reflection strength is diminished; causing a dimming effect on what was a strong response from the top of the reservoir. The strong seismic response from the oil-water contact (OWC) in 1985 has also been dimmed, owing to the production of oil. The smaller oil accumulation to the left of the fault was drained by 1999, whereas much of the oil was still to be recovered from the larger trap, to the right of the fault.

1.4.3 Ekofisk

The examples from Magnus, Gullfaks and Draugen in the late 1990s triggered the "real" business for 4D seismic projects. Several Joint Industry programs were started to progress the technology, an example being ETLP at the Herriot Watt University, started in 1999, and the RCP program at the Colorado School of Mines, which started including 4D as a part of their program. In Norway, the state-led Force program that started in 1995, had 4D seismic as one as one of its core activities and hosted a number of informal and very open workshops, where the oil companies shared their experiences but also concerns like "4D seismic does not work in carbonates". This led Conoco Phillips, the operator of the important Ekofisk field, to do a number of dedicated rock physics studies, eventually leading to the recognition that time-shifts were by far the largest time-lapse effects when matching time-lapse seismic data across a compacting reservoir, figure 1.9.

The Ekofisk oil field is located in the southern part of the North Sea, in water depths of 70–75 meters. The Ekofisk field produces from the naturally fractured chalk of the Ekofisk and Tor Formations of Early Paleocene and Late Cretaceous ages. The reservoir rocks have high porosity but low permeability.

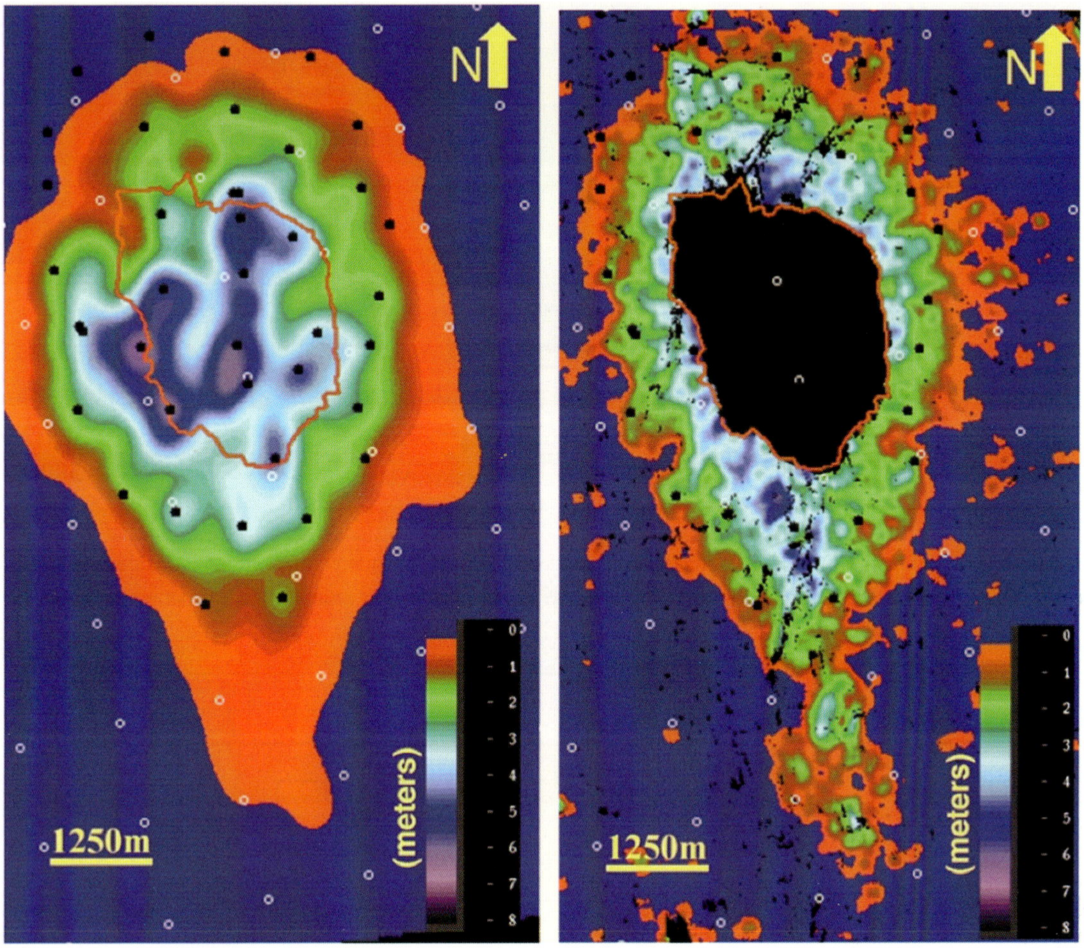

Figure 1.9 Left: Ekofisk field: Compaction from geomechanical modeling, with water injecting wells displayed as black dots. Right: Seismic-derived compaction map. From Guilbot and Smith, (2002).

The reservoir has an oil column of more than 300 meters and lies 2 900–3 250 meters below sea level. Ekofisk has been producing oil since 1971. The field was originally developed by pressure depletion and had an expected recovery factor of 17 percent. Since then, limited gas injection and comprehensive water injection have contributed to a substantial increase in oil recovery. Large-scale water injection started in 1987 and in subsequent years, the water injection area has been extended in several phases. Experience has proven that water displacement of the oil is more effective than predicted and the expected recovery factor for Ekofisk is now approximately 50 percent. In addition to the water injection, compaction of the soft chalk provides extra force to the drainage of the field. Reservoir compaction has resulted in the subsidence of the seabed, which is now more than 9 meters in the central part of the field. It is expected that the subsidence will continue for many years but at a slower rate. Source: NPD Fact pages (2011).

1.4.4 Valhall

The observations of surprisingly large production induced changes in seismic velocity above the reservoir at Ekofisk was followed by similar observations at the Valhall field, figure 1.10 Barkved et al. (2002) and from depleting a North Sea gas field, Hatchell et al. (SEG 2003).

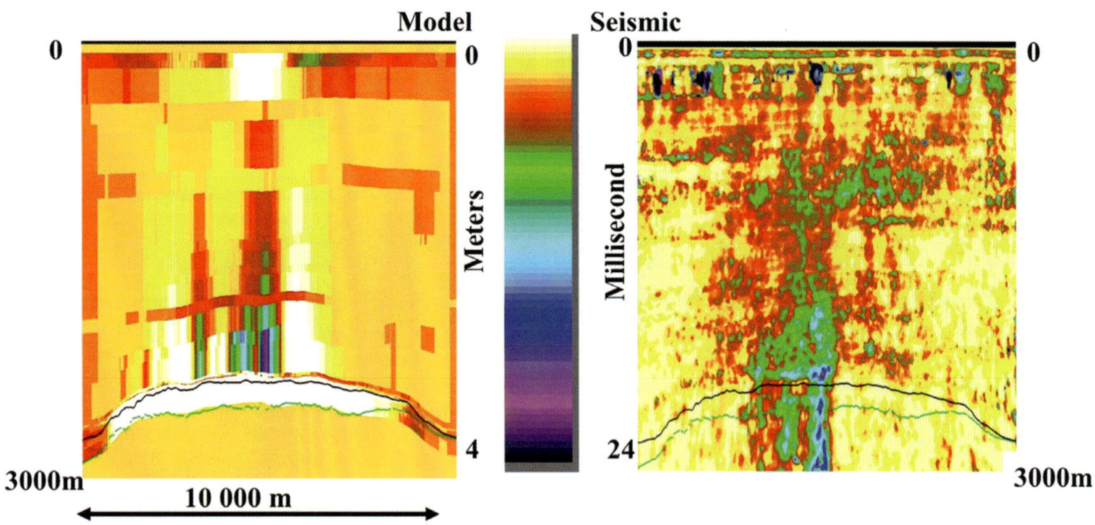

Figure 1.10: The small volumetric strains in the overburden (left), accumulated in meters of subsidence at the reservoir level. The effect is enhanced by changes in the seismic velocities; time-shifts up to 48 msec have been measured from streamer 4D seismic data (right). Modified from Barkved et al. (2003).

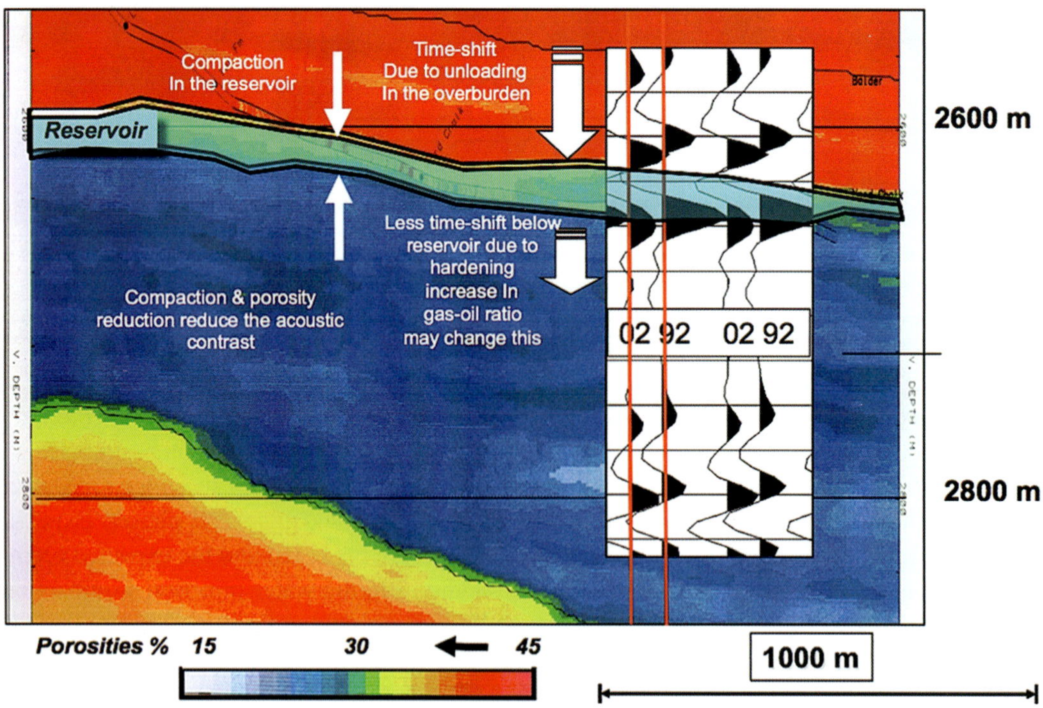

Figure 1.11: Production induced seismic time-lapse effects at Valhall. The subsidence induced traveltime delay above the reservoir (A) is about twice as large as the pull up due to reservoir compaction. Below the reservoir the time-shift (B) has been reduced to about 6 ms.

The time-shifts above the reservoir appear to be at least as much as the production-induced effects in the reservoir, as shown in figure 1.11.

1.4.5 Halfdan

The Halfdan Field is located in the Danish part of the North Sea Central Graben, 250 km off the West Coast of Denmark, figure 1.12a. The field is operated by Mærsk Olie og Gas AS on behalf of the Danish Underground Consortium (DUC). The Halfdan oil and gas accumulation was discovered in December 1998. The field is laterally extensive and comprises a high porosity (25–35%), low permeability (0.5–2 mD) chalk reservoir of Maastrichtian (Upper Cretaceous) and Danian (Lower Paleocene) age, located without structural closure, Albrechtsen et al. (2001).

The depletion strategy was to drill alternating 10–15,000 ft. long horizontal production and water injection wells at a spacing of 600 ft between the wells to develop the field.

Productivity from the tight chalk is achieved by acid stimulation of the wells. To maintain voidage replacement and pressure maintenance in the low permeable reservoir, water needs to be injected above the fracture propagation pressure, from Jørgensen (2002).

Rock and fluid properties in the Halfdan field are favorable for displacement of oil with water due to the lower mobility of water compared to oil. The dense well pattern facilitates an efficient line drive water flood. However, an enhanced reservoir contact area is required to achieve adequate injection rates for voidage replacement and pressure maintenance at attractive production rates. In non-fractured Danish North Sea chalks this may only be realized by injecting water above the fracture propagation pressure. The risk of premature water breakthrough associated with injection at fracturing conditions is mitigated by actively controlling the induced stresses during injection and production. Inducing and monitoring the stresses are a key element in managing the reservoir. Maersk has developed a technique named "Fracture aligned Sweep Technology" (FAST), which utilize that the fact that fluid flow in low permeable rocks affects the reservoir stresses. Before propagating a fracture from a well, the prevailing pressure is manipulated for a period of time. This is achieved by injecting below the fracturing pressure in the actual well, while producing from the neighboring wells. By actively controlling the injection and production, they manage to align the induced fractures with the injector wells, Rod and Jørgensen (2005).

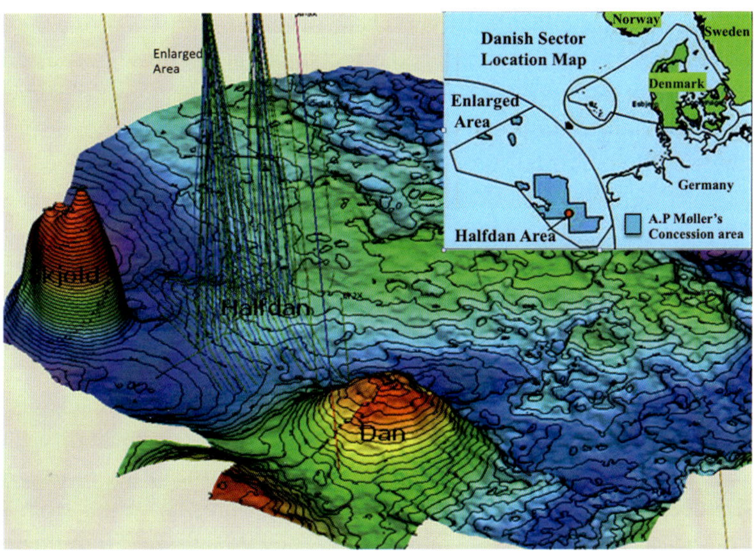

Figure 1.12a: The Halfdan field is located in the Danish part of the North Sea. The trap is dynamic, structural and stratigraphic. The map shows the top Maastrichtian. Dons et al. (2007)

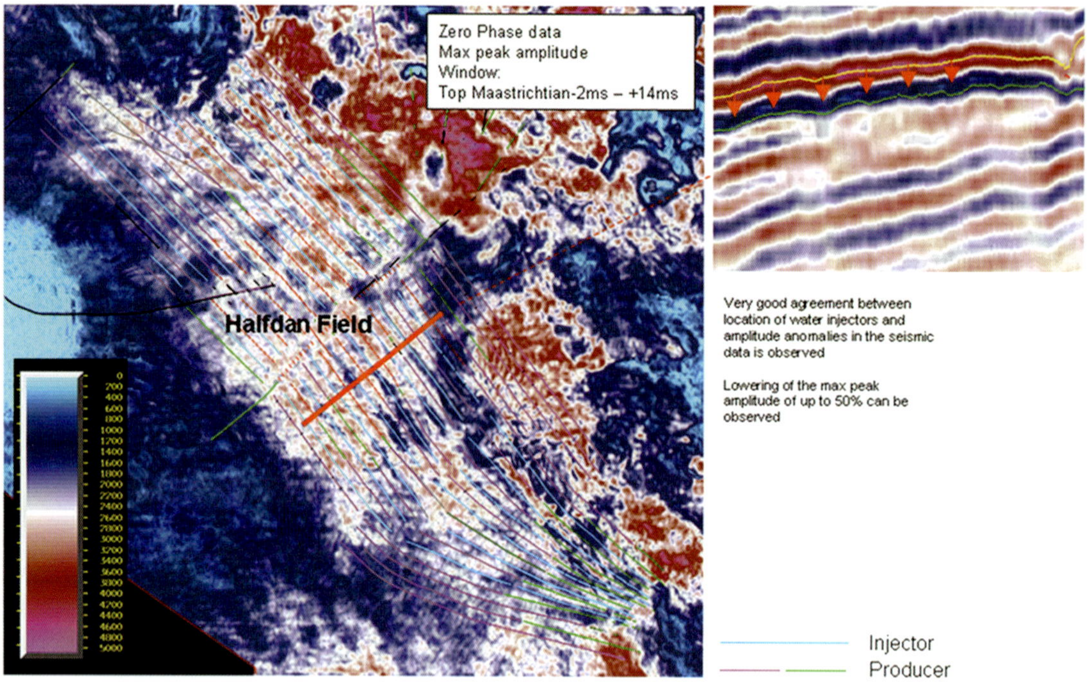

Figure 1.12b: Seismic attribute map showing the presence of the injectors and planned and existing producers. The caption to the right shows how the pairs of injectors and producers are affecting the seismic response, from Dons et al. (2007).

A new seismic survey was acquired across the field in 2005, with the intention of using it for 4D imaging by matching to the previous seismic survey acquired prior to production/injection starting. It turned out that the new 3D seismic reflected the sweep pattern without any 4D processing. The amplitude difference between a production and an injection well was as high as 50%, figure 1.12 a.

The Halfdan case proves an injection/production management strategy and demonstrate how such strategies are related to fluid and rock mechanics considerations. These observations also prove that it is possible to use seismic techniques to monitor production-induced effects in relatively low porous chalk. (The porosity is less than 30% and the permeability is 0.5–2.0 mD). The seismic response observed in figure 1.12b is a combined effect of fluid substitution and compaction of the rock, both increasing the acoustic impedance, and from increasing the pore pressure leading to lower acoustic impedance.

1.4.6 Carbonates

There are few examples of the successful use of 4D seismic applications on hard carbonates reservoirs in the Middle East, there may be several reasons for this. First of all, the stress sensitivity of the rock is normally significantly less than in clastic rock and the softer carbonates fields we find in the North Sea. Secondly, surface conditions may be such that repeatability is difficult, i.e., due to sand dunes.

Al-Jenaibi et al. (2006) describe a successful 4D seismic pilot from an onshore upper cretaceous carbonate field in Abu Dhabi. One of the challenges with this field is the large contrast in permeability, due to differences in rock type and digenesis. In this case the upper reservoir zone has permeability in the Darcy ranges, while the permeability of the deeper zone is in the tens of mD range. Porosity ranges from 20% in the water leg to 35% in the best oil bearing crestal areas , figure 1.13. They are around 50 m thick and extend 12–30 km, with a relief of 180 m. Water injection is used to maintain pore pressure but the water

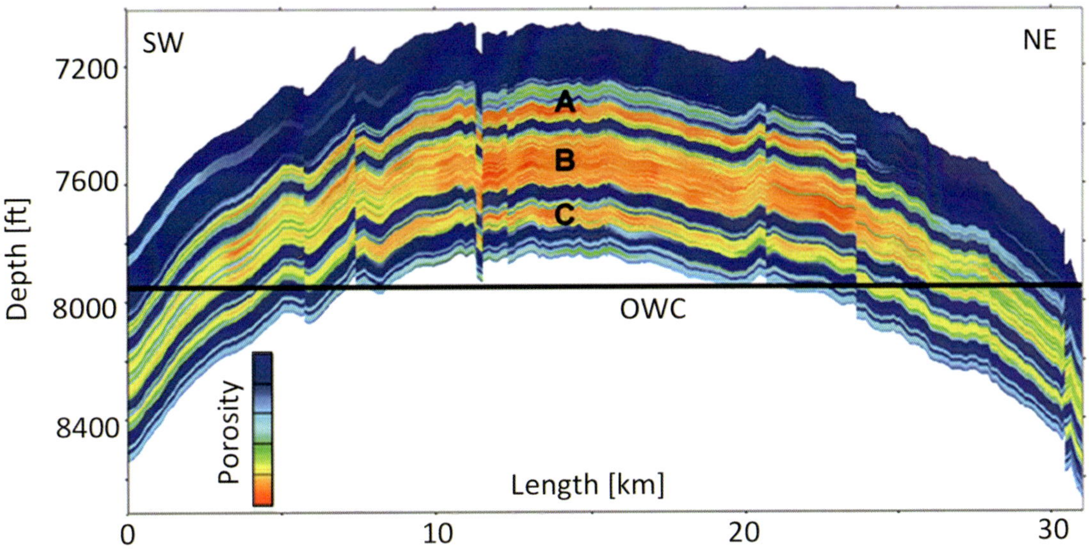

Figure 1.13: A crosssection of the geologic model showing porosity of the reservoir zones. Note the decreasing porosity down flank with an increase in dense bodies in the middel of Zone B. OWC = oil-water contact. Melville et al. (2004).

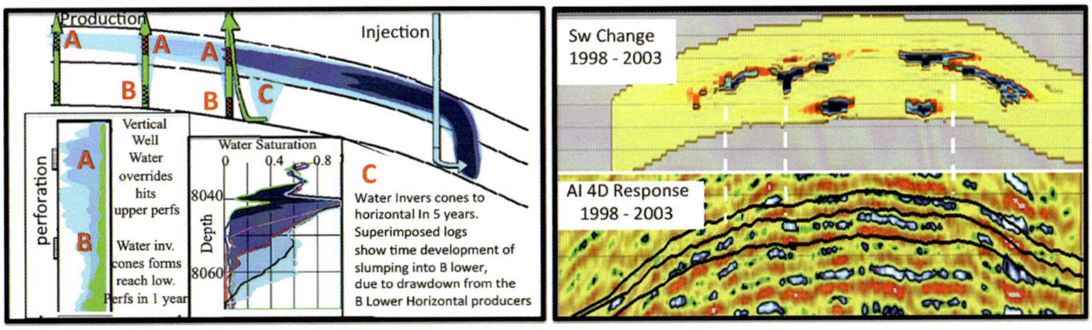

Figure 1.14: 4D response (bottom) compared with the saturation changes in the simulation model, showing some examples of good agreement. Model update and pressure changes in the crest account for most of the discrepancies. From Al-Jenaibi et al., (2006).

has a preference for moving into the highest permeability in the upper zones. As a result, the water does not sweep the lower layer effectively, leaving bypassed oil. The draw-down of vertical wells tends to cause water coning, so horizontal wells are used to more effectively drain a wider area, figure 1.14.

Al-Jenaibi and co-worker's results show that 4D can work in hard carbonates and that reasonable repeatability can be achieved onshore in a sand dune environment. The 4D results correlated very well with available well data. It was found that the 4D results were especially useful for monitoring the upper level, where the porosity could be as high as 35%, which is consistent with findings from the North Sea carbonate (chalk) fields.

1.5 Permanent systems

Permanently installed receivers allow for constant position and coupling are more stable than re-deployable systems. A fixed receiver system also has advantages in that it can be designed to provide additional measurements on random or systematic noise. Permanent receivers were in fact used in some of the first seismic time-lapse studies on land, but did not gain traction. The next wave in permanently installed systems at scale came in the 1990s, and in the following section, we will discuss some of the key installations.

1.5.1 The Foinaven benchmark

To assess the value of seismic reservoir management, BP initiated the Foinaven Active Reservoir Management (FARM) project in 1995 (Cooper et al. 1999a). Foinaven is located 190 km west of the Shetland Islands, figure 1.15. Following the successful use of detailed 3D seismic data for rapid field appraisal and development drilling, the partners (BP and Shell) made an early investment in 4D seismic in 1995, in the expectation that this new technique could further improve the economic return on investment in the field.

The FARM project included towed streamer surveys and surveys acquired using permanently installed cables to evaluate which technology was "fit for purpose". The project was a joint venture between BP, Shell and Schlumberger. The objective of the project was to assist the asset team in managing the reservoir through imaging reservoir changes following production. Seismic repeatability was seen as fundamental to this goal and to test this, the partners installed the first permanently emplaced, deep-water, seismic recording system.

The seabed recording system included six parallel sensor cables, trenched into the seabed in the configuration shown in Figure 1.16. The cables contain hydrophone sensors; four per group and the distance between group centers is 50 m. The cables are 300 m apart. Baseline surveys were acquired in 1995, prior to the start of production, using conventional towed streamer and seabed recordings. The data from

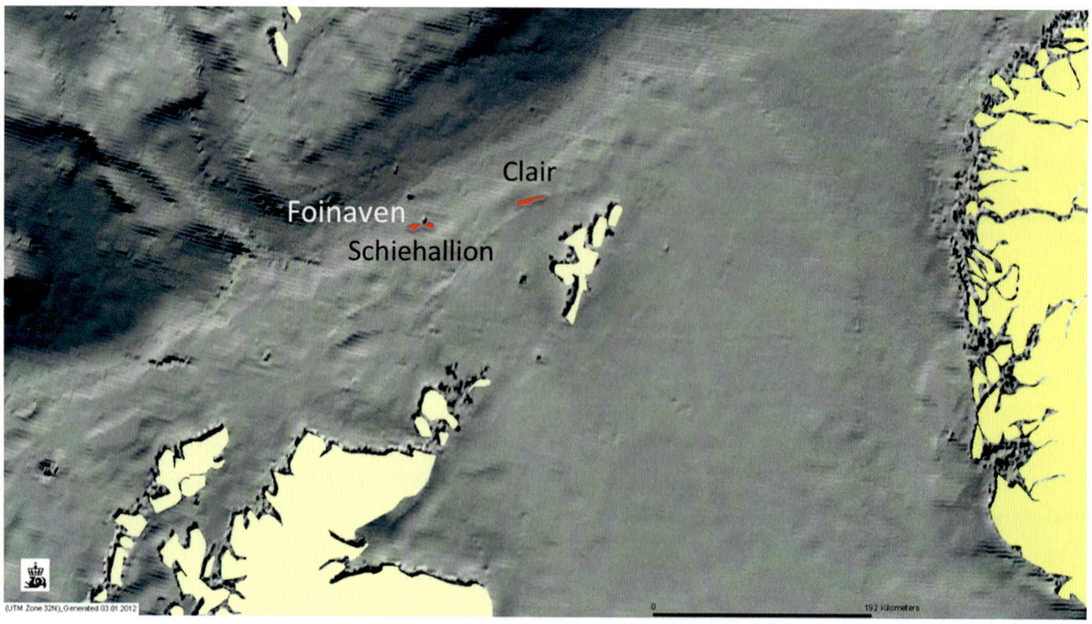

Figure 1.15: Map showing the location of the Foinaven field.

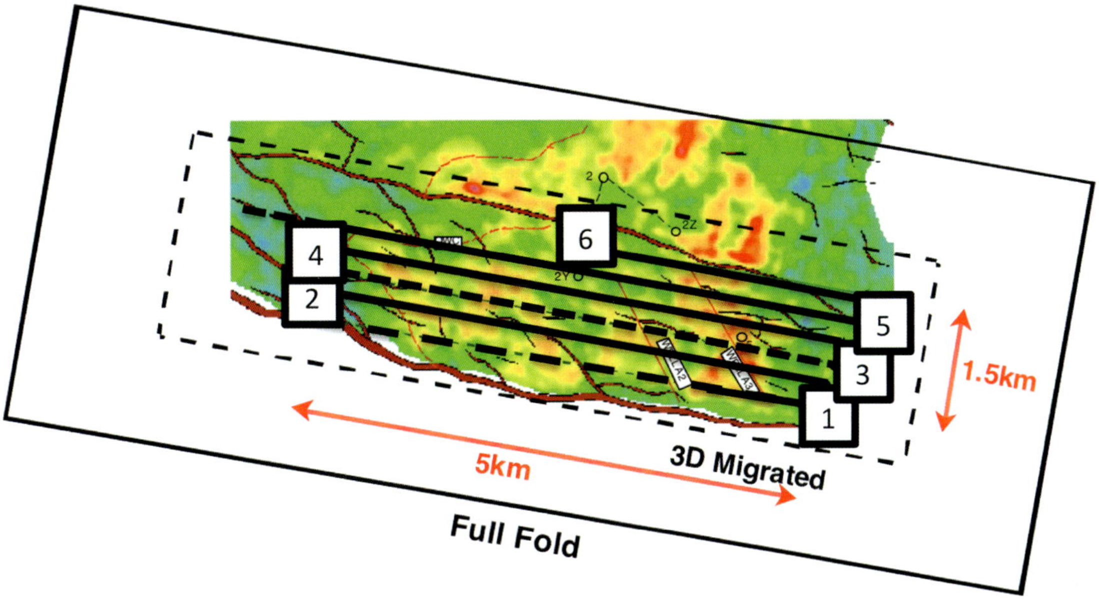

Figure1.16: configuration of seabed cables over Panel 4 of the Foinaven field (shown as an amplitude map). Solid lines indicate sensors that were active for both the 1995 and 1998 surveys. Modified from Cooper et al. 1999.

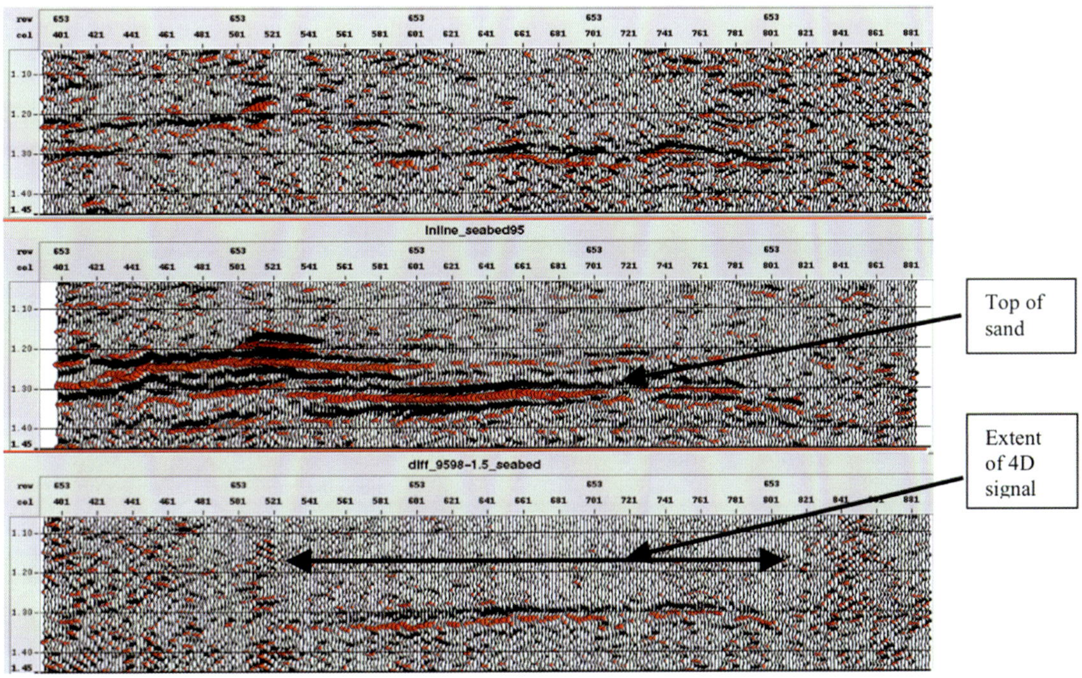

Figure 1.17: In-lines at reservoir target levels. Top: difference from the towed streamer data, middle data from the 1995 survey and the bottom differences for the seabed recordings. From Kristiansen (2000).

the seabed array was acquired with a recording vessel connected to the sensor spread and a source vessel traversing source lines parallel to the seabed cables. A 3D vessel, towing four streamers and a single source array, acquired the towed-streamer surveys.

In 1998, after 10 months of production, both the towed-streamer and seabed surveys were repeated. To maximize the repeatability of the data, the acquisition conditions were matched as closely as possible, including using identical spread, recording settings, identical source, same sail line orientation and shot point position. Due to feathering, re-occupying the receiver locations for the streamers was not an exact match, but the positioning errors were typically on the order of the positioning accuracy error. For further details see Cooper et al. 1999.

Four seismic datasets have been acquired: The baseline (1995) and repeat (1998) surveys, recorded on the seabed spread and with towed streamers. Figure 1.17 shows a comparison of the results from the two methods.

1.5.2 Teal south 4C/4D PRMS test

"In late 1996, Texaco and Input/Output embarked on a novel experiment to test a low-cost 4-C/4-D permanent reservoir monitoring system (PRMS). Western Geophysical and Digicourse were brought into the project to provide data acquisition and positioning services. In July and August 1997, four-component (4-C) data acquisition was performed over 9 km^2 in Eugene Island Block 354 (Teal South), using a dense shot grid (25×25 m). In late 1997, Texaco turned the control of the project over to the Energy Research Clearing House and invited industry participation"; Ebrom et al. (1998).

The Teal South Field is in shallow water (85 m). Its multiple reservoirs (unconsolidated Tertiary sands) produce both oil and gas from depths of 4000 to 8000 ft. A sparse receiver grid (four lines of six 4-C receiver stations) was deployed, the station interval was 200 m and the line separation was 400 m. A 25×25 m grid of seismic shots was acquired; the shot grid extended 1000 m beyond the receiver grid. This produced a full-fold, fully migrated *P*-wave image of the 4500-ft sand. The system used dual - axis gimbaled sensors and sandbags were used to ensure good coupling and reduce noise. Orientation deter-

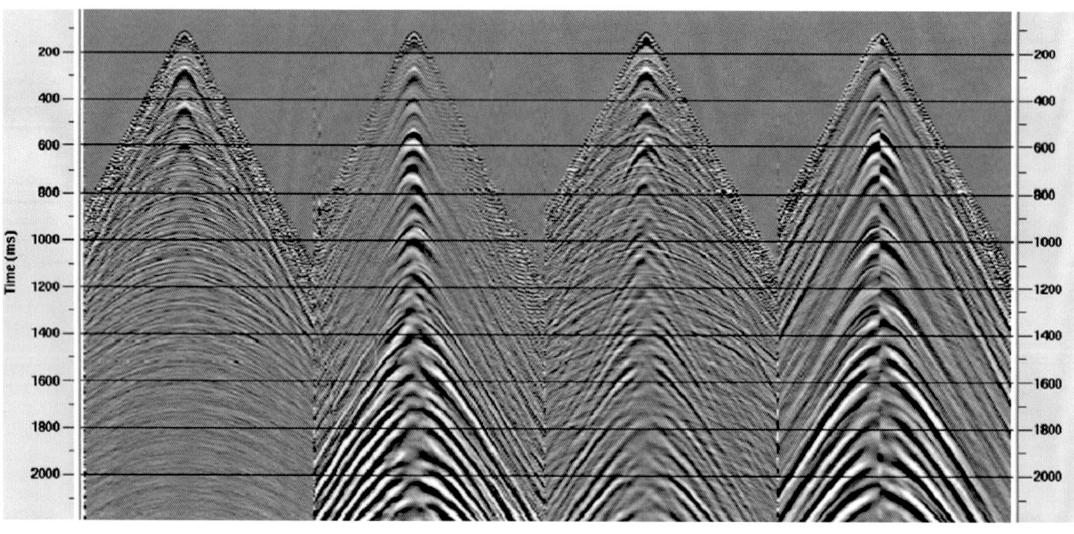

Figure 1.18: Common receiver gathers: a) hydrophone, b) cross-line geophone, c) vertical geophone, and d) in-line geophone. From Ebrom et al. (2002).

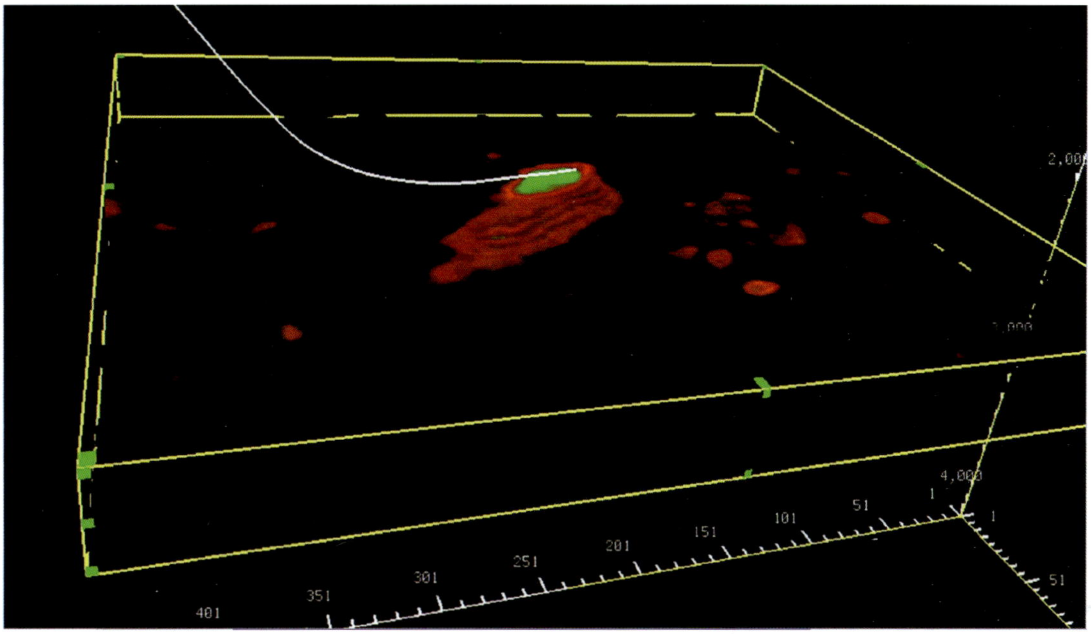

Figure 1.19: Visualization of difference cube from the Teal South PRM. Well track of D10 well is also shown intersecting with the southern end of the difference anomaly. From Ebrom et al. (2002).

mination was made from analysis of first breaks. An analog riser cable connected the ocean bottom sensors to a temporary recording system at surface. The recording equipment was housed in a buoy. Figures 1.18 and 1.19 show that the system delivered data of excellent quality.

1.5.3 The Valhall permanent installation

In 2002, BP and the Valhall License partners (BP, Hess, Shell and Total), decided to install a permanent seismic recording array at Valhall. The primary objective was to monitor the future Valhall Water Injection program. Initially, the Life of Field Seismic, LoFS, program included 6 seismic surveys. By the end of 2011, 14 surveys have been acquired and the current program includes 2 surveys/year for the next two years. The Valhall License is now looking into an extension of the permanent array into West and South Valhall and further into the Hod license. BP Norge AS operates both the Valhall and the Hod fields and the current partner is Hess Norge AS

In addition to conducting repeated 3D sourced surveys, the nearly 10,000 channel subsea seismic system can be used for passive monitoring of seismic activity in the vicinity of the reservoir in real time. This reservoir scale seismic system is composed of more than 120 km of GEOSPACE Subsea cables, covering 45 Km^2 of the Valhall Field. The armored cable segments incorporate 4-component (x, y, z, P) seismic sensor stations at 50-meter intervals. The cables were deployed in parallel lines over most of the Valhall Field's producing formations, see figure 1.20 and figure 1.22.

The seismic cables are coupled to a GeoRes recording system at the Valhall IP platform. The recording systems have a capacity of 40 TBytes of data. Recent upgrades to the high-speed fiber-optic communications link to the BP office in Stavanger now allow for transfer of real-time recordings to shore at a sampling rate of 4 msec. Secure network communications have been set up to allows for year-round remote supervision and operation from the shore. The system is run without the need for an operator on the platform.

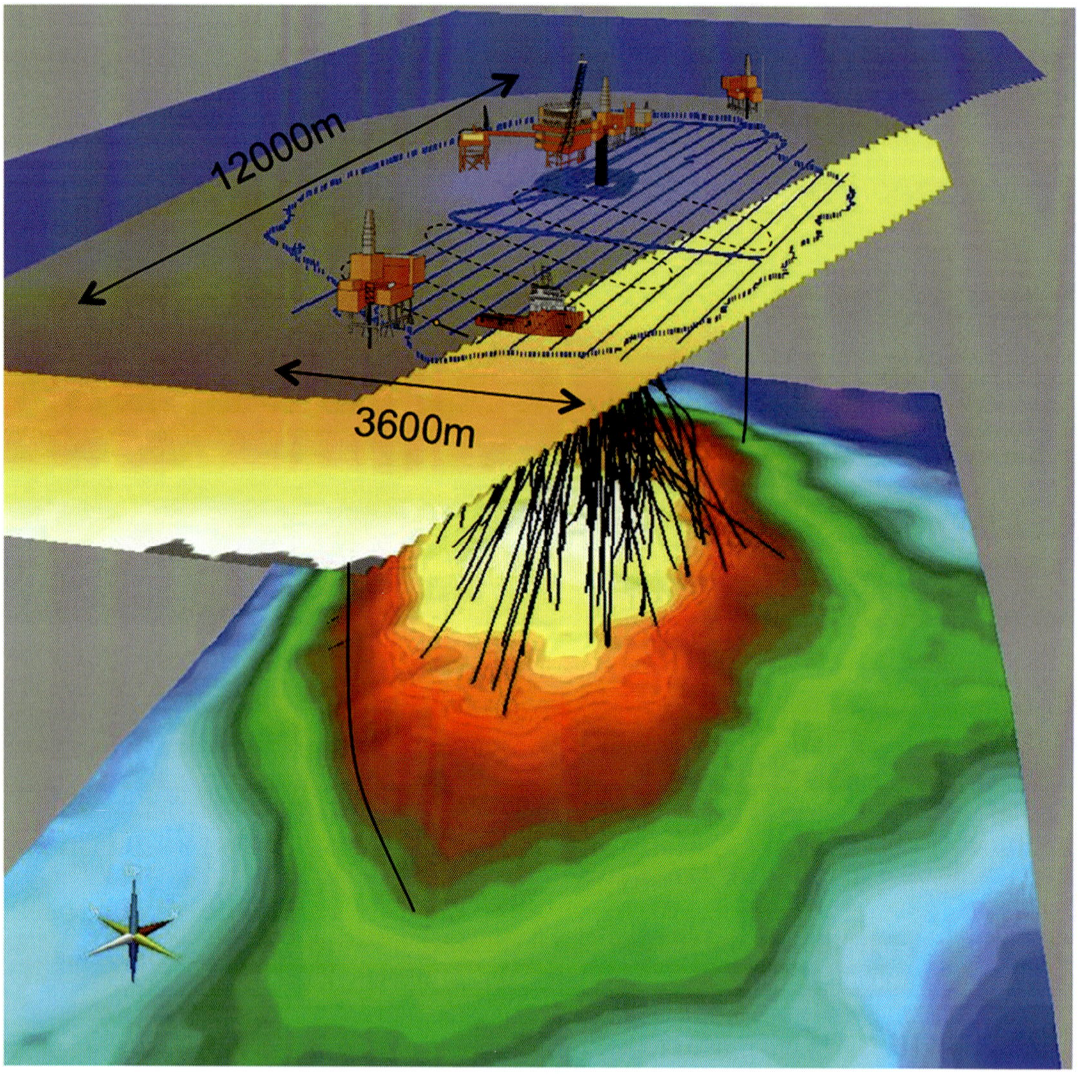

Figure 1.20: The Valhall LoFS array.

Station spacing	50 m
Stations	2508 Three mutually orthogonal (non-gimbaled) geophones per station One hydrophone per station
Pear signal input	250 millibars / 10 mm/sec
Noise floor	Noise floor 0.1 microbar RMS / 4×10^{-6} mm/sec RMS
Electrical cross-feed isolation	>80 dB
Total Harmonic Distortion	< 0.2%

Table 1. 1 System characteristics.

The permanent placement of the recording system makes acquisition simpler and more accurate than the conventional marine tow system in many ways since:

- Receiver position is the same for each survey
- Survey design is repeated for every survey
- Receiver signal and noise characteristics are constant
- Source repeatability is optimum as the same source is used for all surveys
- Data management is simpler

The system includes long period seismometers, tide-gauges and continuous recordings of barometric pressure at the surface. Figure 1.21 shows some of the elements of the system and installation process. We will discuss the acquisition, processing workflows and the results from the Valhall LoFS array in some more detail in the following chapters.

Figure 1.21: Key elements of the Valhall LoFS array.

1.5.4 The Clair permanent installation

In 2006, a small permanent trenched seabed receiver array (40 km of cable covering ~11 km^2) was installed over the up-dip core area of the Clair Field's Phase 1 development, Foster et al. (2008). The Clair LoFS array was installed with the objective of confirming whether a useable 4D response could be obtained from the moderate to low porosity reservoir. The LoFS array covers just over 11 km^2, which is only part of the Phase 1 Development area. Seismic surveys have been acquired in 2006, 2007, 2008 and 2010. Ricketts and Barkved (2011).

Clair (discovered in 1977) is a large oil reservoir of Devonian age, located 75 km west of Shetland, figure 1.23. The Clair development is to be a phased development and Phase 1 started producing in 2005. The installation of the LoFS array was difficult, due to an extremely hard seabed and was only partially successful in burying the cables.

The LoFS data, acquired with dense shot spacing of 25 m×25 m has delivered the highest quality 3D images of the field, improving on the previous OBC survey acquired in 2002, which had a 250 m shot-line spacing, figure 1.23 Davies et al. (2011). Data from both ocean bottom surveys were superior to the previous streamer data, acquired in 1993 and processed in 1999.

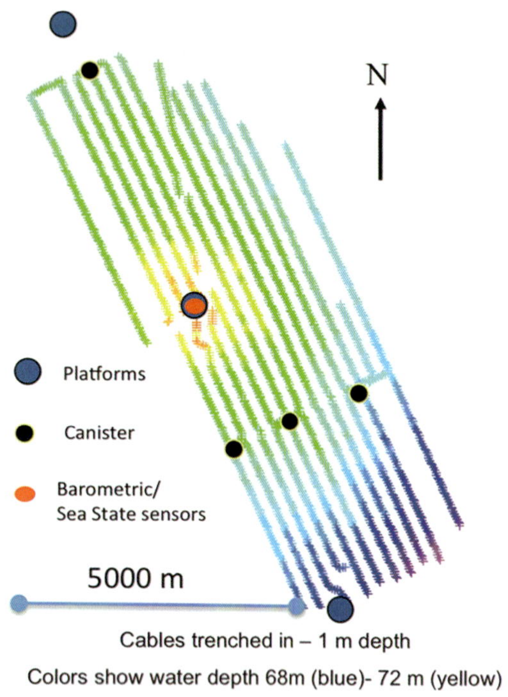

Figure 1.22: Valhall LoFS System: The cables were trenched 1 meter into the seabed. The cables end in seismic array terminators (SATs), which are located in 4 canisters that have been dug into the seabed. The SATs are connected to the Valhall IP platform via umbilicals. The umbilicals deliver power and optical fibers for data transfer.

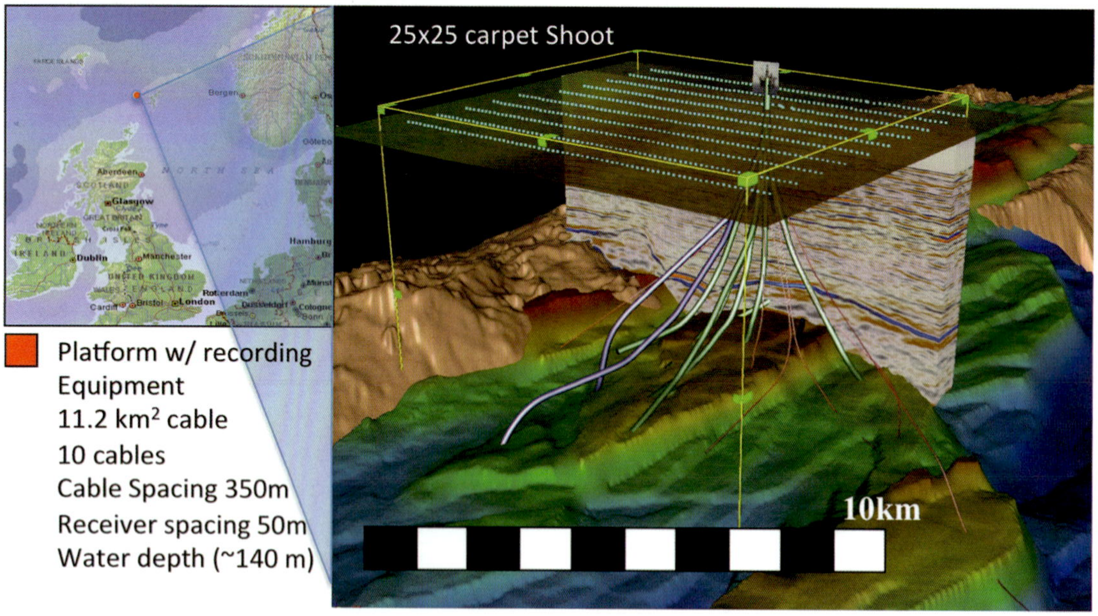

Figure 1.23: the Clair LoFS array. From Ricketts and Barkved (2011).

The 4D data from the LoFS array has demonstrated that it is possible to observe a 4D response from the Clair reservoir. Figure 1.24 shows the significant uplift observed due to high density, wide azimuth data.

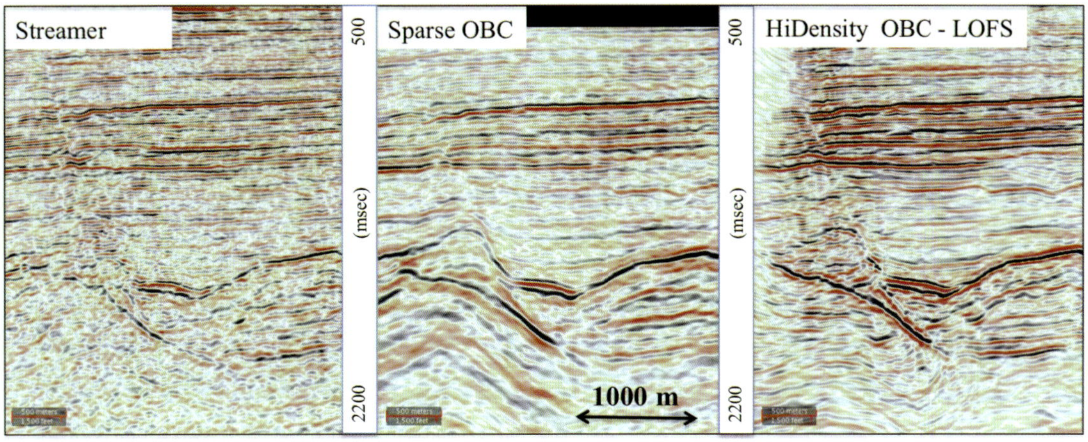

Figure 1.24: Comparisons of towed streamer, sparse OBC and High-density OBC (Davies, et al. (SPE 2011).

1.5.5 The Ekofisk permanent reservoir monitoring system

Installation of the Ekofisk PRM system was started in April and completed in October 2010. The system covers an area of 60 sq. km., Folstad (2011). In addition to trenching the seismic cables 1.5 m into the seafloor, the installation also included tieback to a containerized topside recording system on the Ekofisk M-platform. Figure 1.25 shows the design of the seismic array. Thc Ekofisk License opted for a system based on Bragg-grating optical sensing elements, the Optowave system from Optoplan/CGGVeritas. The Optowave system is made up of fiber optic seismic array cables, lead-in cables and laser interrogation instrumentation.

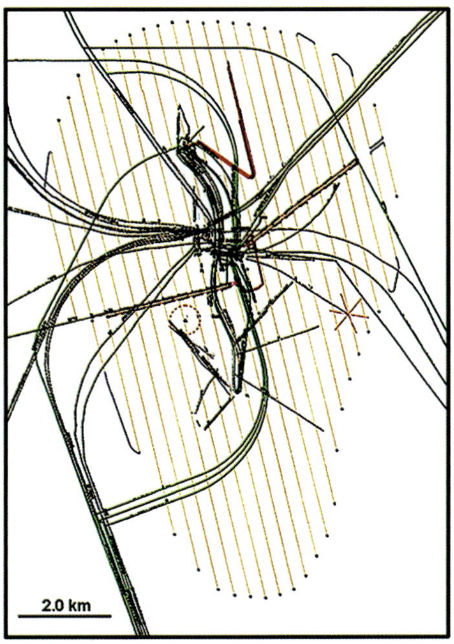

Ekofisk Life-of-Field Seismic Design Overlain on Ekofisk Facilities

300 m seismic array cable separation
50 m sensor station interval
200 km seismic array cable
40 km umbilical cable
24 cable lines
~ 4000 sensor stations
60 km^2 sensor station area

Sensor cables extend 500 to 1000 meters beyond intended time-lapse monitoring Area.

Purple – LoFS backbone cable
Tan – LoFS seismic array cable
Blue – LoFS jumper cable
Red – LoFS system test cable

Green active pipeline
Black (dashed) – abandoned pipeline

Figure 1.25: The Ekofisk Life of Field Seismic design overlain on existing Ekofisk facilities. Folstad (2011).

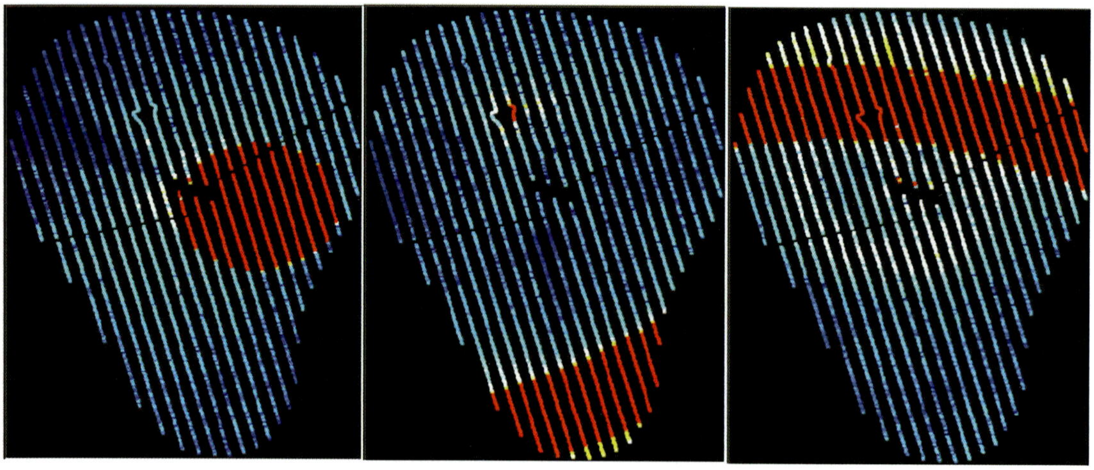

Figure 1.26: Location map and Screen dumps of GIS-display, showing the wave propagation and response of the 3966 hydrophones installed at Ekofisk from different shots. Red color is high energy level and blue is low energy level. From Nakstad et al. (2011).

Seismic acquisition is performed with a containerized source system operated from one of the Ekofisk supply vessels. Time stamps and navigation headers are transmitted from the source vessel to the recording system via a satellite connection. The first Ekofisk survey was completed in January 2011. So far, three surveys have been acquired. Figure 1.26 shows a snapshot of the response of the 3996 hydrophones from 3 different shots.

A separate, smaller recording system for passive seismic activity monitoring was added in early 2011. This system will continuously record and transmit data to shore from receiver stations at seven different seafloor locations outside active data acquisition periods.

1.5.6 Cere La Ronde gas storage

Meunier et al. (2001) describe a system using permanent in-place receivers and sources. Several seismic monitoring experiments have been carried out over the Cere La Ronde Gas storage reservoir. The brine-filled reservoir is used to store gas through the summer to be used in the winter. Conventional surface seismic recording was not sensitive enough to measure pressure and fluid saturation changes, Meunier and Huguet (1998), major issues where permits to use the large seismic sources in an agricultural region and variation in humidity in the weathering zone caused large changes. Vertical seismic arrays were therefore installed, clamped on the well casing and regular VSP walkaways were recorded. This improved the repeatability but coverage was limited to near the well bore, Dubois et al. (1998). Subsequent studies and analysis led to a detailed understanding of how to mitigate the effect from the weathering layers and to the design and installation of a system based on permanent sources and vertical receivers arrays installed in wells, see Muenier (2001) for more details. In figure 1.27 we see the results from real-time monitoring of the pressure variations in the reservoir.

Low-energy stationary seismic sources were used to prove the ultimate sensitivity of a seismic monitoring method. Measurements allowed of significant differences in the order of 1/100 ms. Traveltime difference of 0.1 msec corresponded to a 4-bar pressure increase measured in the reservoir.

The methodology has been developed to include the use of piezo-electric sources. Schissele et al. (2009) describe the use of buried sources and sensors and compare this to surface sources and recordings. The source they used was a 58-second long frequency range sweep (12–300) Hz, repeated every 60 seconds.

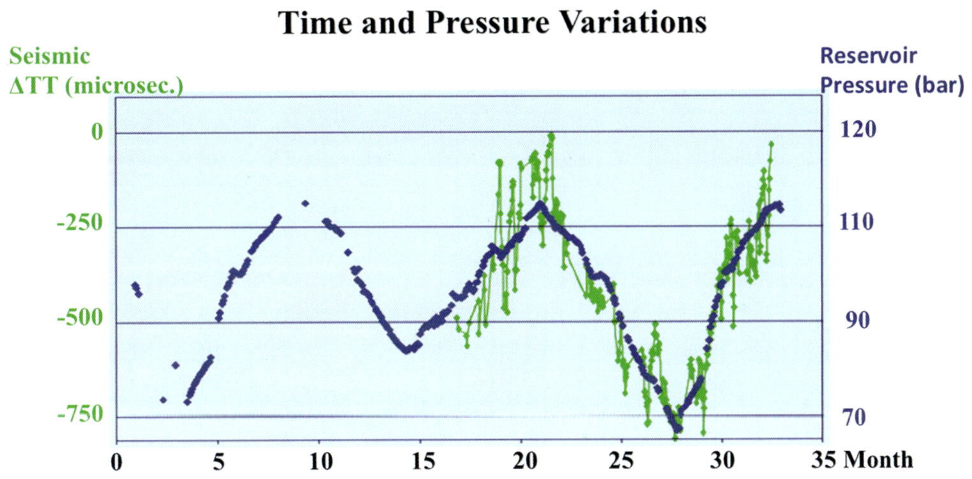

Figure 1.27: Continuous monitoring of the Cere-la-Ronde Gas storage reservoir over 36 months. The changes in traveltimes (green) were obtained by picking horizons above and below the reservoir and following closely the seasonal variations in reservoir pressure Courtesy of CGGVeritas/ J. Meunier.

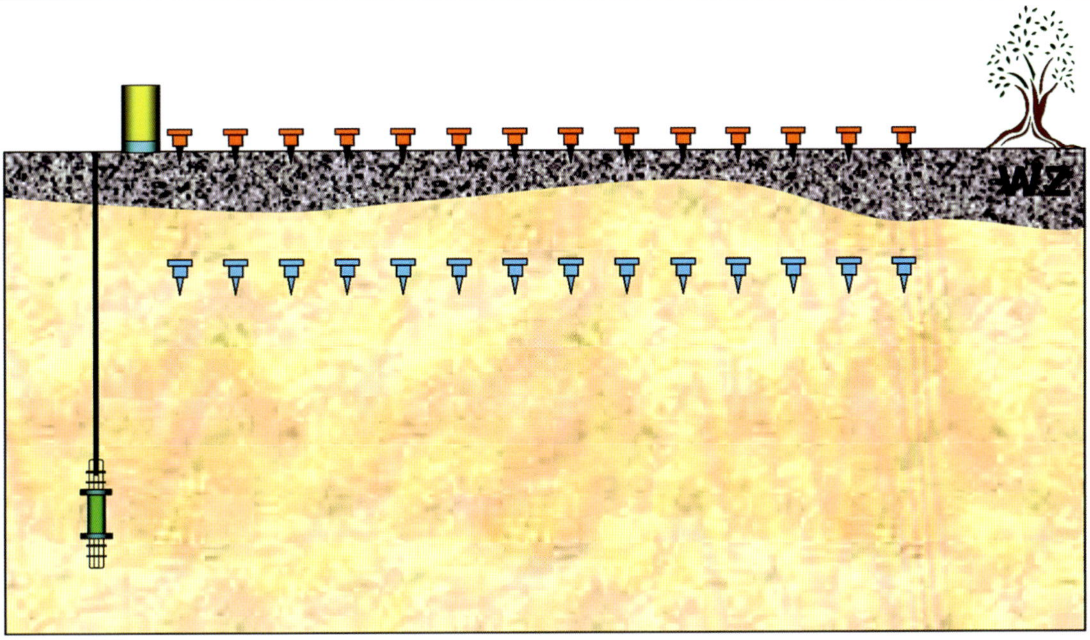

Figure 1.28: Acquisition setup used to record the data shown in the figure below. From Schissele (2009).

To enhance the signal-to-noise ratio, they average the signal continuously over a day. Figure 1.28 shows the experimental setup. The combination of buried sources and buried geophones gave a RMS Time variation in 6 micro seconds and a 0.5% change in RMS amplitude, which is both six times better than can be achieved by using the source and the sensors at the surface. Figure 1.29 shows the impressive repeatability they achieved.

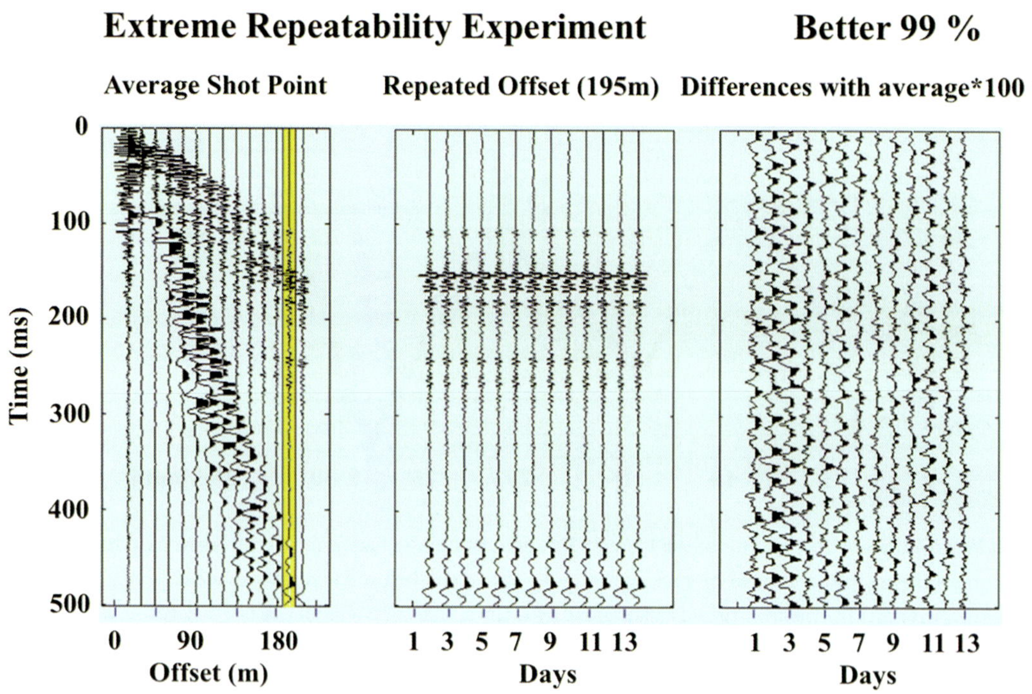

Figure 1.29: Seismic repeatability for the buried source. Left panel: averaging of all shots recorded during the experiment. Middle panel: daily averages of the trace highlighted in yellow in the left panel. Right panel: differences between the daily average and the full average, multiplied by 100. From Schissele (2009).

1.6 Monitoring hydro fracturing

Shale gas has become an increasingly important source of natural gas. Some experts think that shale gases may supply more than 50% of the gas supply in the US by 2020. Shale can contain a high volume of gas, as the porosity can be high, however, it will be hard to produce due to low permeability. To be able to produce large quantities of gas requires fractures; these could be natural but the recent shale gas boom has been due to developments in horizontal drilling and hydraulic fracturing, to create a massive amount of induced fractures around the well. Pumping fluid under high pressure, normally water with some additives, generates the hydraulic fractures in the reservoir rock. Proppant, grains of sands or ceramics, may be injected with the fluid to keep the fractures open. Injection pressure and injection rates are important parameters for "designing" a frack. The pressure can be as high as 100 MPa and fluid rates might reach 250 liter/s. Optimizing the pressure and fluid is important for making the treatment more cost effective, to minimize the possible environmental footprint but also to avoid pumping too much water, which might be trapped in the pore space and effectively reduce the gas flow.

To guide the hydraulic fracturing, pressure rates, additives and advanced numerical models are used. The various types of formation might benefit from different treatment, creating short, wide fractures or long, thin ones. With the right parameterization, some help can be obtained from a simulator model in achieving this. Microseismic monitoring has turned out be useful in understanding the performance of a frack job and in optimizing the parameters for a specific play or field. Concerns about fracturing into the rock outside the reservoir, and possibly introducing an environmental risk, have increased with the increasing activity. This has led to requests for more documentation and maps of where the frack has been going on.

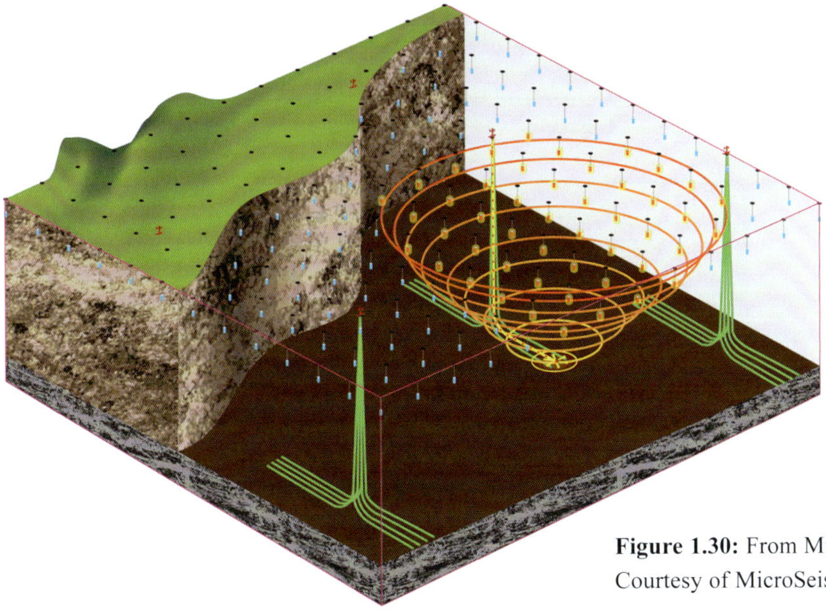

Figure 1.30: From Microseismic Acquisition. Courtesy of MicroSeismic, Inc.

To monitor hydrofrack, onshore-dedicated observation wells or instruments in the treatment well have been the common approaches. However, this may be expensive and also results in biased observations. Recording from the surface will often give the best coverage and be the most effective approach. Recently, it has become more common to view the hydro-frack monitoring in the light of the longer-term needs and systems have been permanently emplaced to monitor acoustic emissions associated with production.

1.7 Conclusion

The early seismic experiences concerned the registration and characterization of large-sized earthquakes. The 4D seismic technology started out by using a permanent seismic array onshore; today we have seen the first full-field permanent systems covering up to 80 sq. km being in placed on the sea-bottom, returning high-quality dynamic images of the changes in the reservoir and its surroundings. As illustrated in figure 1.31, the use and focus started out by aiming at understanding complex enhanced recovery schemes on shore. The technique became very popular during the 1990s, when the technology first of all demonstrated a big economic impact by pointing to bypassed oil or new infill well candidates due to compartmentalization. Since then, the focus has become more important as a tool to understand the depletion mechanism and for water management. Lately, due to the continuous improvement in data quality, seismic time-lapse has become also an important tool for optimizing production and injections in individual wells. The seismic industry was founded on the applied learnings from earthquake seismology. Today, the passive recording of micro-seismicity associated with hydraulic fracturing of oil wells has become a multimillion-dollar industry.

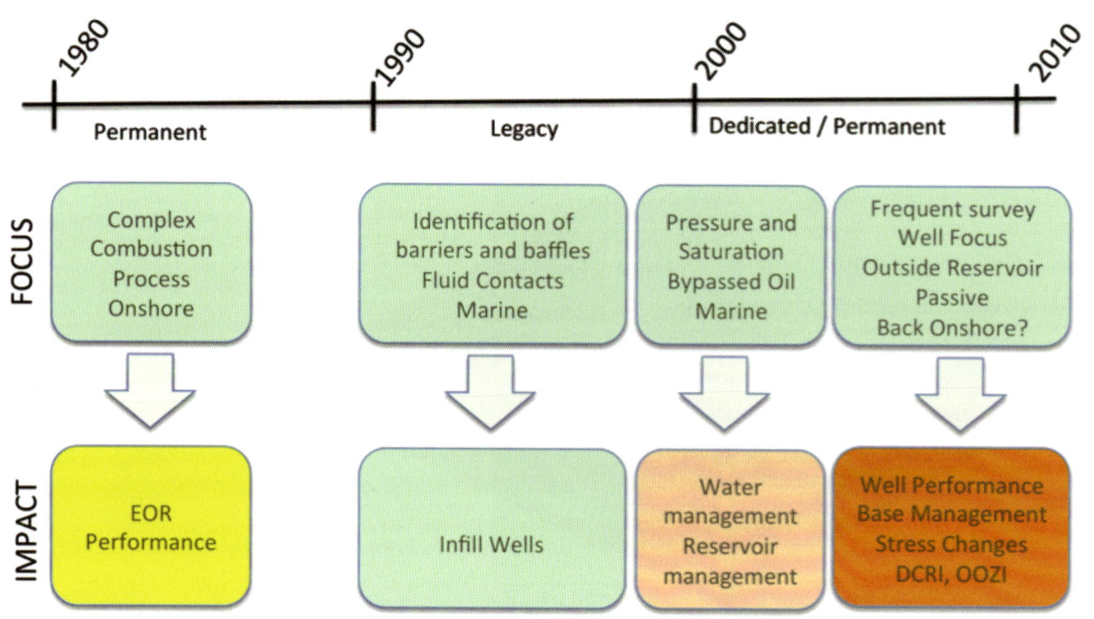

Figure 1.31: Development of Seismic Time-lapse projects.

2 Introduction to the case history

"I want minimum information given with maximum politeness." - *Jackie Kennedy*

We will be using data examples from the Valhall field for the remainder of the book. In this chapter I will provide some relevant background information. The Valhall field is an oil-bearing carbonate reservoir; and is a reservoir that compacts as oil is produced from it. Some of the production-induced changes observed at Valhall would be considered second-order effects in other fields, however since the geomechanical effects are so strong, this is an ideal case to learn from. The reservoir is also fractured, and difficult to image using conventional seismic. The general view, prior to obtaining the first 4D images across a compacting chalk field, was that these types of field did not qualify for time-lapse seismics. Today we know that a number of fields actually fall into the category of compacting reservoirs, e.g., unconsolidated sandstones and HPHT fields. Another aspect of Valhall, which makes it interesting from a 4D perspective is the low permeability, which "allows us more time" to study the production-induced seismic response. Models are key to predicting and understanding the 4D seismic response and we will briefly discuss the role of seismic and the key models we have for the field: the geological model, the reservoir flow model and the geomechanical model.

The first 4D seismic dataset from Valhall was acquired in 2002, and revealed details even down to the individual perforation levels in horizontal wells.

2.1 North Sea chalk fields

The chalk fields of the North Sea Central Graben have a geologic history that in many ways makes them unique in the world. The sequence of geological events combined with conditions that led to the prolific producers of hydrocarbons we find today. "The happy cojoining of high porosity, overpressure, extremely high saturation of a live oil and fracture permeability makes the fields economic to produce"; Andersen (1995).

The chalk fields of the North Sea are mainly located in the Norwegian and Danish Sectors in the Central Graben area, figure 2.1, with a few fields also located on the UK continental shelf. The fields have generally been characterized as fractured, over-pressured, high porosity chalk. However, subtle changes in deposition processes and rates and burial history can, in some cases, result in a reservoir, but in others, a perfect seal. The awareness that producible hydrocarbon can also be available in lower porosity chalk has grown, as operators have been testing out more and more of the deeper part of their fields. The use of horizontal wells, which dominates as the preferred depletion strategy, has allowed using an extended toe section to appraise the upside of the lower flanks of existing fields and push the presence of the reservoir deeper. Chalk is a carbonate and shares most carbonate characteristics, including that carbonates can function both as a reservoir and a seal, which introduces a range of exploration opportunities.

An engineer used to clastic reservoir behaviors will find that working with chalk reservoirs challenges many of his basic assumptions. First of all, porosity can range up to 50% or more, while permeability could be only a few millidarcies. The grain size of the particles making up the rock is also much smaller, 0.5–10 microns, figure 2.2.

For a long time the common view has been that the 4D seismic methods do not work in carbonates. Although the rock mechanics model clearly showed a reduced porosity during pressure drawdown

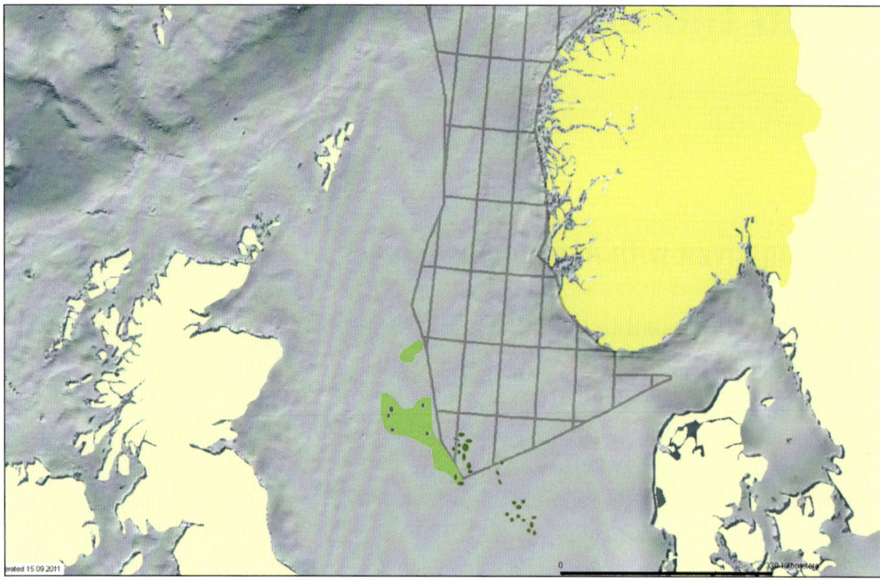

Figure 2.1: North Sea Chalk Field is located in the Central Graben. (Map for NPD.no, 1909.2011).

Figure 2.2: Chalk Grains (right) compared with Sand Grains (left). (Copyright BP, used by permission).

and gas coming out of solution when pressure is drawn down below the bubble point, chalk was still a carbonate. When the first 4D images of chalk showed some stunning 4D effects, the common view changed. Now the explanation is that the 4D seismic techniques works so well in chalk because it behaves like an unconsolidated sandstone. The examples presented in the previous sessions underpin that none of these simplifications are very relevant, as the chalk reservoirs with a proven 4D response cover a wide range of static and dynamic properties. A much more constructive approach would be to group the reservoirs according to the nature of the production induced-changes. In chalk we observe compaction as in unconsolidated clastic rock and we observe the effect of closing and opening fractures like in carbonates. A key element is the low matrix permeability; this implies that the production-induced effects move slowly, which allows for better capture of certain phenomena and the possibilities for studying the effects in more detail.

2.2 Reservoir and depletion plan

The Valhall field is an over-pressured, under-saturated Upper Cretaceous chalk reservoir located in the southern part of the North Sea, 290 km offshore southern Norway in 69 m of water, figure 2.3, Barkved et al. (2003c).

The field produces from chalk in the Tor and Hod Formations of Late Cretaceous age. The reservoir depth is approximately 2 400 meters. The chalk in the Tor Formation is fine-grained and soft, with pervasive fractures allowing oil and water to flow more easily. The field was discovered in 1975 and production started in 1982. The field was initially developed with three platforms, for accommodation, drilling and production. In 1996, a wellhead facility, WP, with 19 slots for additional wells, was installed. Bridges connect the four facilities. A water injection facility, IP with 19 more slots, was installed centrally on the field in the summer of 2003 and connected by bridge to Valhall WP. Valhall WP will also use the drilling rig on this facility. The flank development consists of two wellhead facilities positioned in the north and south of the field. The southern facility came on stream in 2003 and the northern facility in 2004. Valhall processes production from Hod and delivers gas for gas lift in Hod. A plan for development and operations, PDO, for Valhall redevelopment was approved in 2007. In 2010 a new combined process and hotel platform was installed. The platforms are now run by electrical power from shore. Figure 2.4 shows a cartoon of the Valhall and Hod Structure and current platforms.

Recovery originally took place by pressure depletion with compaction drive. As a result of pressure depletion from production, compaction of the chalk has caused subsidence of the seabed, presently exceeding six meters at the central part of the field. Water injection in the center of the field started in January 2004, the strategy being to expand water injection to new parts of the field for more pressure support and better displacement of the oil. Gas lift is also important to optimize production in most of the wells. (NPD Fact pages, 11.09.2011).

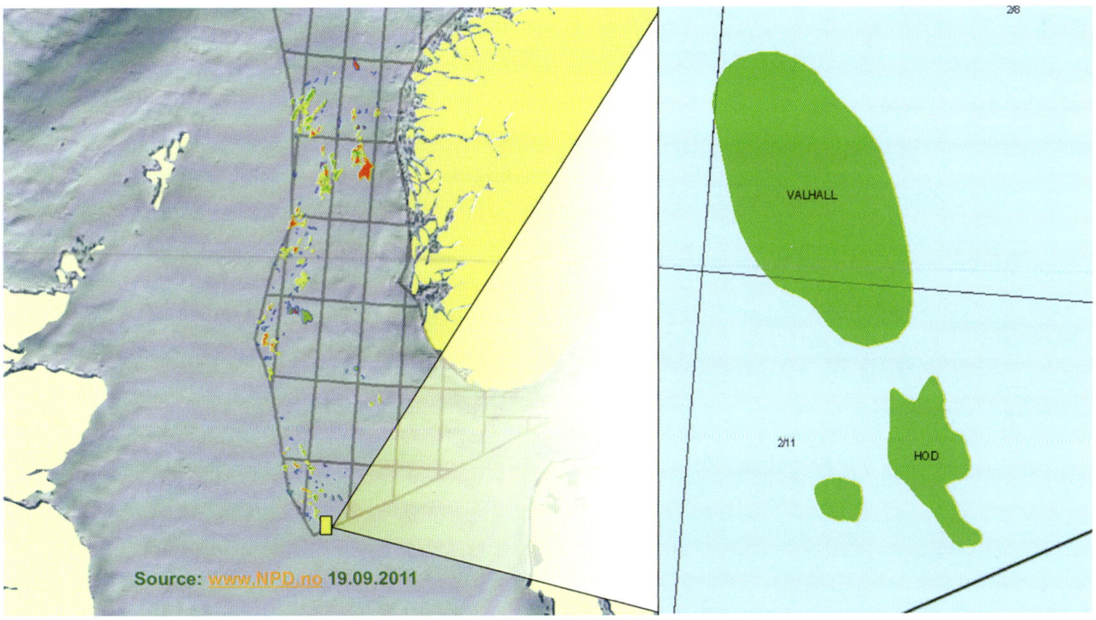

Figure 2.3: The Valhall Field is located in the southern part of the Norwegian North Sea.

The estimated recoverable reserves in 1982 were 247 million barrels, mmboe, which has later been upgraded and is currently 1272 mmboe; this corresponds to an estimated recovery factor of 42%. The estimated production in 2011 is 37 500 boe from 45 producers. The field is expected to produce until 2050.

The application of new technology has been fundamental in bringing the field up to current performance. New completion techniques combined with dedicated draw-down schedules have been established to reduce influx of the weak chalk formations and reduce well failures. Special well design for a compacting and subsiding environment has also been implemented. Drilling of the overburden has been a key challenge. This has been partly mitigated by detailed geological mapping of problem zones *versus* structure, lithology, overpressure and well-bore stability analysis, but has also resulted in replacing extended reach wells by the installation of new flank drilling and production facilities. Since 2006, the water flood scheme has been in place.

Due to the compaction in the reservoir and subsidence in the overburden, geo-mechanical modeling is a fundamental tool supporting long-term depletion and development strategies. Modelling of overburden stress has had a significant effect on overburden drilling successes.

Seismic imaging has generally been a challenge at Valhall due to the presence of hydrocarbons in the overburden. Use of 4D seismic technology has improved this. In addition, the 4D seismic response appears to be very sensitive to production-induced depletion, compaction of the reservoir and subsidence of the overburden and seafloor. Seismic techniques will be critical in future development.

The Hod field is a satellite chalk field 13 kilometers south of Valhall that was discovered in 1974 and put on stream in 1990. The field was developed with an unmanned production facility, which

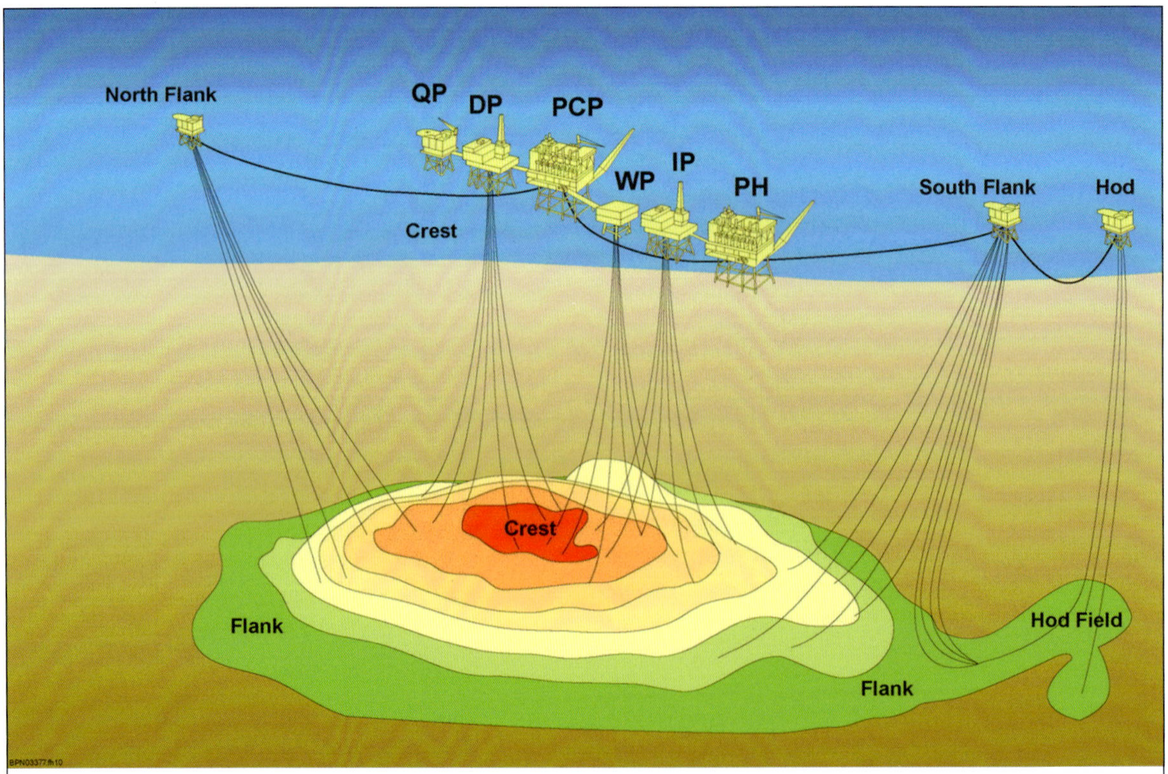

Figure 2.4: The Valhall Field has been developed by a field center, currently consisting of 6 platforms, a northern and southern Well Head platform and a separate platform producing from the Hod field.

is remotely controlled from the Valhall field. The reservoir at Hod consists of chalk in the Ekofisk, Tor and Hod Formations of the Early Paleocene and Late Cretaceous ages. The reservoir depth is approximately 2 700 meters. The field consists of the three structures Hod West, Hod East and Hod Saddle. Hod Saddle connects Hod to Valhall and is producing through four wells drilled from Valhall. Recovery at Hod takes place through pressure depletion. Gas lift is used in two wells to increase production. A water injection pilot program was started at Hod in 2011.

The Valhall and Hod Fields are both operated by BP Norge AS. BP Norge AS and Hess Norge AS are the licensees.

2.3 Geological setting

The Valhall Field is a double plunging NNW-SSE trending anticline, figure 2.4. A simplified reservoir zonation is shown in figure 2.6. The primary reservoir is the Tor Formation with a secondary reservoir from a unit within the Hod Formation. The thickness of the Tor reservoir varies from 0 to 80 m and the lateral thickness changes can be abrupt. Porosities vary considerably and could be as high as 50% in the thickest areas. Permeability is generally between 1-10 mD. Hardgrounds or dense zones are commonly found in the more central area of the field and may act as a conduit for water during water injection.

The Valhall Field is at the highest structural position along the Lindesnes Ridge in both the strike and dip directions. The Lindesnes Ridge is a major post-Jurassic structural element, which is characterized by tectonic inversion, Wride (1994). A cross-section across the field is shown in figure 2.5, and a typical log from a well at the crestal position of the field is shown in figure 2.6. The field has the basic characteristic of an inverted structure, where the antecedent structural low has been transformed into a relatively positive feature and *vice versa*. The strong variations in Tor and Hod formation thickness across the structure reflect the various stages of the development of a growing structure. See Farmer and Barkved (1999) for a more detailed discussion.

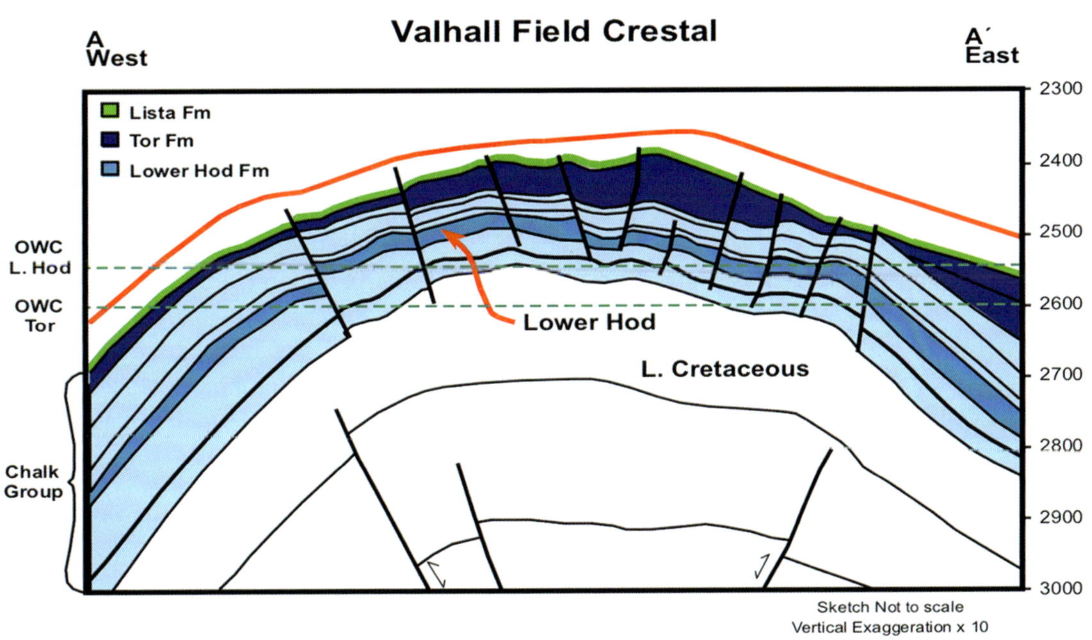

Figure 2.5 Structural Cross/section from Valhall.

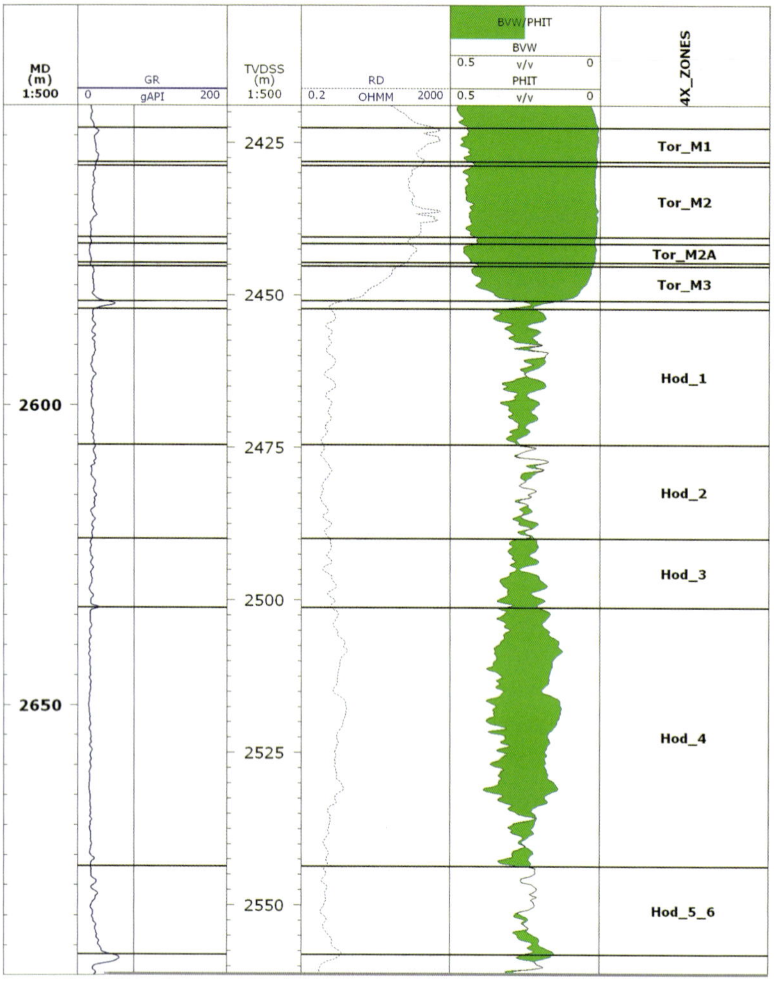

Figure 2.6: Typical well from the crest of Valhall.

Understanding the detail is important for the zonation of the reservoir. Combining graphic correlation techniques, detailed analysis of the biostratigraphy with sequence stratigraphic techniques, the age and the relations between the formations has been established, Mann and Lane (1995). The most important stage is the forming of horsts and grabens in the Campanian. This is a result of extension across the top of the growing structure and subsequent collapses (syn-depositional faulting), forming local highs and lows.

During the Maastrichtian, Tor formation chalk sediments filled the crestal grabens with a combination of remobilized and pelagic chalk. The local highs commonly have a condensed section developed owing to low sediment accumulation. Occasionally with hard ground developing or with a pronounced hiatus in the depositional record, they are considerably more extensive than on the flanks and in deeper parts of the basin.

By the end of the Maastrichtian, most of the grabens had been filled; the structure remained structurally high and provided less space for the younger Paleocene formations. Only a thin condensed section of shale, equivalent to the Ekofisk and Våle Formations, drapes the crest of the structure. Regional thinning of the Paleocene, Eocene and Oligocene intervals across the Lindesnes Ridge indicates that the structural high was present until the early Miocene.

Chalk is mainly built up of coccoliths - fragments of coccospheres from a calcareous planktonic algae, with minor contents of planktonic foraminfera and fragments of larger fossils like bryozoa. The coccospheres are complex structures 2–20 microns across and are composed of individual calcite plates or coccoliths, a typical volume of 0.3 to 1 micron. Andersen, (1995).

Chalk would be referred to as a mudstone in a clastic classification. The non-reservoir quality chalk may contain more than twenty percent volume of clay and clay-sized quartz particles. The amount and the distribution of these particles combined with the packing and preservation of the coccolith fragments (texture), figure 2.2, have an important impact on the dynamic properties of the chalk.

The Tor Formation has been divided into four reservoir zones based on detailed biostratigraphy; Tor-M0, Tor-M1, Tor-M2 and Tor -M3. The Tor Formation consists of chalk deposited in the Maastrichtian periods. The Hod Formation is Santonian, to Turonian age. Between Tor and Hod we might find the Magne formation, which is a relatively clean, reworked Hod age microfossils of Campanian age. The Tor Fm. is bounded by unconformities at the top and base. The most complete stratigraphic section is found at the flanks or in local structural lows.

The source rock for oil at Valhall Field is from the Upper Jurassic Mandal Formation (Kimmeridge Clay). Oil generation occurred during the Early Miocene to the present time.

Above the Chalk, a 1000 m thick Tertiary claystone section of Paleocene, Eocene and Miocene age provides the top seal of the Valhall reservoir, Figure 2.7. Hydrocarbons that have escaped from the reservoir penetrate and permeate micro fractures in some of these claystones and have led to the low seismic velocity "gas clouds" overlying the central parts of Valhall Field. A detailed geomechanical description the Valhall overburden can be found in Kristiansen 1998.

Table 1 summarizes the pertinent reservoir rock and fluid property data for Valhall. The early exploration wells were drilled with water-based mud and development wells were drilled with oil-based mud. For the exploration wells, porosity are calculated directly from the density logs, assuming a limestone matrix of 2.71 g/cc and a fluid density of 1.0 g/cc. For wells drilled with oil-based mud the calculations have been changed accordingly.

Discovery	1975
First Production	1982
Water depth, (m)	69
Reservoir depth, (m)	2400–2600
Initial Reservoir Pressure (psi)	6550
Reservoir Temperature (deg C)	90
Oil gravity, (API)	36
Oil viscosity, (cp)	0.40
Original oil-in-place, (MM stb)	2600
Cumulative oil prod. 31.12.2010, MM stb)	980
Average thickness, (m)	27
Matrix Permeability range, (mD)	1–10
Total Permeability range,(mD)	1 300
Connate water saturation, (%)	5
Porosity, (%)	33–50
Bubble point pressure, (psi)	3000–4000
Solution GOR (scf/stb)	800–1400
Rock Compressibility (10-6 psi-1)	10–100

Table 1: Pertinent Data Sheet.

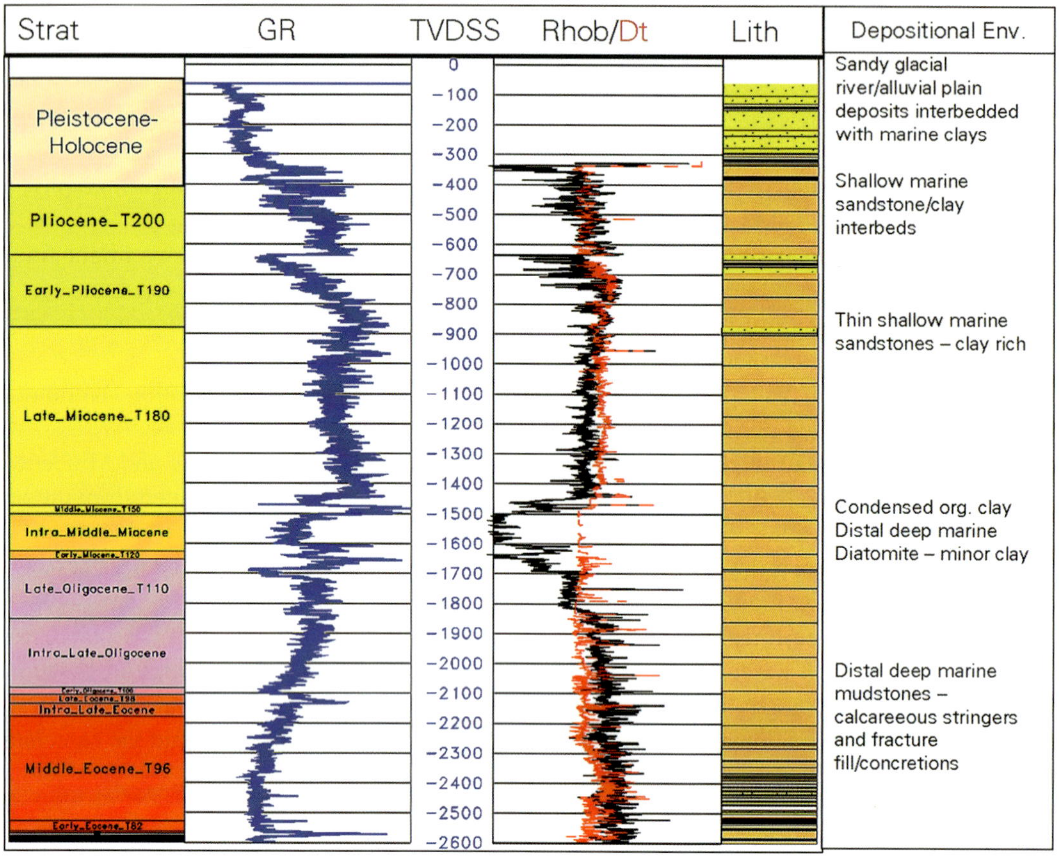

Figure 2.7: Overburden description.

2.4 Reservoir characteristics

The high porosity and low matrix permeability has resulted in relatively low gas mobility. This has been observed in field data, figure 2.8, and confirmed by gas-oil relative permeability measurements. Seismic data are used to guide the prediction of porosity away from the wells. Integrating log with capillary pressure data helped in understanding the distribution of hydrocarbons within the field. As a result of improved porosity and saturation models, new well locations have been identified and drilled.

Early production, rates and decline from the central area of the field reflected that this part of the field was naturally fractured initially, Ali and Alcock (1994). Production tests showed that total permeability was more than ten times matrix permeability. In some parts of the field this initial permeability declined during production and pressure depletion, as rapid decline in rate was seen in build-up response and pressure transient analysis (PTA), Powley et al. 1992. This permeability decline is not uniform across the field. In some areas of the field, or more precisely in the hard ground and the dense zones, fractures tend to stay open.

Resolution and definition of the fracture system using logs was initially very difficult in the poorly consolidated Tor formations. But the development of advanced image logs has changed this dramatically. The modern azimuthal density tools provide vital details, which are especially useful during the drilling of horizontal wells, Barkved et al. (2003c). Reservoir compaction is a major source of reservoir energy and has provided a large boost during pressure depletion. The structurally highest parts of the Tor

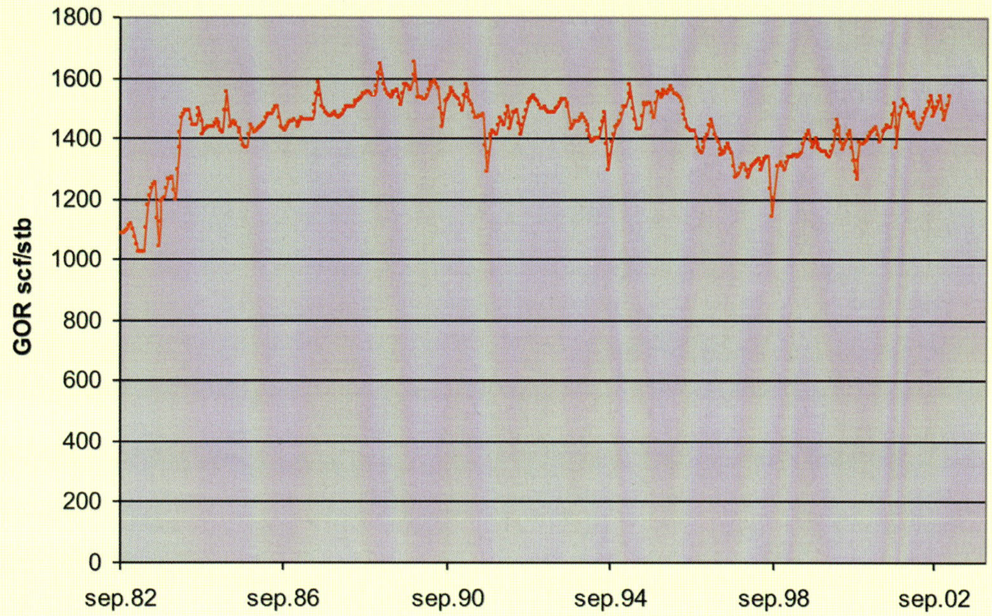

Figure 2.8: Field GOR over 20 years of production history. The average solution GOR for the field is around 1100 scf/stb, with an increase in field GOR of only 35% after 20 years of depletion. From Tjetland et al. (2007).

formation reservoir were around 2900 psi over-pressured at discovery. The pore pressure was initially 6550 psi and with an overburden pressure of around 7200 psi, the effective stress is 650 psi. This is comparable to a burial depth of only 700–800 meters, Barkved et al. (2003c).

Reservoir compaction has been observed by porosity reduction in infill wells and has also resulted in reservoir thickness reduction. Porosity and thickness reduction complicated the geological and rock property modeling. An example of porosity reduction is seen in figure 2.9. The porosity reduction is not 'uniform' and matching porosity reduction with the rock compaction curve has not been straightforward, Halvorsen (1997). Geomechanical modeling has been useful to assess the impact of stress arching and areas with additional strain and hence, higher compaction energy.

Water injection from dedicated injector wells started during 2006 and will initially focus on the crestal parts of the field. Exact incremental recovery from the field is difficult to predict. The early observations are revealing a potentially varying water-flood response in different parts of the crest, with a fractured region having an influence on sweep in the area of one injector and a more matrix-dominated response in another area. Understanding these variations will influence the location and design of future injectors and the expansion of the water injection area, Tjetland et al. (2007).

2.5 Wells and completion

A variety of completion methods have been used on Valhall. Since 1995 the preferred completion method has been horizontal wells that are proppant fractured in the Tor formation and acid fractured in the Lower Hod formation. An exception to this completion technique has been for thin lower porosity Tor Flank wells, where some wells are completed with an un-cemented liner and no stimulation. See Barkved et al. (2003c) for a summary of historic completion techniques.

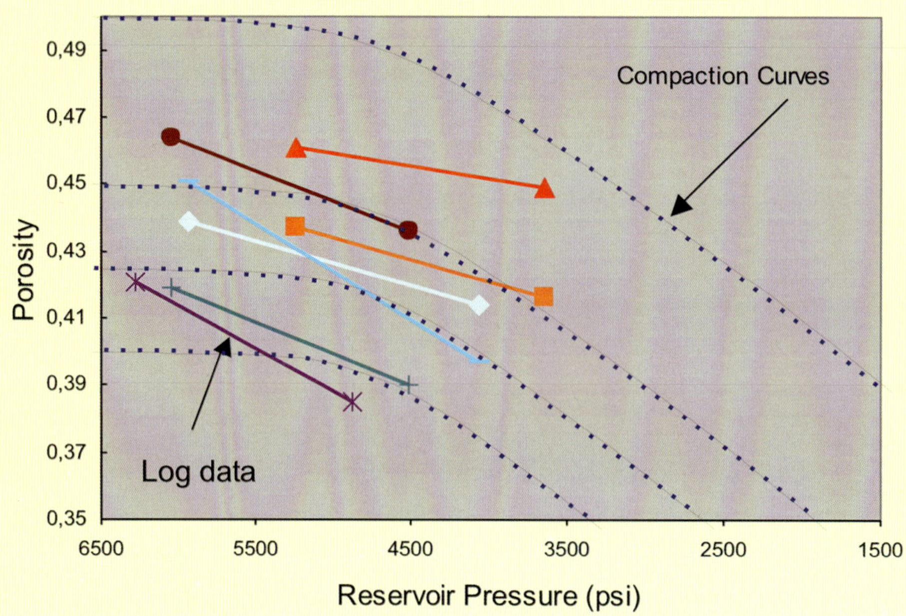

Figure 2.9: Porosity reduction observed when drilling a "replacement well" close to a previously "failed" producer. There is a clear trend of reduced porosity as a function of pressure (or net stress). Modified from Kristiansen et al. (2005).

2.6 Compaction and subsidence

After only three years of production at Valhall, the seafloor had subsided 0.5 meters. Since then the subsidence of the platforms at Valhall has been measured with pressure gauges at the seafloor and infrared sensing, single frequency GPS and currently, dual frequency GPS on the platform.

Figure 2.10 shows the subsidence data at Valhall from the initial measurements until today. Until 2002 a relatively constant subsidence rate of around 25 cm/year was measured. Since then, the observed compaction has followed a lower trend, currently around 11 cm/year. The subsidence rate has partly been arrested by the injection of water.

Laboratory tests show that the chalk will compact as the pressure decreases. The compaction results in subsidence in the overburden and eventually of the seafloor.

Compaction has been measured in the reservoir using wells instrumented with radioactive bullets. A baseline log was run before start of production and monitoring logs have been run during the depletion of the area. Compaction is a result of deformation of the fine-grained structure. Due to their complex microstructure, chalks are very strain rate sensitive materials, but this strain rate can also change contingent on the stress path the chalk is exposed to. Andersen (1995) provides more details and references for this complex behavior.

The impact from compaction at Valhall is both positive and negative. The compaction provides reservoir drive energy, which significantly helps in improving the recovery from the field. The value of the added production far outweighs the negative effects, which are chalk production, well failures and impact on infrastructure.

Mitigating for the negative effects of compaction has been a major activity since production started at Valhall. The initial focus was to complete the wells such that they produced oil and not chalk. Highly

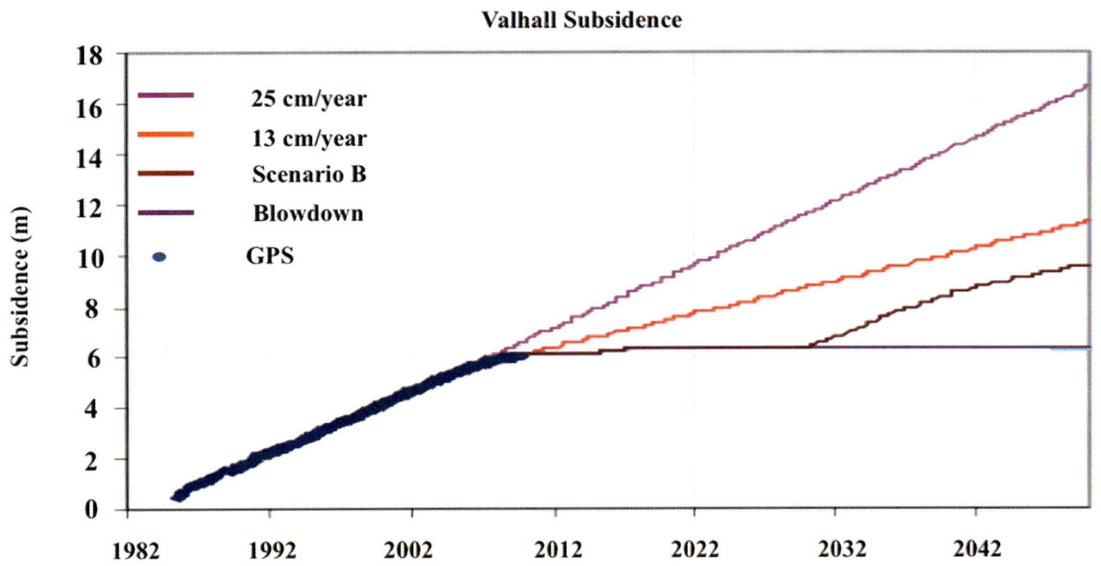

Figure 2.10: Subsidence: actual and predicted subsidence rates for various scenarios. Modified from Tjetland et al. (2007).

innovative completion methods were tested and implemented and these turned out to be essential for making Valhall a prolific field.

Soon, the need for being able to predict sea-level subsidence drove the need for geomechanical modeling at Valhall. Similar effects were also observed in other producing chalk fields in the North Sea. Already in 1980, the chalk field operating companies strongly encouraged by the Norwegian Petroleum Directorate and the Danish Energy Agency had agreed to start up a joint industry program aimed at understanding chalk behavior during production. Rock Mechanics became a major element in this program and the chalk fields became important rock mechanical "laboratories" for the oil industry.

2.7 Reservoir modeling

The reservoir model is the key to developing the field. The model has been continuously upgraded since the first was built in 1981 to incorporate more details and updated as new wells were drilled, Powley et al. (1992). In 2003, prior to the start of the water flooding, a three-phase model consisting of 44×84×12 (44,352) cells with a grid size of 150×150 m was used as the base model. The model included production data from the start of production in 1982 and is based on information from more than 200 well penetrations, of which 130 are being used in history matching. The layer is based on a geological zonation, where layers 1-6 represent the key reservoir interval, the Tor Formation.

Parts of the crestal area of the field showed initially 3–10 times enhancement of the matrix permeability, probably due to pervasive hairline fracturing, Powley et al. (1992). To match the performance of the producing wells, all reservoir models since the late 1980s have had an elongated "engineering zone" of high permeability in the crest of the field. In the preparation for water flooding, the nature and the presence of these enhanced systems needed some basis. A number of different geological fracture realization models were built and tested in the simulator and the most likely scenario appears to be a combination of fault induced fracturing combined with the presence of extensive dense zones, which occurs at the base and in the middle of the Tor formation, figure 2.11. Buer and Trice (2004); Olsen et al. (2005).

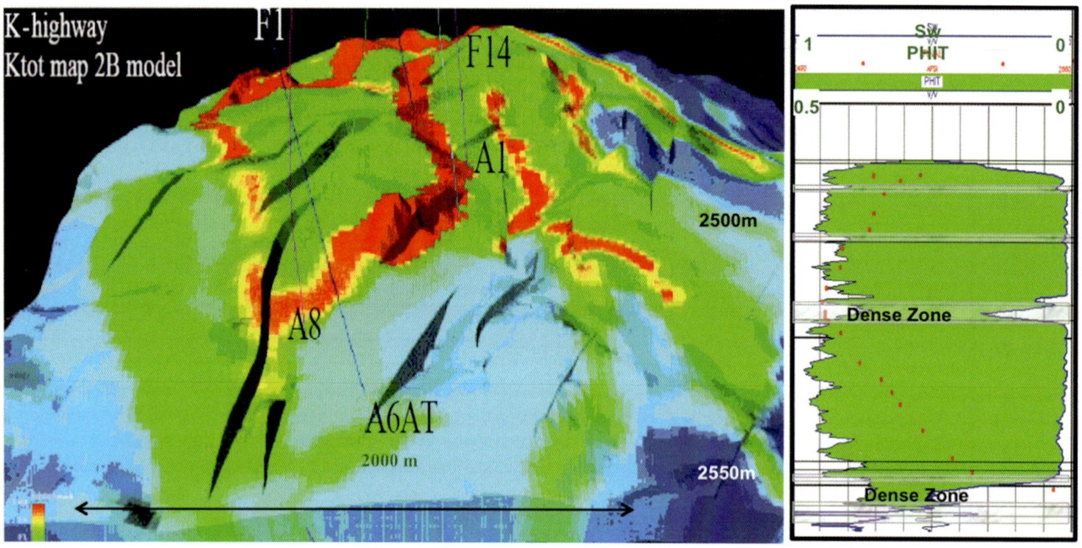

Figure 2.11: 3D model showing the significance of the permeability highway in the crest of Valhall. Red-orange color indicates an area of high permeability and blue color areas of low permeability. Right: Open hole logs, the pressure measurements (red dots) indicate influence of a dense zone as a high conductive layer. Modified from Tjentland et al. (2007).

Mechanical compaction represents an important source of energy in the model. Rock-compaction tables in terms of pore volume multiplier as a function of pressure, determined from the porosity-dependent vertical stress curves, are used to simulate compaction in the flow model. The model has also incorporated functionality for chalk weakening by water. The hydrocarbon fluid is modeled as a single oil type but there are variations in the bubble point between the regions. Special core analysis laboratory measurements, SCAL data, from a number of wells are the basis for the relative permeability of the model. Pseudo curves generated by a fine-scale dual porosity model are used for the highly fractured and the intermediate fractured area. The curves for the non-fractured area are up-scaled matrix curves.

Even with the high number of well penetrations, the uncertainty of the model has been considerable. One reason for this could be the lack of correct representation of the additional energy from compaction-induced overburden loading effects in the up-scaled compaction curves. The more recent modeling is done incorporating the geomechanical effects though explicit modeling of the overburden.

Laterally large variations in thickness could also easily create baffling or compartmentalization effects, which are difficult to nail down precisely without, e.g., 4D images. As more information becomes available, it will be natural to revisit the fracture models.

2.8 Geomechanical modeling

Various suites of finite element software applications have been used to help the modeling of compaction. In the reservoir fluid flow models; the compaction effect has typically been addressed using pore volume multipliers. These multipliers are based on stress-strain measurements on cores in the lab and adjusted for the strain rate differences between a laboratory experiment and what the rock is exposed to in the field. These models have often been run using the pressure history from the reservoir flow model to evaluate the effects of the deformation. The drawback is that in reality, the effective stress is impacted by the compaction induced subsidence effects from the surrounding rocks and it has therefore become more popular to couple the flow model and the geomechanical modeling to obtain a more correct pressure prediction.

Getting the compaction energy right has been a major challenge, partly due to up-scaling of laboratory core data to reservoir cell size. Other issues are related to the nature and size of the deformation in the reservoir and that a standard reservoir model has not been able to handle the complex load/unload transfers from one area to another, which are strongly influenced by the overburden behavior (stress arching). Figure 2.12 shows the calculated changes in vertical stress between production start-up in 1982 to June 1999 for one of the reservoir layers, Tor M1. The differences easily translated to errors in the pressure predictions of +/– 450 psi over short distances.

Kristiansen and Plischke (2010) describe the development of a full field finite element for the Valhall Field. Herwanger and Koutsabeloulis (2011) also based their example model on a Valhall "look-alike" model. In the following we will have a closer look at what is involved in the process of developing a model.

The Valhall geo-mechanical model serves a number of purposes and is a key tool for almost all field activities that include the subsurface. Geomechanical modeling impacts well design, production, drilling, reservoir and field management. While the impact of rock compaction on reservoir energy might be not fully appreciated, the need for predicting subsidence and its impact on the infrastructure is obvious. The Valhall license partners are just about to complete a major redevelopment, which includes building new platform facilities. One of the reasons for this is the diminishing air-gap between the surface and the lower platform decks. Due to the subsidence, the older platforms have to demobilize personnel during the most severe winter-storms. The old platform will have to be phased out at some point in time. This requires long lead planning and understanding how the subsidence will develop is one of the factors influencing this.

In addition to using the model for seabed subsidence predictions, the model will be used to:

- Guide the interpretation and capture the insights gained from the seismic surveillance data.
- Support well bore stability analysis
- Assist in history matching of the reservoir flow model
- Help to improve porosity modeling from logs run in depleted reservoirs.

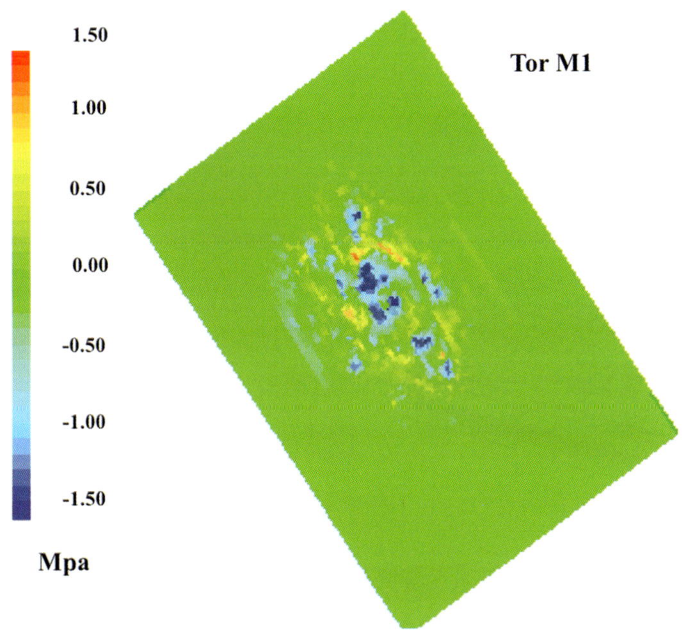

Figure 2.12: How the overburden weight or the vertical load is modified across the field from the initial conditions prior to production start-up in 1982 and up to June 1999. From Barkved et al. (2003c).

The new finite element model consists of 30 layers and covers an area of 16 by 28 km. The 2400 m thick overburden is divided into 11 layers; there are then 13 layers in the reservoir and the under-burden consists of 4 layers and ends at a depth of 4200 m. The overburden definition is guided by a combination of log measurements and seismic observations. The reservoir layers are virtually the same, as is the latest full field reservoir flow model, apart from the finite element model requiring continuous lateral layers. Where the reservoir layers are missing, due to erosion or non-deposition, a 30 cm thick layer is used in the geomechanical model. The model includes 30 of the most important faults. Unlike the flow model, where faults may be "verticalized", the fault dip needs to be accurate in the finite element model, to model possible shear displacements. Figure 2.13 shows an overview of the resulting 3D mesh. The model consists of 176,460 higher-order elements, with 645,631 nodes. See Kristiansen and Plischke for further details of the model building process, parameterization and boundary conditions used.

History matching is an essential part of qualifying the model for future predictions. This includes testing a significant set of parameter variations. The prediction has been matched against measured platform subsidence, actual seabed bathymetry, 4D seismic, measured strain across radioactive markers in wells and stresses in the reservoir. A GPS sensor was installed at one of the central platforms as early as 1984. This was upgraded in 1994 and is now using a dual frequency GPS. The match, as can be seen in figure 2.14, is impressive.

Results of bathymetric surveys are available for 1978, 1992, 2001, 2002, 2003, 2004 (3), 2005(2), 2006 and 2007. These surveys do not have a local accuracy such as platform subsidence measured by GPS, due to varying wave height, higher order tidal variations, etc., but reflect their areal variation, which is interesting. Again, figure 2.15 shows very good comparison between observed and modeled results. The slight deviation in shape will probably be corrected by a refinement in reservoir faulting, which has recently been revealed.

From seismic time-lapse data we can derive an estimate of the compaction at top reservoir level. The central part of the field suffers from poor imaging owing to gas in the over-burden. For improved quantitative analysis in the future, a shale rock physics approach will be used, as presented by Fjær and Kristiansen, 2009. For this comparison we have used a linear (R-factor) approach, which might not be

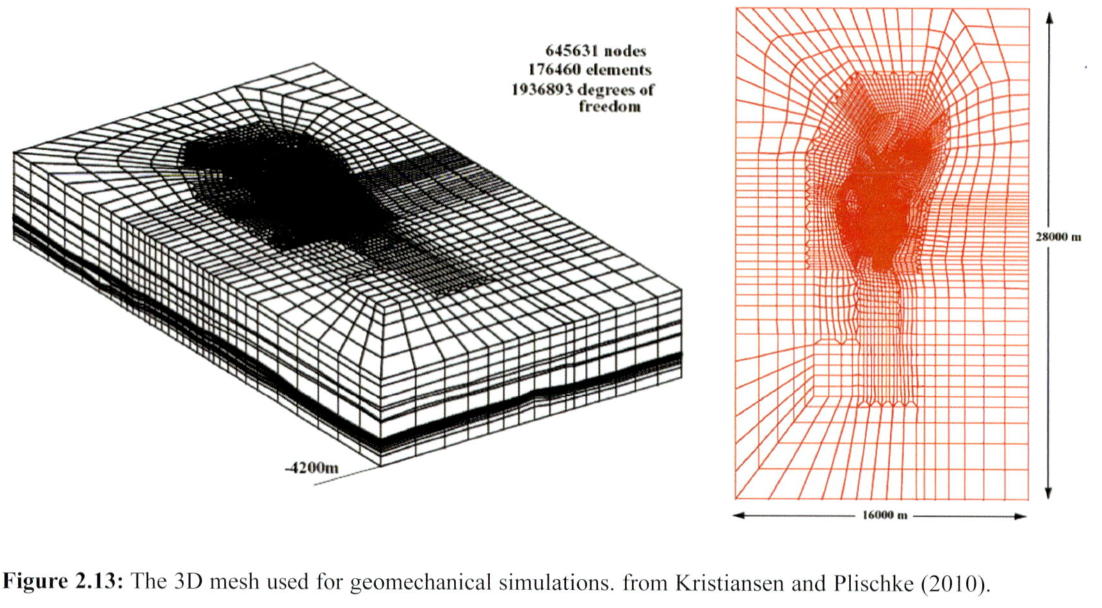

Figure 2.13: The 3D mesh used for geomechanical simulations. from Kristiansen and Plischke (2010).

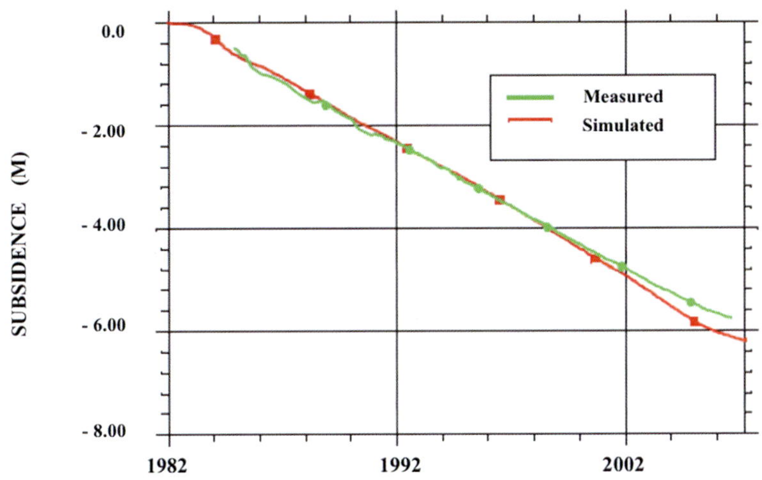

Figure 2.14: Measured versus simulated platform height, from Kristiansen and Plischke (2010).

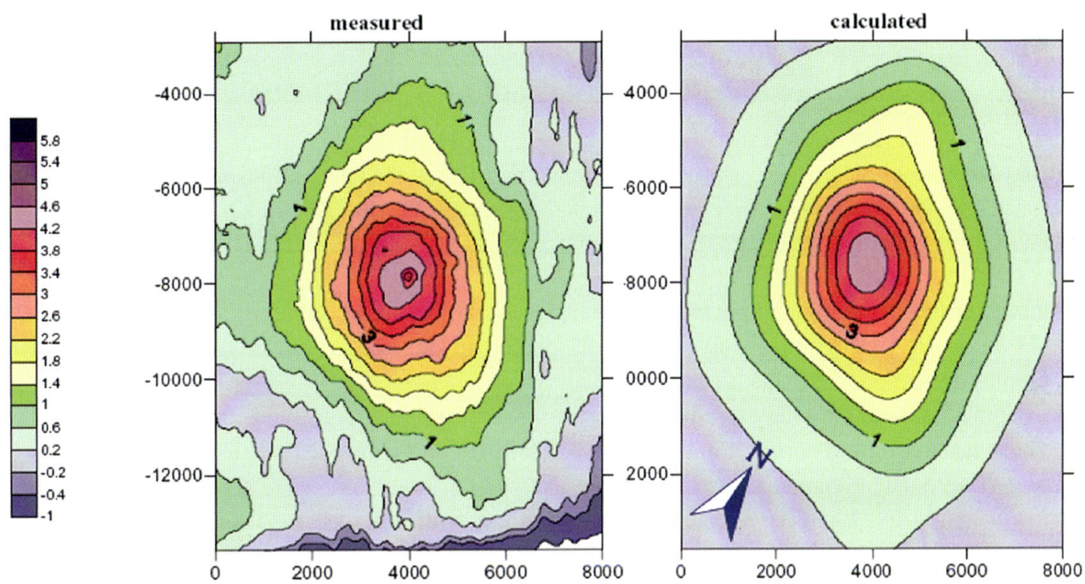

Figure 2.15: Comparison between observed and modeled seabed subsidence, from Kristiansen and Plischke (2010).

valid at the edges of the thickness comparison. Still, the overall match is very good, Figure 2.16. The details of the crest have been ignored due to poor seismic imaging, however with new, improved processing and a velocity estimate, we have been able to refine the dynamic response. This indicates that there are structural elements that have not been included in the models. Most likely, with these included the match would have been even better.

In a couple of near-vertical wells, radioactive markers were shot into the formation before installing the casing. A baseline log was run; subsequently monitor logs were run after a period of production. These markers were installed in and above the reservoir. Especially across the reservoir section there is very good agreement between model and prediction. The level of detail built in the models most likely helps with this match. The reservoir flow model will not predict the compacting volume correctly in

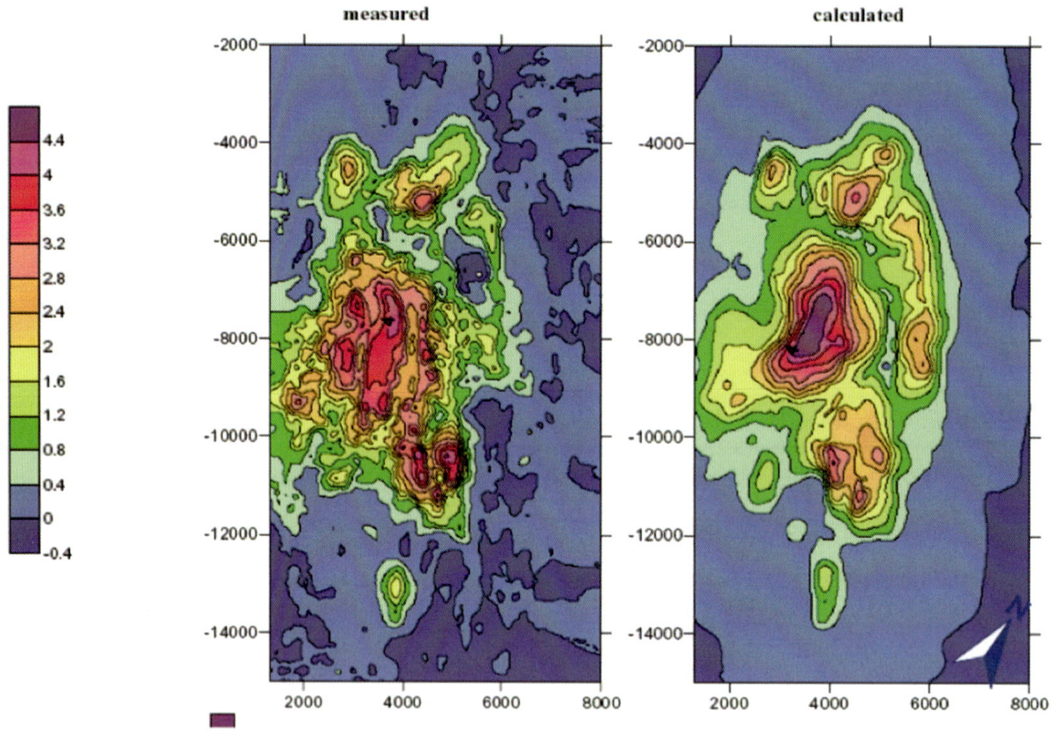

Figure 2.16: Comparison between compaction estimated from 4D seismic (1992–2002) and vertical strain calculated from the model for the same period, from Kristiansen and Plischke (2010).

detail; however there are some expected consistencies in the variation, which may be used to check the model. The porosity reduction observed in wells drilled in depleted areas is another possible history matching parameter.

The geomechanical modeling predicts interesting changes in magnitude and location of vertical stress. There is a very good correlation between well failures and where the vertical stress anomalies due to stress arching occur, Kristiansen (2007).

Various sensitivities have been tested using this model. Introducing anisotropy into the overburden provides a slightly better match, while possible drainage from the overburden shale will not make significant enough changes to conclude based on prediction *versus* observations. Comparing the synthetic seismic, using the more rigorous approach discussed by Fjær and Kristiansen (2009), they are also in agreement that including anisotropy is the right approach.

Drainage of the cap-rock has also been evaluated. At around 30 m above the reservoir there is very little change in pore pressure. Still, the pressure depletion in overburden shale could impact drilling and indications that there are changes impacting the fracture gradient and the mud weight needed to avoid borehole collapse has been observed when drilling, Kristiansen and Plischke (2010).

2.9 Role of seismic data

The seismic data provide the structural framework; thickness variations, fault definitions, subsurface topography and porosity trend needed in reservoir modeling. Due to the abrupt lateral variations in thickness of a relatively thin reservoir, seismic data are critical for detailed well planning and the safe drilling of horizontal wells needed to drain the field economically.

Weak rocks and the presence of gas limits the operational "mud" window in the overburden. This is further complicated because the reservoir is compacting during production, creating stress redistributions above the reservoir. Seismic data are important in defining the complex fault patterns in the overburden, which, if poorly understood and combined with well angle and the issues mentioned above, could increase the cost of drilling a well significantly.

The presence of the low velocity gas charged tertiary overburden and especially the highly porous diatomaceous Miocene interval creates a "gas-cloud" effect across the central area of the field that distorts the seismic images and makes it difficult to image the top of the reservoir in the crest with conventional techniques. Figure 2.17.

Outside the "gas cloud" area the seismic quality is very good and the seismic data were vital in developing the flank areas. Barkved, (1996).

Advanced processing of conventional 2D seismic combined with the extensive use of VSP to undershoot the gas cloud were the seismic tools used during the first ten years of production, Leonard and Munns (1987). When the horizontal drilling program started in 1990, the need for detailed lateral seismic coverage became urgent. Horizontal wells made the thinner flank area of the field economic. In this area the seismic data quality was very good and the first 3D seismic data set was acquired in 1992. The 3D seismic data provided a more detailed definition of structural elements and helped in identifying the upside area by the given alternative context for earlier well observations. For just one of the areas of the field, this resulted in 40 MMstb oil in added reserves. The increase in reserves was identified by the refined 3D seismic images and proven by drilling and later production for new horizontal wells drilled into the area.

The step change in detailed reservoir characterization from 3D seismic coverage stimulated an appreciation of the value of seismic data and raised the need for technology that could provide seismic images of the reservoir beneath the gas cloud. An advanced pre-stack imaging scheme was tested and alterna-

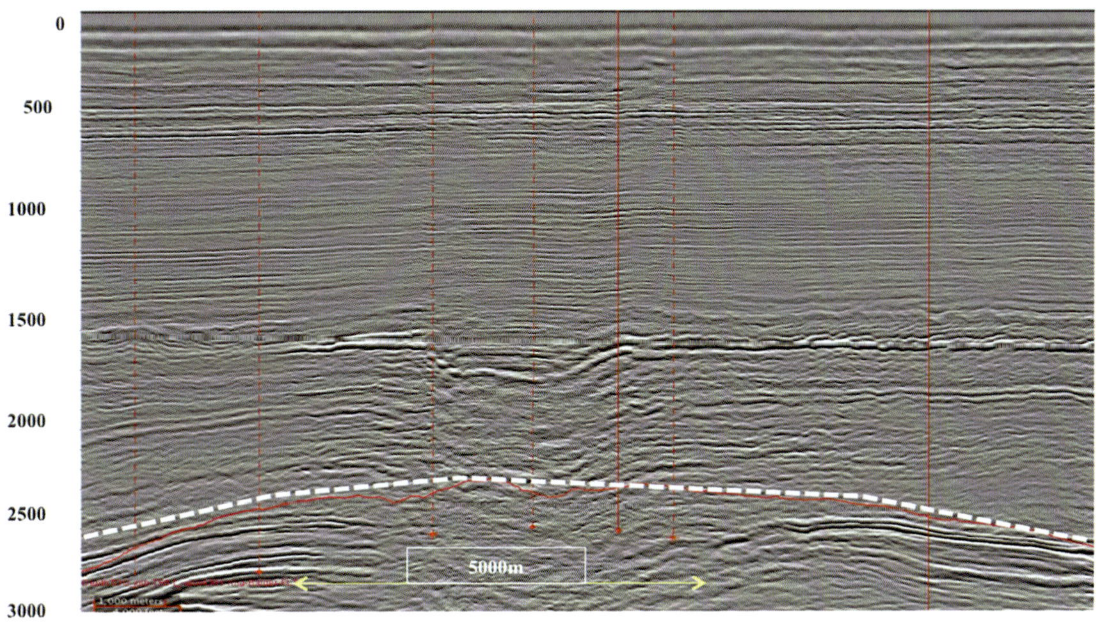

Figure 2.17: The presence of the low velocity gas charged tertiary overburden creates a "gas-cloud" effect across the central area of the field, which distorts the seismic images and makes it difficult to image the top of the reservoir in the crest with conventional techniques.

tive 3D seismic acquisition, placing the receivers in the wells, 3D VSP, was also considered. In the end, the partnership agreed to look more closely into the possibilities inherent in the emerging 4-component ocean bottom seismic (4C OBS), which also allowed for imaging using shear waves, which are "not" impacted by the fluid in the rocks.

The Valhall field became one of the key assets for demonstrating the viability of imaging below gas using converted shear waves recorded by 4C sensors on the seafloor, Thomsen et al. (1997). Converted (shear) waves are reflected shear energy from down-going compressional (p-wave) seismic energy, at the acoustic interfaces in the subsurface. One of the world's first 3D 4C seismic datasets was acquired at Valhall in 1997, Rosland et al. (1997), Barkved et al. (1997). Images constructed from this dataset were critical for planning and drilling several of the subsequent wells drilled in the field, Figure 2.18.

The 3D/4C OBS data acquisition also provided a wide azimuth P-wave data set and was analyzed for Amplitude *versus* offset and azimuth, AVOA, effects. These analyses revealed azimuthal anisotropy effects, which showed some correlations to existing fault patterns, Hall et al. (2002a).

The first attempt to produce a 4D image from time-lapse 3D data at Valhall was done in 2001, Hall et al. (2003). The two 3D datasets considered in that study were the 1992 Towed streamer (TS) 3D seismic volume and the 1998 3D/4C OBS data. The two datasets were almost as different as they could be. The 1992 data were acquired using two streamers and a single source. The OBC data were acquired on a grid rotated by 25 degrees to the TS grid. There was no attempt to match the surveys during the pre-stack processing. The 3D OBC data were acquired using patch shooting with orthogonal source and receiver lines with 600-meter line spacing and 25 m in-line source/receiver interval. The same velocity field has been used in the migration of both data sets and was conditioned on both volumes. However, the different acquisition apertures and the different azimuths relative to the subsurface structure remained as

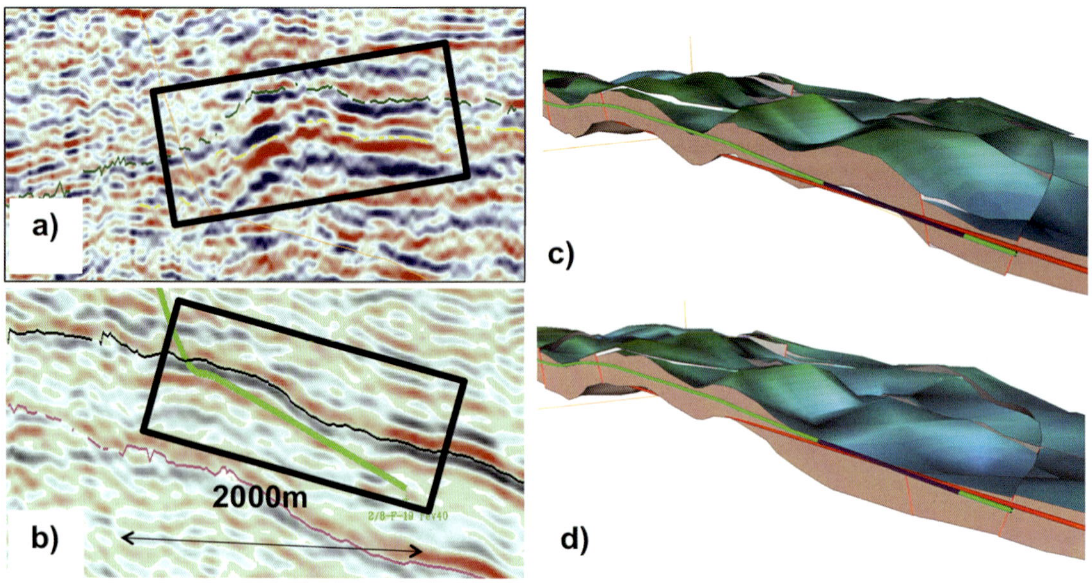

Figure 2.18: Converted wave images were critical in drilling several wells. a) P-wave section in P-wave time, the data was of little help in defining the entry and the first 500 m of the well (black box) b) C-wave in C-wave time provided enough confidence to drill the well. As we see, the prognosis from the C-wave in c), matched the final trajectory in d fairly well). Barkved et al. (2004).

a difference between the two 3D volumes. A cross-matching scheme following the approach of Rickett and Lumley (2001) was used. This included survey re-alignment, bandwidth and phase equalization and amplitude balancing to the same energy level.

To resolve the spatial time-shift between the two surveys, we used a 3D warping technique. The resulting warping vector was then utilized to interpret areas of compaction and induced stress state changes in the overburden. The resulting vertical vector (time-shifts) appears to be consistent with the production data, figure 2.19. The red color could reflect stress pillar effects but appears too high and most likely this is a datum calibration problem. It is interesting to see that we recovered a consistent signal inside the gas area (central blue area). The guiding surface is almost impossible to interpret, so we used well control and velocity data to construct the surface in this area.

In 2002, a second towed marine 3D streamer data set was acquired. This was matched to the 1992 3D and the resulting 4D images demonstrated "stunning" 4D responses, clearly defining how wells are draining the field, Figure 2.20, Barkved et al. [2003b]. The observations from the 4D seismic soon become essential for planning and drilling new wells.

Drill cutting from new wells being drilled is normally slurrified and injected in the formation above the reservoir. Regular waste disposal started up in 1992 and more than nine MMbbls were injected down hole at Valhall, prior to the 2002 TS 3D seismic being acquired. The DCRI is injected in batches of 2–3,000 bbls at rates of 5–8 BPM. Close monitoring and detailed operational guidelines are critical for managing the DCRI wells, which are needed to support drilling and production activities.

In Figure 2.21, we have an example from a non-productive area of the field, where a well drilled to test reservoir presence was converted into a drill cutting re/injection well (DCRI). Pressure and temperature monitoring are used to ensure that the drill cuttings are confined at the planned interval. The

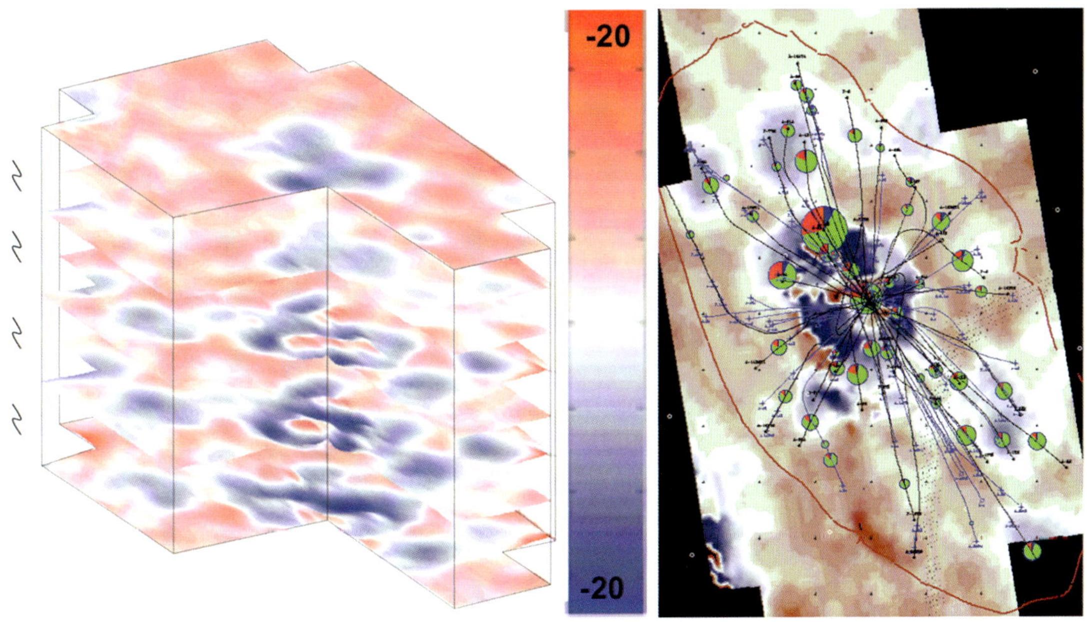

Figure 2.19: Left: stack of warped surface with vertical time-shift posted. Middle: red indicates lower traveltime down to surface from monitor, blue larger. Right: Time-shift from warping vector. Blue indicates lower velocity /unloading in the overburden. The Green-red-blue bubble indicates production; the size of the bubble indicates volume, color fraction of oil (green), gas (red) and water (blue). From Hall et al.(2003).

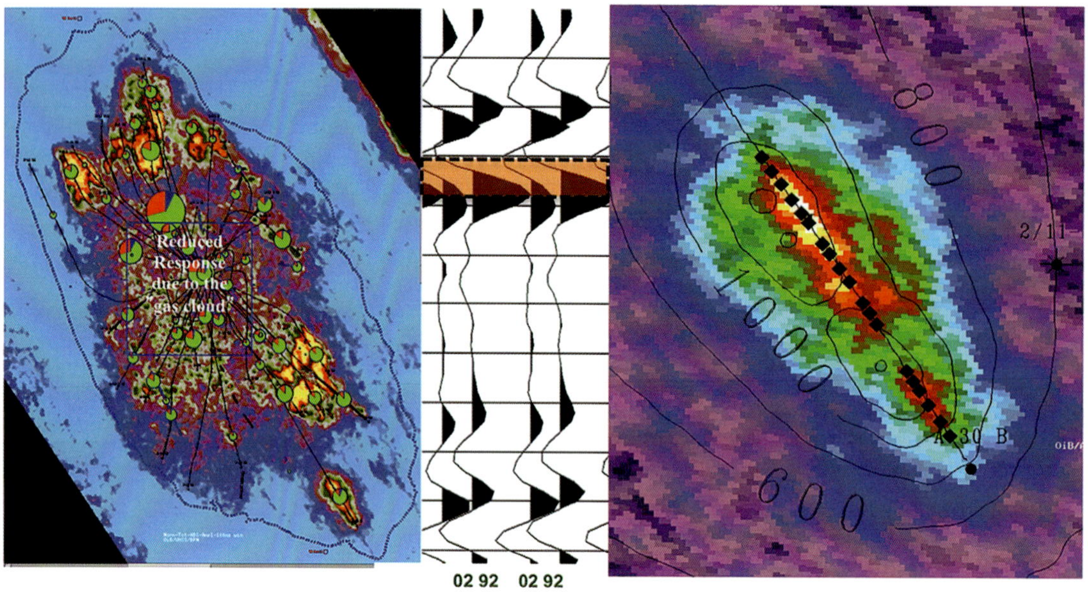

Figure 2.20: Left: map showing the lateral variation in 4D responses. In the middle, interlaced traces from 1992 and 2002 show the effects at the reservoir, before and after six years of production.The reservoir is shown by the transparent orange box. The image to the right shows the details from one well. The black dots represents the perforation points and the contours refer to drawdown from the reservoir model. Both maps show the combined response of a time-shift and an amplitude change. From Barkved et al. (2005).

4D response from this area reflected that the changes have been confined within the planned area. At a later stage, when more detailed 4D images become available from the LoFS array, we observed that the DCRI area was in fact confined in compartments by lateral heterogeneity and the lateral extension was more complex than the initial model. This model also happened to be confirmed by a new well, which was drilled though the area *en route* to a sweet-spot northwest of this area.

In 2002 the Valhall partnership approved the installation of the first field-wide permanent seismic installation at Valhall. More than 2500 four-component sensors were put on the seafloor and the partnership committed a work program of six surveys to be acquired. The primary objective of the Valhall Life of Field Seismic (LoFS) program was to monitor the water injection program, but the first surveys were to be used to establish the "technical" limits of such a system for future acquisition. See own section on LoFS installation and operation.

Other seismic techniques such as coherency processing and spectral-decomposition have been used to extensively define faults in the reservoir. Today, these techniques are complemented by curvature analysis. Dedicated attribute analysis technology helps in fracture characterization and in articulating structural details, which are important for detailed well planning, when a rigorous and detailed 3D model of the subsurface is constructed to guide optimal drilling of a well. Curvature analysis of time-lapse surveys has lately been suggested in various technical forms, but no results have been published so far. Combining 4D seismic images with curvature coherency maps, to verify structural consistency is commonly used, but the "edge" attribute map is then represented as static information.

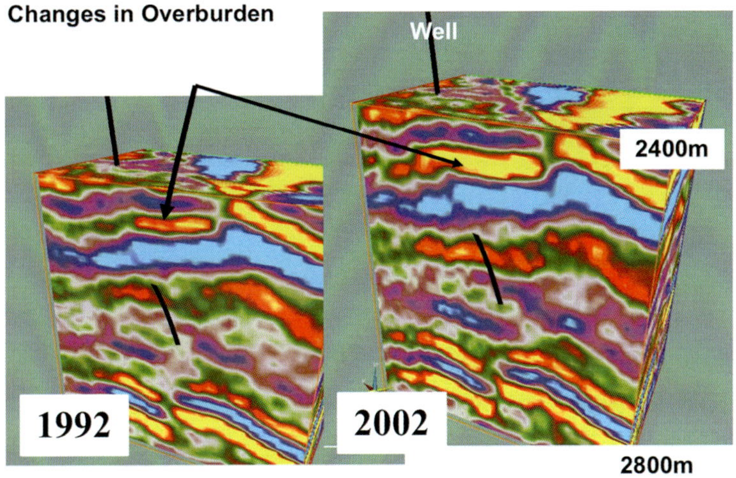

Figure 2.21: Apparent brightening of the reflector above the reservoir supports the interpretation that the "slurrified" drill cutting is injected into the cap rock, Paleocene shales; in a non-productive area of the field. From Barkved et al. (2005).

2.10 The Valhall imaging challenge

Pre-Stack Depth Migration (PSDM) is required to obtain optimal quality seismic images in a structurally complex subsurface. Gas charging in the overburden, of the types we see across several fields of the North Sea Central Graben, typically results in strong lateral variations, which are addressed more correctly with depth processing. Before the millennium, 4D processing was almost without exception based on time processing (PSTM). Depth processing was clearly more expensive and most of the fields studied were not obvious candidates for 4D depth processing. With the commitment to install a permanent array, with the primary objective to monitor a water flood program, where most of the wells were in seismically poorly defined areas, Barkved et al. (2003a), proposed to test whether PSDM would add value in the poorly imaged waterflood areas.

With a 4D towed streamer data set available, a new monitor survey in 2002 and a legacy data set from 1992, we initiated a project to test the viability of the PSDM, Askim et al. (2004). The pre-processing for both surveys was the same, de-signature, standard editing, tidal corrections and tau-p predictive deconvolution. The PSTM workflow was done in an in-house processing center operated by CGG and included more rigorous cross matching between the surveys. For the PSDM processing, a new densely sampled (25×25 m), gridded velocity model was established using tomographic inversion.

Figure 2.22 shows a comparison of the PSDM and the PSTM amplitude difference results. The displays show RMS amplitude extracted along a horizon at the top of the reservoir. The general noise level in the PSDM image seems slightly higher than in the PSTM image. The primary reason for this is that the PSTM workflow also included a more rigorous post migration and a dedicated post stack matching process, which was simplified for the PSDM version. Looking in more detail at figure 2.22 and 2.23, the area marked (1) shows a large production-related 4D response. The anomaly has a similar shape and magnitude in both versions. However, the anomaly extends somewhat further south in the PSDM image, reducing the effect of the gas cloud. Area (2) shows an interesting 4D response. Significant production in this area took place before the first survey in 1992, which pre-depleted the central area, leading to the doughnut-like shape in the difference map. The anomalies are very similar in both maps but if anything, there is possibly a better consistency between amplitude anomaly and structure for the PSDM version, as seen in Figure 2.22, area 2, which shows a better alignment between anomaly and well location, most likely again due to depth migration better handling lateral

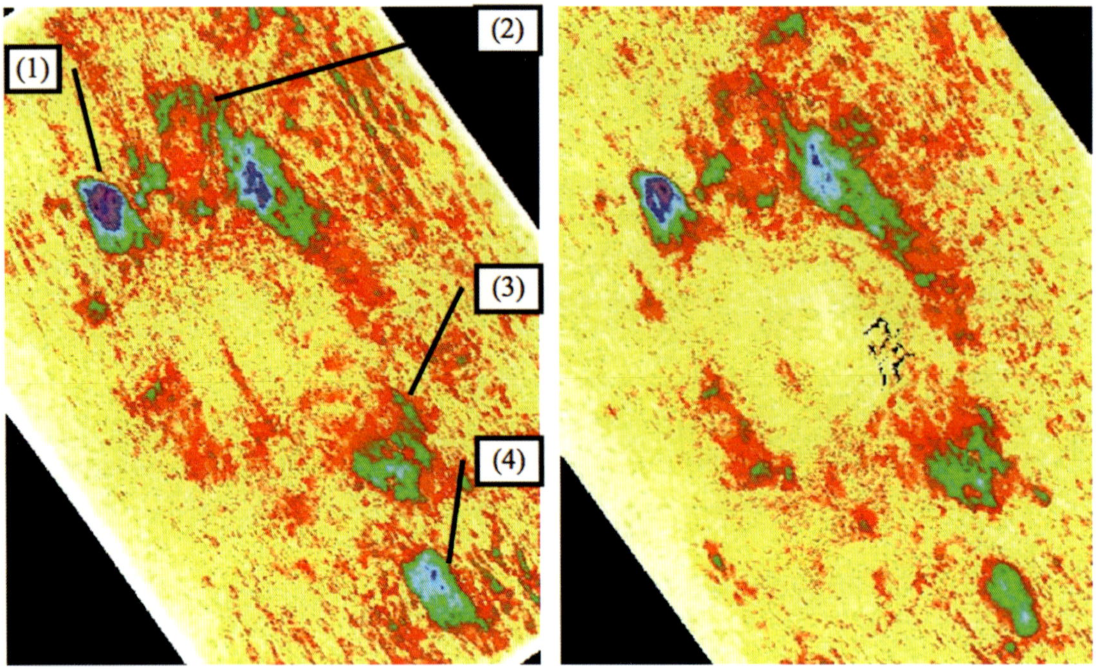

Figure 2.22: Amplitude difference from PSDM processing (left) and PSTM processing (right). From Askim et al. (2005).

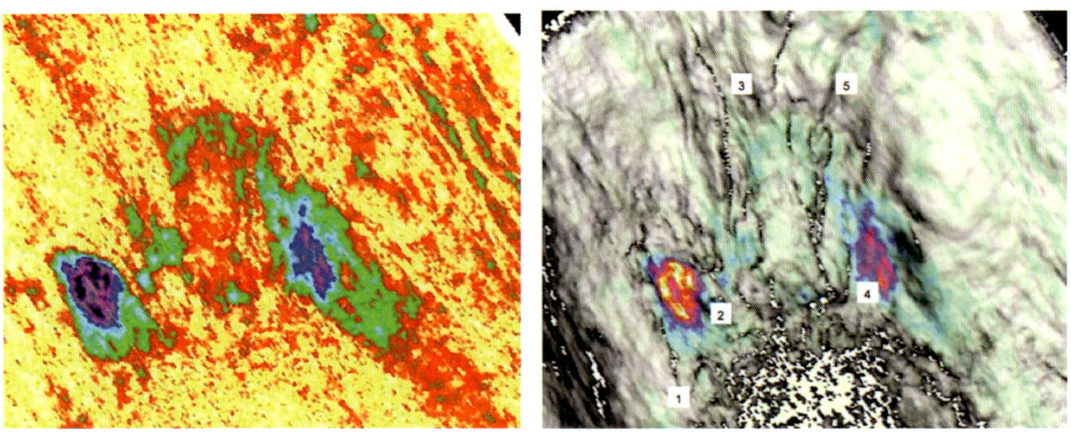

Figure 2.23: Left Amplitude difference from PSDM showing the north flank of Valhall, Right: same 4D attribute co-visualized with coherency. Askim et al. (2005).

velocity changes. In area 4, the PSTM version was clearly more consistent with the structural details and further investigation showed some residual move-out issues for this area in the PSDM version.

Based on these results, we concluded that PSDM, including a slightly less sophisticated matching process, preserves more of the 4D signal. Askim et al. (2004).

Figure 2.22 showed some minor improvement in pushing the 4D signal "under" the gas cloud, but it is obvious that a significant part is still very poorly defined. We believe that a permanent array would allow for better undershooting of the gas, as shown by the 1997 3D4C OBC survey. Use of converted waves (C-waves) was the second candidate for delivering data in the central area. C-wave data had been used successfully in delivering new wells, although the resolution was not optimal.

Advanced finite difference modeling has taught us that significant uplift in seismic image quality is possible if the detailed and accurate velocity distribution is known, see figure 2.24 More details may be found in Brandsberg-Dahl et al. (1999). This study was limited to elastic simulation and an-elastic attenuation is not addressed. The key observation is that the poor imaging in the presence of gas over the reservoir may be due to inaccurate or "too smooth" velocity fields and not necessarily be due to attenuation. This is consistent with observation in the wells, where we have clearly recorded down-going seismic energy in wells in the most severe data degradation areas.

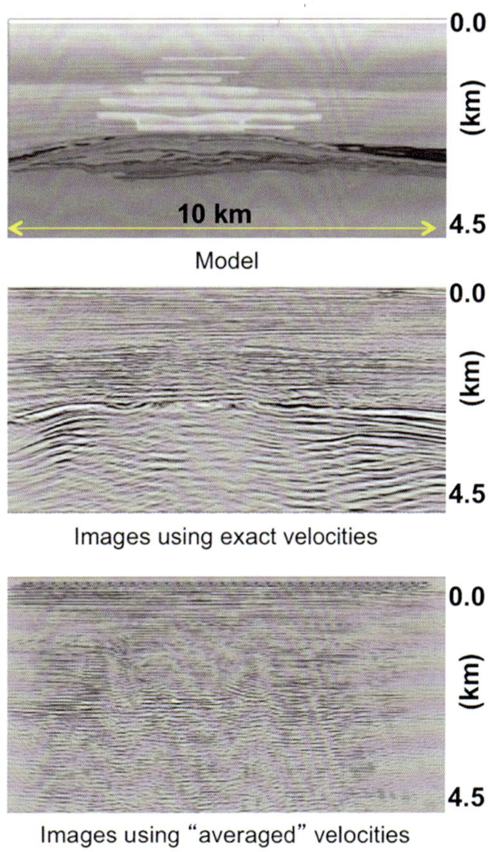

Figure 2.24: Top: model used for full elastic modeling of a line across the Valhall field, below the images from migrating with the exact velocities field. Bottom: migrated with a smoothed velocity field. From Brandsberg-Dahl et al. (1999).

Following the insights from these numerical studies resulted in a strong focus on defining the best possible velocity definition across the field. A real breakthrough came with the use of Full Wave Form inversion, Sirgue et al. (2009). Figure 2.25 shows a cross-section through the Valhall field of the best tomographic velocities and the resulting seismic images when migrated with the best possible velocity field from tomography, compared to what was achieved using the Full Waveform Inversion (FWI) velocity. The inherent properties of the data acquired across the permanent seismic array at Valhall, low frequency and large offsets, made these results possible.

The detailed seismic velocity volumes resulting from the FWI could also be used for 4D seismic analysis, however, the methodology is less mature for detailed 4D analysis as yet, Zheng et al. (2001) offer some interesting alternatives. FWI allows us to work with the details of the velocities, which obviously complements working with amplitude or the derived acoustic impedance estimates. The level of detail that can be observed in a FWI velocity model across Valhall is seen in figure 2.26.

These results, as seen in figure 2.27, demonstrate that considerable P-wave energy does make it through the "cloud". Full-waveform inversion was able to find a velocity model that could produce a significantly better image than we had been able to do with velocities from tomography analysis.

There does appear to be some loss of amplitude and higher frequencies and it is still difficult to make a unique "static" interpretation of the top and base of the reservoir looking at a single vintage of data. However, the dynamic images appear to offer some help. The FWI velocity model has significantly improved the S/N ratio in the poor data area and the reservoir may be uniquely defined based on the 4D response, Barkved et al. (2010). Figure 2.27 shows the improvement in 4D response in the gas-affected area.

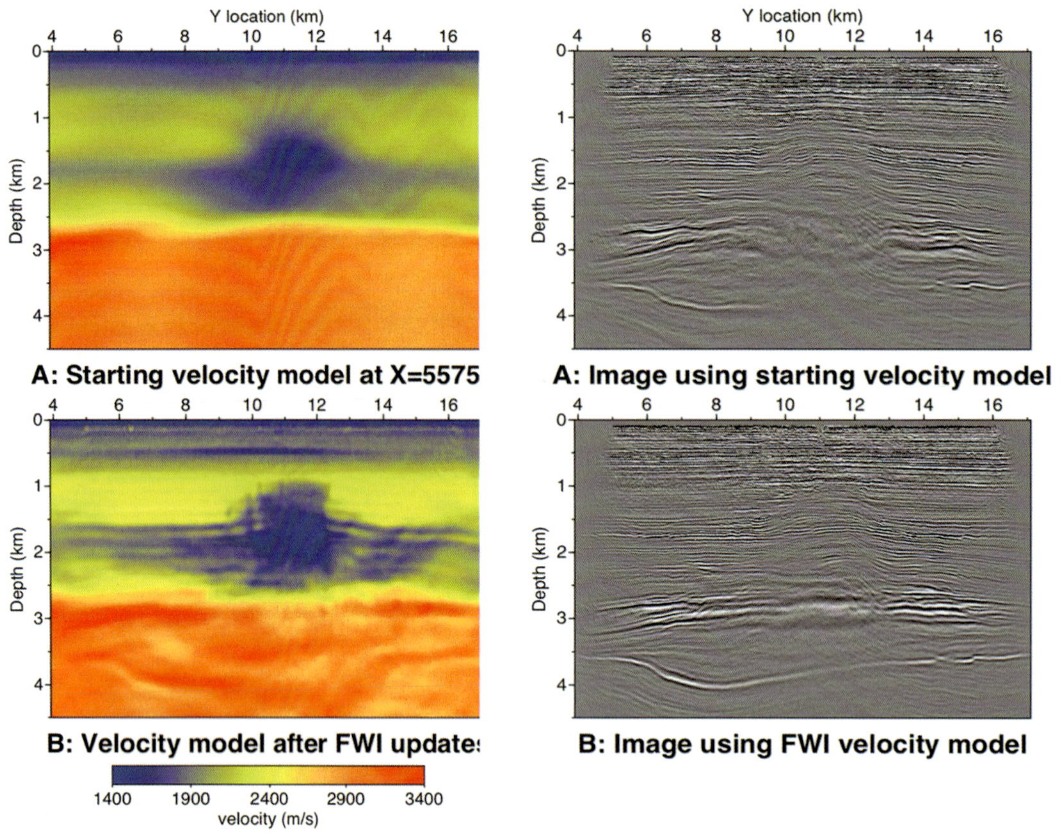

Figure 2.25: Left: Section through the starting 2007 velocity model from conventional tomography and (B), the corresponding section through the final 2009 velocity model from full-waveform inversion. Right: the pre-stack depth-migrated images corresponding to the velocity models A and B. From Sirgue et al. (2009).

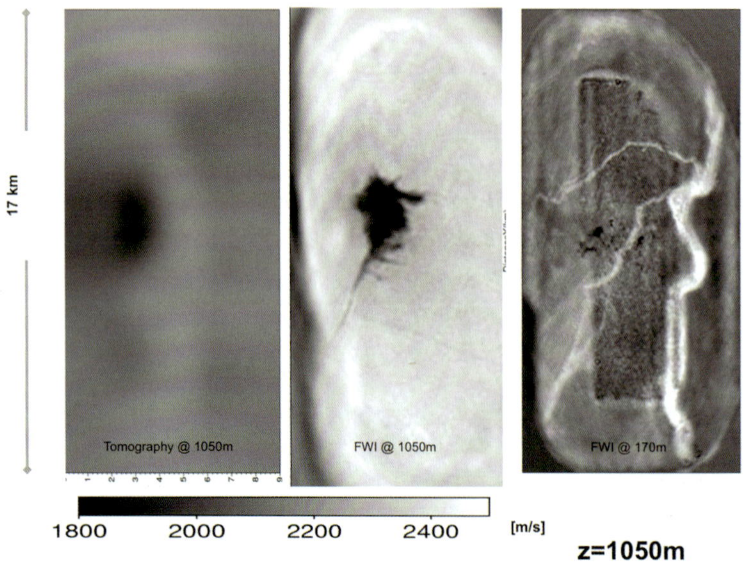

Figure 2.26: Right and center: details from the FWI data at 1050 m and 170 m below seabed. Left-hand image shows the resolution commonly seen in the tomographic-based velocities at 1050 m. Data from the LoFS array. From Sirgue et al. (2009).

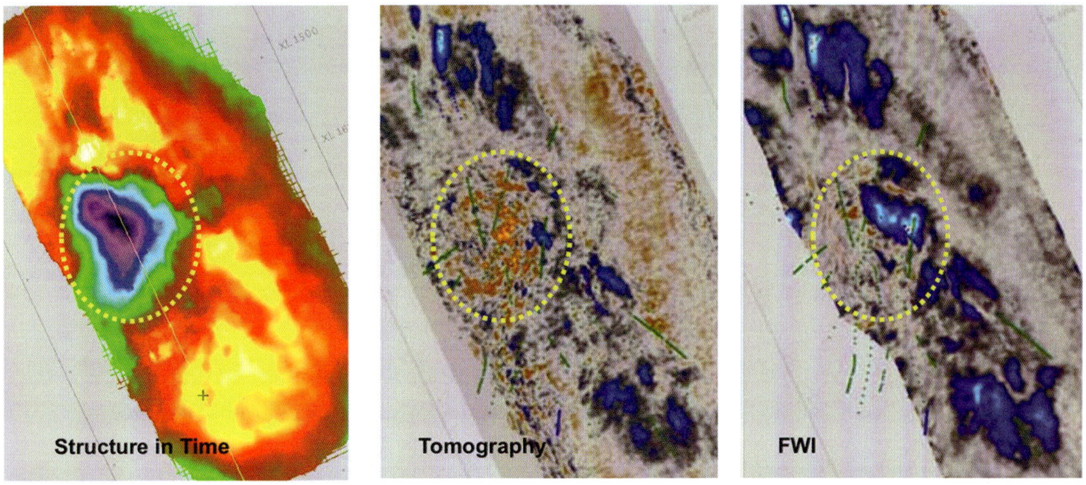

Figure 2.27: Left top reservoir structure map, blue-purple colors shows delay of 300 msec in the central area (outlined by yellow circle) of the field due to gas in the overburden. Center and Right: NRMS map, 250 msec windows, centered on top reservoir, shows the production-induced effects in blue colors. The improvement in the 4D response by using a more detailed velocity model is simply stunning. From Barkved et al. (2010).

2.11 Conclusion

In this chapter we have provided some field-specific information from the Valhall field. The intention is to provide enough context for the subsequent case history examples. The main messages are that the Valhall field is a complex field, with the compaction properties of a weak, unconsolidated or HT/HP clastic field and the permeability and fracturing most often associated with a dense carbonate field. The low permeability and field life makes it an ideal candidate for monitoring depletion induced effects over time.

Part of the Valhall field is in principle ill-conditioned for conventional seismic imaging. We have however had some success, reducing the impact of the shallow gas on imaging the reservoir, using techniques that should work in theory and actually do. When getting the velocities right we are in fact able to see the reservoir in the dynamic images, even though it is virtually impossible to nail down the reservoir in the static images.

3 Basic Rock Physics, Geomechanics and Modeling

"Personally, I look forward to better tools for analyzing C++ source code."
- Bjarne Stroustrup

Rock and fluid physics coupled with geomechanics provide the framework for modeling the impact of production-induced changes on the seismic response. The rock mechanics effects were initially not considered in seismic time-lapse studies, but over the last ten years it has been demonstrated that these effects can dominate the 4D seismic response in some cases. The current rock and geomechanics tools can provide a stronger basis for understanding the key controls on the seismic responses and hence a more definitive interpretation. In this chapter we will briefly revisit some of the key concepts and see how these are wrapped into a tool for use in feasibility studies and for use in qualitative and quantitative analysis of seismic time-lapse observations.

3.1 Introduction

When we analyze the seismic data from porous media we are looking at the velocity, amplitude and frequency of the reflected or transmitted seismic signal. The amplitude is linked to the changes in elastic properties across an interface. The elastic properties are compressional and shearwave velocities, V_p and V_s, and density is often denoted as ρ. Frequency is linked to attenuation, which is in general less well understood than density and velocity. Attenuation is a potentially interesting seismic time-lapse parameter but is outside the scope of this work.

In this section we will briefly introduce the fundamentals of rock physics and geomechanics that we will need to translate production induced changes into changes in the subsurface elastic properties and eventually their impact on the seismic response. We will also explain how this is implemented in our modeling.

3.2 The elastic equations and some relevant concepts

The elastic equations express the velocities as:

$$V_p = \sqrt{\frac{K+4/3\mu}{\rho}} \tag{3.1}$$

and

$$V_s = \sqrt{\frac{\mu}{\rho}} \tag{3.2}$$

Where K is the effective bulk modulus, μ is the corresponding shear modulus and ρ is the bulk density of the rock, which may be expressed as a volume average of the mineral and fluid density:

$$\rho_{sat} = \varphi\rho_{fluid} + (1-\varphi)\rho_{matrix} \tag{3.3}$$

where φ is the porosity. The porosity is the ratio of pore volume to the total volume of the rock, and is a function of grain size distribution, packing and chemical and physical deformation.

The dynamic frame moduli are the bulk and shear modulus. The bulk modulus (K) of a rock describes the rock's volumetric elasticity and is a measure of the rock's resistance to uniform compression. The units are giga-pascals (GPa) or millions of pounds per square inch. The inverse of bulk modulus is compressibility. The shear modulus (G or μ) of a rock describes the rocks' tendency to resist shear stress when opposing forces are acting upon them.

Gassmann (1951) and Biot (1956) showed, based on continuum mechanics approximation, that seismic velocities in porous media at low frequencies (~< 500 Hz) are linked to the sediment fabric (microstructure) through porosity and the dynamic frame moduli. This has provided us with a means of transforming the elastic moduli of one fluid composition to another.

Other sets of models are referred to as inclusion models and are based on perturbing an elastic solid with a cavity or inclusion. Reviews of such theories are given by Zimmerman (1991) and Wang and Nur (1992).

Contact models are based on viewing the rock as a collection of grains and stiffness and deformability are based on the contact between the grains and the elastic moduli of each individual grain. Most of them are based on Hertz-Mindlin's theory of contact mechanics, and are useful for gaining insights, i.e., from including cement minerals between the grains. It is difficult to extend this model to realistic rocks.

Digital rock or modeling is a fairly new development. These models may be based on deriving the geometry of the grain and pore space from detailed CT scans of thin sections of the rock and use numerical modeling to calculate the response of an external impact.

Mixing laws or bounds are another important tool for use in estimating the elastic properties of a saturated rock. They provide robust limits for the properties and offer an intuitive basis for the validation of estimated or measured properties.

Avseth, Mukerji and Mavko (2005) provide an excellent overview of rock physics models, strengths and limitations and I refer you to this book for more in-depth discussions.

Material	P-wave Velocity [m/sec]	S-wave Velocity [m/sec]	Bulk Density [g/cc]
Dry Sand	300-700	100-300	1.7-2.4
Wet Sand	1500-2000	400-600	1.5-1.7
Saturated Shales/Clays	1100-2500	200-800	2.0-2.4
Porous Saturated Sand	2000-3000	750-1500	2.1-2.6
Chalk	2300-3500	1100-1900	2.4-2.8
Limestone	3500-6000	2000-3300	1.8-2.7
Salt	4500-5500	2500-3100	2.1-2.3
Dolomite	3500-6500	1900-3600	2.5-2.9
Granite	4500-6000	2500-3300	2.5-2.7
Basalt	5000-6000	2800-3400	2.7-3.1
Coal	2200-2700	1000-1400	1.3-1.8
Water	1450-1500	--------------	1.0
ICE	3400-3800	1700-1900	0.9
Oil	1200-1250	--------------	0.6-0.9
Quartz	5900	3750	2.65
Calcite	6260 - 6640	3270 - 3440	2.71
Halite	4550	2630	2.16

Table 3.1 Examples of elastic properties and minerals, modified from Bourbie et al. (1987).

3.3 Empirical rock physics models

An extensive summary of various measurements of elasticity of selected rock and minerals can be found in Woeber et al. (1963).

An empirical model is generally based on assuming a functional relation between two or more parameters and calculating representative coefficients using regression analysis. The most famous one is Gardner, Gardner & Gregory (1974), who state the compressional velocity, V_p, as a function of ρ as:

$$V_p = (\frac{(1-\varphi)\rho_{dry+}\varphi\rho_{fluid}}{0.23})^4 \tag{3.4}$$

where V_p is given in feet/sec and ρ in g/cc.

Other well-known relations are Han's (1986) for velocity-porosity-clay behavior in sandstone, and Greenberg-Castagna (1992) relations for Vp and Vs.

Dvorkin and Nur (1998) compared three of the most well-know porosity/velocity relations Raymer, Hunt and Gardner (1980) (RHG), Wyllie, Gregory and Gardner (1956) (WGG) with Gardener, Gardner and Gregory (1974) (GGG) to some real observations and showed good correlation with (fast) cemented sandstone and the RHG equations.

WGG is a heuristic model based on an idea that the time it takes for a P-wave to travel through the rock is equal to the time it takes to travel through pore-fluid plus the time it takes to travel through the solid phase. This may be written as:

$$1/V_{Psat} = \frac{1-\varphi}{V_{Pdry}} + \frac{\varphi}{V_{Pfluid}} \tag{3.5}$$

Where V_{Psat} is the measured P-wave velocity for the saturated rock, while V_{Pdry} and V_{Pfluid} is the P-wave velocities for the solid and for the fluid and φ > is the fractional porosity.

RHG is expressed as:

$$V_{Psat} = (1-\varphi)^2 V_{Pdry} + \varphi V_{Pfluid} \tag{3.6}$$

As we can see this is similar to WGG but explicitly requires that $\varphi < 0.37$.

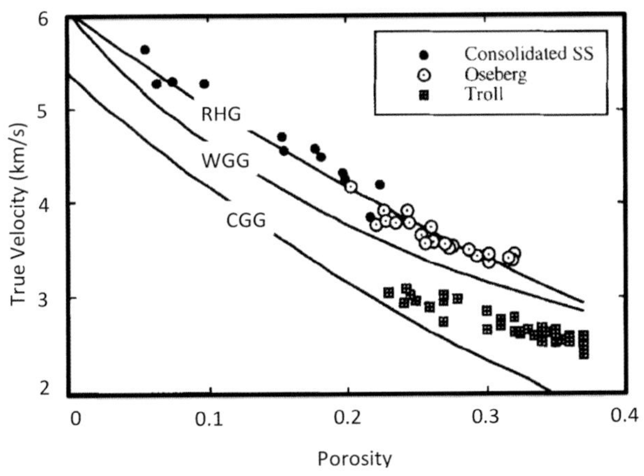

Figure 3.1: True velocities *versus* porosity in clay-free water-saturated sandstone. The symbols represent three data sets. The solid lines are the theoretical predictions. From Dvorkin and Nur (1998).

Gassmann's (1951) equations provide us with the tools for estimating the bulk and shear moduli of saturated rocks at low frequencies (<500–1000Hz) That is to say, when there is ample time for the fluid to move such that there is no pressure gradient introduced into the fluid by the wave.

This provides us with an approach for modeling velocity and density changes, which are based on first principles. The bulk modulus of the saturated rocks is the sum of the modulus of the frame and the modulus of the fluid-filled pores:

$$K_{sat} = K_{dry} + K_{pore} \tag{3.7}$$

and

$$\mu_{sat} = \mu_{dry} + \mu_{pore} = \mu_{dry} \tag{3.8}$$

Where K and μ are the bulk and shear moduli. The subscripts sat, dry and pore refer to saturated (undrained) rock, the dry (drained) framework and the fluid-filled pore spaces. To derive the elastic moduli of a mixture of grains and pore spaces we need information about the relative volume fractions of the phases (minerals, fluid), the elastic moduli of each of them and their geometric configurations. We will typically not have enough data about the geometric configurations nor the model to predict the exact impact. But we can calculate the response based on simple geometric models, which represent end members, usually referred to as bounds, figure 3.2.

The Voigt upper bound is based on a iso-strain model, where the impact of passing waves are carried by the strongest member of the constituents of the composite

$$M_v = \sum_{i=1}^{N} f_i M_i \tag{3.9}$$

The subscript v refers to Voigt, f_i and M_i refer to the volume fraction and modulus of the i-th member of N phases. Reuss provides a lower bound of the composite in the case that the soft member dominates the effects of the composite.

$$1/M_R = \sum_{i=1}^{N} f_i M_i \tag{3.10}$$

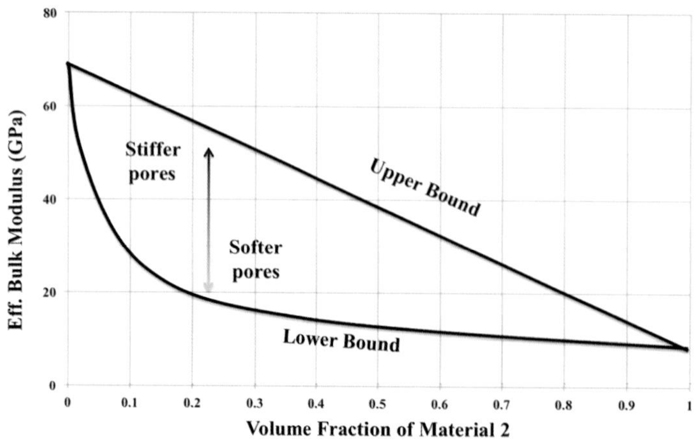

Figure 3.2: Illustration of the bounds concept. Actual effective bulk moduli for the mixture of the two materials will be within the lower and upper bounds.

Avseth et al. (2005), provide further context for the relevance of these bounds. We know that the current velocity of a sediment is linked to its history. Knowing the composition of the rock, we may deduce its development. The sediment will start its life in suspension, which will be close to a Reuss bound, while processes that increase the strength of the material will move its elastic properties closer to a Voigt average, figure 3.3. Knowing the relevant history and ranges of composition of the rock, we can then start compiling a template for the actual geological setting we are using, to help guide our interpretations quantitatively as well as qualitatively, figure 3.4, Avseth et al. (2006).

The other often-used bounds in rock-physics are those defined by Hashin-Shtrikman (1963). The upper and lower bounds are calculated by interchanging the two constituents of the composite.

$$K^{HS\pm} = K_1 + \frac{f_2}{(K_2 - K_1)^{-1} + f_1 (K_1 + 4\mu_1/3)^{-1}} \tag{3.11}$$

$$\mu^{HS\pm} = \mu_1 + \frac{f_2}{(\mu_2 - \mu_1)^{-1} + \frac{2 f_1 (K_1 + 2\mu_1)}{5\mu_1 (K_1 + 4\mu_1/3)}} \tag{3.12}$$

K_1 and K_2, $\mu 1$ and μ_2; f_1 and f_2 are the bulk and shear moduli and the fraction of the individual phases. A more general form, applying to more than two phases might be found in Berryman (1999). One advantage of H-S is that there are no assumptions about geometries. The physical arrangement of the 'H-S' material is shown in figure 3.5. A lighter rim surrounds the darker spheres, the sizes of the combined spheres may vary but the relative volume factions are constant. If the moduli of each constituent are similar, the distance between upper and lower bound becomes small, on the other hand, if they are very different like fluid *versus* mineral, the distance remains large and offers wider ranges for estimation of the rock.

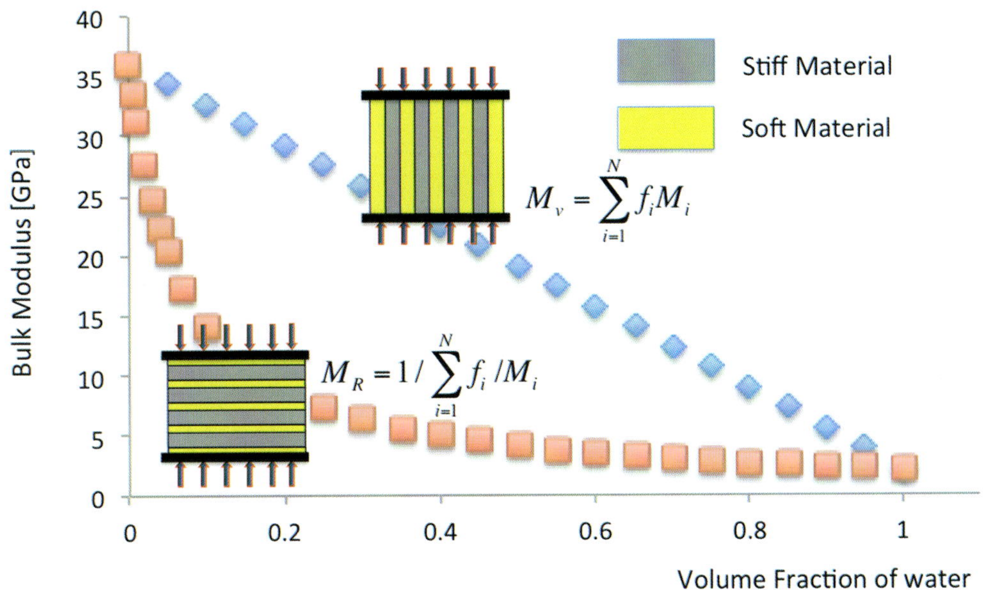

Figure 3.3: Diagram showing the upper and lower bound of the bulk moduli (M=K) of a mixture of quartz grain and water. The bulk modulus for water is 2.2 GPa and for quartz 36 GPa.

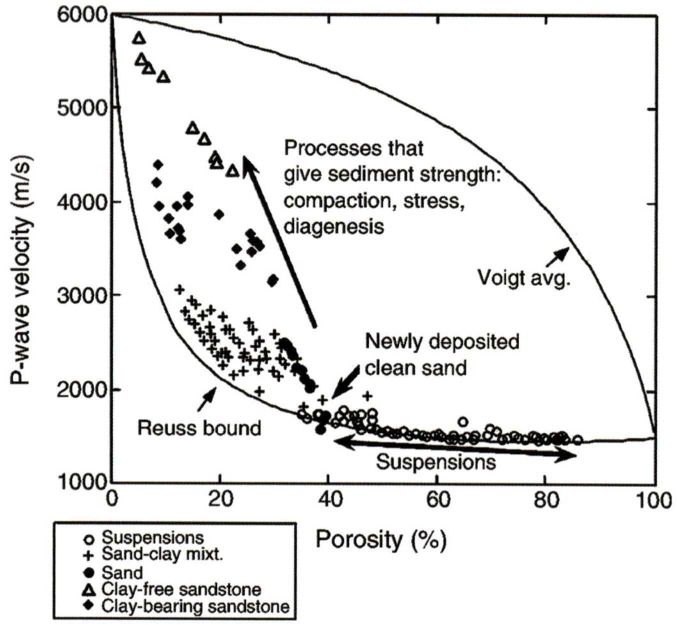

Figure 3.4: P-wave velocities *versus* porosity for a variety of water saturated sediments, compared with the Voigt-Reuss Bounds. From Avseth et al. (2005).

Figure 3.5: Physical interpretation of the Hashin-Shtrikman bounds. The two fabrics are for the bulk moduli of two materials with a different bulk modulus. We obtain the upper and lower bounds if we use the "softer" material in the core and/or around the core, respectively.

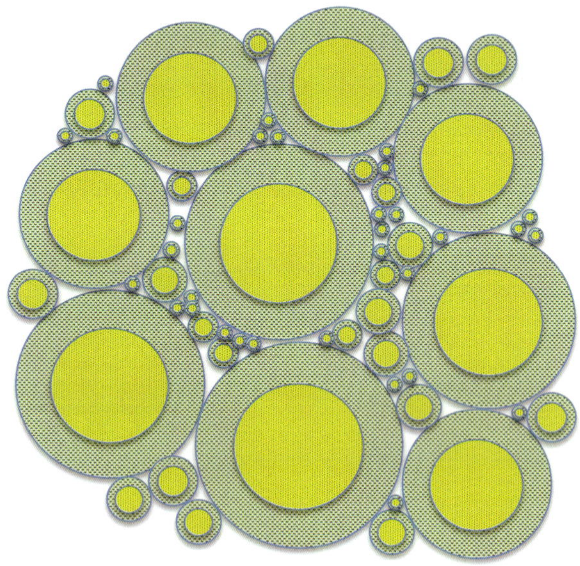

There are several modified versions of these equations. One case is that beyond a critical porosity the fluid will be the supporting phase, as there are no longer any cohesive effects between the mineral grains. Beyond critical porosity the K_{dry} and μ_{dry} frame will become zero. Using this concept, we simply replace φ with φ/φ_c in the above equations.

Gassmann's theory assumes a homogeneous mineral modulus and isotropic pore properties but does not assume anything about the pore geometries. The pore properties may be included through the frame moduli by multiplying the functional dependency with the frame moduli of porosity, with the moduli of the mineral grain:

$$K_{dry} = K_{grain} \cdot f_K(\varphi) \tag{3.13}$$

and

$$\mu_{dry} = \mu_{grain} \cdot f_\mu(\varphi) \tag{3.14}$$

Mavko et al. (1998) provide us with a range of alternative ways to express the moduli, with respect to their interdependencies.

3.4 Brief on rock mechanics

As the sedimentary rocks in a basin are steadily being buried deeper and deeper in the Earth, the stresses acting on the sediments increase because of the increasing weight of the overlying masses and due to the effects of tectonic forces. The changes in the stress impact the properties of the sediments, porosity and permeability are generally decreased and the seismic velocities increase. If the rate of deposition is high, the sediments may be buried faster than the fluid manages to escape. This transfers more of the load from the grain to the fluid and we may experience overpressure. The sediments continue to be subject to physical and chemical changes during burial; temperature, hydrodynamics and subsequent burial and uplift all influence the end results. An understanding of how these effects interact with the morphologies of the sediments is essential for understanding today's distribution of sedimentary properties and stresses, Bjørlykke et al. (2010 in Petroleum Geology, Bjørlykke and Avseth).

Geomechanics is addressing the mechanical responses of all geological materials and includes soils mechanics and rock-mechanics. Rock mechanics originated from mining and were used to predict behavior in solid rock. The term has also been used for the engineering principle applied to the design of oil and gas wells, drilling and reservoir production in mining, tunneling, etc. Soil mechanics is a geotechnical discipline that is used to describe the behaviors of soils or softer rocks. In soil mechanics the grain-to-grain contact is negligibly small, in "hard" rocks the pore spaces might be filled with cement, which will increase the contact significantly. In soil mechanics the effective stress concept leads to water taking the external load (under undrained conditions), less so in hard rock, which is significantly less compressible than water.

Today, geomechanics appears to be the term of choice for these activities, while rock mechanics might be used still when working at core sample scale and in the mining industry.

Geomechanics differs from fluid mechanics and solid mechanics in the sense that the rocks consist of heterogeneous mixtures of particles and fluids. The particles are typically clay, silt, sand and gravel but might also include organic material. The source of the grains, the transport mechanism and burial history will all influence what the strengths and weaknesses of the rock will turn out to be. The shear strength of the rock is primarily derived from the friction between the particles and will be influenced by the effective stress acting on the grains.

This short summary is an attempt to provide some context for the 4D seismic examples and applications we will discuss later. For further details the book "Petroleum Related Rock Mechanics" (2nd edition) by Fjær et al. (2008) provides a good starting point. The EET 5 book on Seismic Geomechanics by Herwanger and Koutsabeloulis (2011), is of course very relevant to what we are discussing later and Colin Sayers's DISC on Stress, Sayers (2010), is another practical reference book.

3.4.1 Stress, strain and Hooke's law

Geomechanics is based on well-known physical principles and concepts: stress, strain and elasticity. One of the most important but maybe the simplest to understand is elasticity. We all know of materials that return to their original shape after we have applied a force to them. The classical example is a spring.

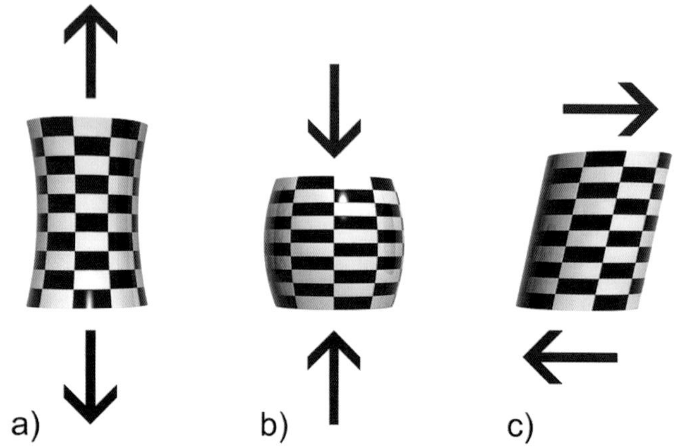

Figure 3.6: a) tensile force, b) compressive force, c) shear force.

If the extension of the spring is linear with the force applied to it, we call this linear. We also know that there is a limit to how far we can pull a spring and still have to go back to its original shape. So if a rock behaves elastically, this will only be true for a very limited range of forces acting on it and a weak rock will probably never be perfectly elastic. In general, elasticity describes materials that return to their rest shape after an applied stress has been relieved. Plasticity describes materials that permanently deform after a sufficient applied stress. A material will often start to behave elastically for lower stress states but when the stress is higher than the yield strength, the rock will deform plastically, resulting in permanent deformation, which is not recoverable through unloading.

The force we apply to a rock can be tensile or compressive and the stretched or compressed distance normalized by its initial length is called strain. If we have a two or three-dimensional body, it may also be deformed by twisting. The forces applied in this case, see figure 3.6 are called shearing forces. If the deformation is proportional to the applied force, the material is called linear elastic and we can use linear elastic equations such as Hooke's law, to describe the relations between deformation (strain) and the applied stress.

Stress σ is defined as force, F, per unit area A,

$$\sigma = F / A \tag{3.15}$$

Rocks may have complex stress states applied to them. In general, the stress is not uniformly distributed over the whole volume of the rock. Therefore, it is necessary to define the stress at the specific point or local area. The local stress at a point or an infinitesimal small part of a surface, i, is defined as the limit when the local area approaches 0:

$$\sigma_i = \lim_{\Delta A_i \to 0} (\Delta F / \Delta A_i) \tag{3.16}$$

A force might be split into its orthogonal components, figure 3.7, and each of them will interact with a volume, the orientation of which might be defined by the orientation of its surface in the three orientations, the sum or the total stress acting on this point (local area is then described by 3×3 equations, which is referred to as a tensor of rank 2. The component of the surface that is normal to the surface is called normal stress, while the forces acting parallel to the forces are called shear stresses, τ. We may use z, y and z or 1,3 and 3 to denote the three orthogonal directions.

This might result in some surprising responses, like simple force acting in one direction might result in a response in the orthogonal orientation. As an example, applying a compressional force to a cube in one direction may result in shearing forces in other directions. As a consequence, the stress state of a point is defined with an infinite number of vectors, associated with a plane through this. Fortunately, according to Cauchy's stress theorem, knowing the stress vectors through three of the planes passing though that point, the stress vector in any other planes might be found.

The other fundamental concept in rock mechanics is strain, ε. When working with rock-mechanics problems we are normally dealing with infinitesimal small changes, which we denote mathematically by *d*.

In its simplest form, Hooke's law states that the relation between stress and strain is constant. This constant is known as Young's modulus - E. A simple relation between infinitesimal small changes in stress, *d*σ, and strain, *d*ε, along the x directions only, may be written as:

$$d\varepsilon_{x=}\tfrac{1}{E}\sigma_x \tag{3.17}$$

If the stress is tensile, the sample will be stretched in the direction of the force, F, figure 3.8. The sample will then become thinner across the direction of stretching in the y and z directions.

These strains are related to the strain along the x-axis by a constant ν, which we know as Poisson's ratio.

$$d\varepsilon_y = d\varepsilon_z = -\nu\varepsilon_x \tag{3.18}$$

This is true when we only have stress along x. In general, we also have stress along y and z, and as we pointed out above, these other stresses will also cause transverse strains and the general result will be:

$$d\varepsilon_i = \tfrac{1}{E}(\sigma_i - \nu(d\sigma_j + d\sigma_k)) \tag{3.19}$$

The subscripts follow a cyclical convention. They denote the principal stress directions (1, 2 and 3) such that (i, j, k) can be (1, 2, 3), (2, 3, 1) or (3, 1, 2). Since principal stress directions are assumed to line up with the axes of the sample, these are often referred to as the vertical and two horizontal directions.

There exists a number of alternative relations that can be used to relate strain and stress in the elastic domains. See Fjær et al. (2008)

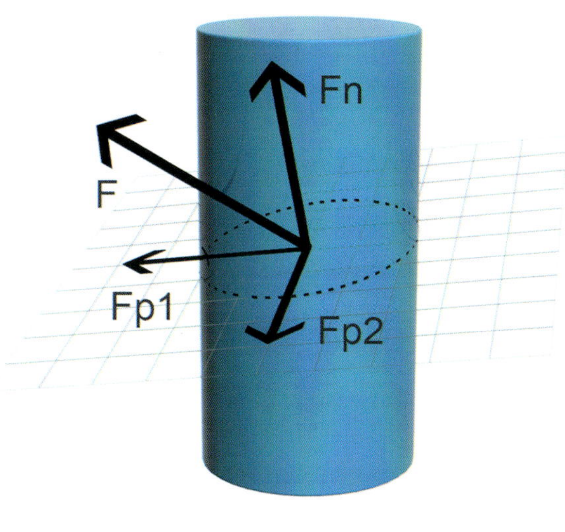

Figure 3.7: A force F, can be split into the three orthogonal components, two parallel, F_{p1}, F_{p2} and one normal, F_n, to a plane through the point where the forces are acting.

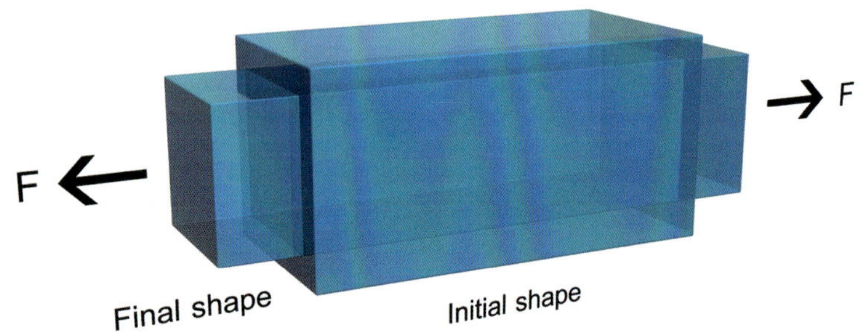

Figure 3.8. If we apply a tensile stress to an elastic sample it will become longer and thinner.

Knowing the principal direction greatly simplifies our analysis and for most basic geomechanics analysis we assume this to be known, and we typically orient the coordinate system to be parallel with the principal axial to avoid having to deal with the shear stresses. In the lab we tend to align the material such that the applied stresses will be along symmetry axis of the material; two directions are typically along the bedding planes and the other orthogonal to the bedding planes

This implies that we end up with nine stress components related to a point:

$$\begin{pmatrix} \sigma_x & \tau_{xy} & \tau_{xz} \\ \tau_{yx} & \sigma_y & \tau_{yz} \\ \tau_{zx} & \tau_{zy} & \sigma_z \end{pmatrix} \tag{3.20}$$

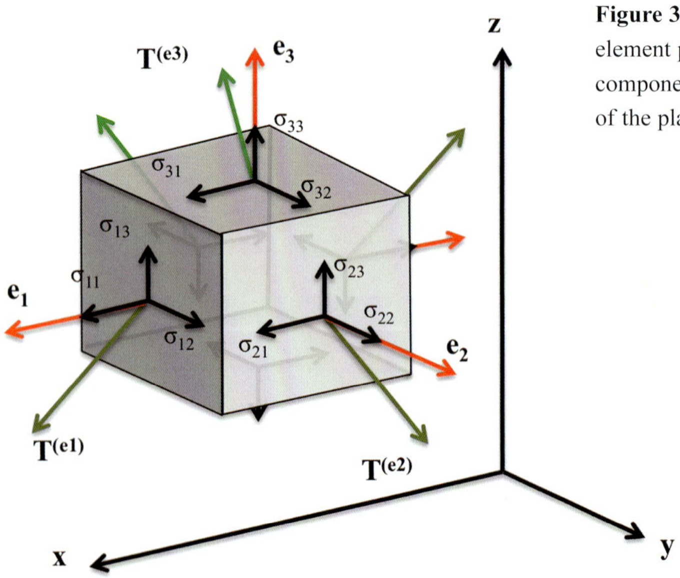

Figure 3.9 The stress vectors T of each of the element planes can be decomposed into a normal component and two shear components, for each of the planes (or axes).

where σ_x defines the normal stresses related to the surface normal to the x-axis, while τ_{xy} and τ_{xz} represent the shear stress in the z and y directions in the same surface, etc. Note also under a static condition $\tau_{xy} = \tau_{yx}$, etc. otherwise the material will start rotating. This means that there will only be six independent stress components.

For advanced calculations it is useful to express Hooke's law in matrix notation. Fortunately, rocks can normally be described by assuming that its properties have three mutually perpendicular planes of symmetry. We make use of the symmetry of the stress and strain tensors and express the six dimensional vectors in an orthonormal coordinate system, we will have the following expression for an orthotropic material:

$$\begin{bmatrix} \sigma_1 \\ \sigma_2 \\ \sigma_3 \\ \sigma_4 \\ \sigma_5 \\ \sigma_6 \end{bmatrix} = \begin{bmatrix} C_{11} & C_{12} & C_{13} & 0 & 0 & 0 \\ C_{12} & C_{22} & C_{23} & 0 & 0 & 0 \\ C_{13} & C_{23} & C_{33} & 0 & 0 & 0 \\ 0 & 0 & 0 & C_{44} & 0 & 0 \\ 0 & 0 & 0 & 0 & C_{55} & 0 \\ 0 & 0 & 0 & 0 & 0 & C_{66} \end{bmatrix} \begin{bmatrix} \varepsilon_1 \\ \varepsilon_2 \\ \varepsilon_3 \\ \varepsilon_4 \\ \varepsilon_5 \\ \varepsilon_6 \end{bmatrix} \tag{3.21}$$

Here σ is the stress tensor and ε is the strain tensor. C is called the stiffness matrix and its component is called the elastic constant. The inverse of the stiffness matrix is called the compliance matrix, **S.** These are not "normal " vectors or matrices, and this notation has specific rotational properties, which has to be taken into account when transferring to a different coordinate system.

Anisotropy tends to complicate geomechanical analysis. The rock properties tend to vary as a function of direction and it easy to accept that the rock properties are different in the vertical directions to those in the horizontal directions because most depositional processes tend to drape the sediments along a horizontal surface. For now, we will ignore anisotropy but we should keep in mind that this could be grossly oversimplifying the issue in some cases.

One often-used parameter in geomechanics is also the bulk modulus K. When all three stresses are equal and the rock is isotropic, the strain will be the same in all directions. The bulk modulus is a measure for how the rock will resist compression when there is a uniform force applied to it. It is the proportionality factor between volumetric strain ε_v and stress σ, defined as:

$$d\varepsilon_v = d\sigma / K \tag{3.22}$$

The inverse of bulk modulus is bulk compressibility C_b. A very important measure for the reservoir engineer is pore volume compressibility C_{porv}. This is a measure for how much the pore volume will change with stress. Pore volume compressibility is important as it represents one of the energies that will drive the hydrocarbon out of the rock.

$$c_{pv} = \frac{dV_{por} / V_{por}}{d\sigma} \tag{3.23}$$

Where V_{por} is the initial pore volume and dV_{por} is the change in pore volume.

We introduced the dynamic moduli early in this section. Their static counterparts have a physical definition, however, among other things, the strain rate is different, rocks are a strain rate dependent medium and the exact relations between them appear to be complex and appear to be linked to porosity, Fjær et al. (2005), and Thomsen (1996). Still, the dynamic elastic moduli generally provide

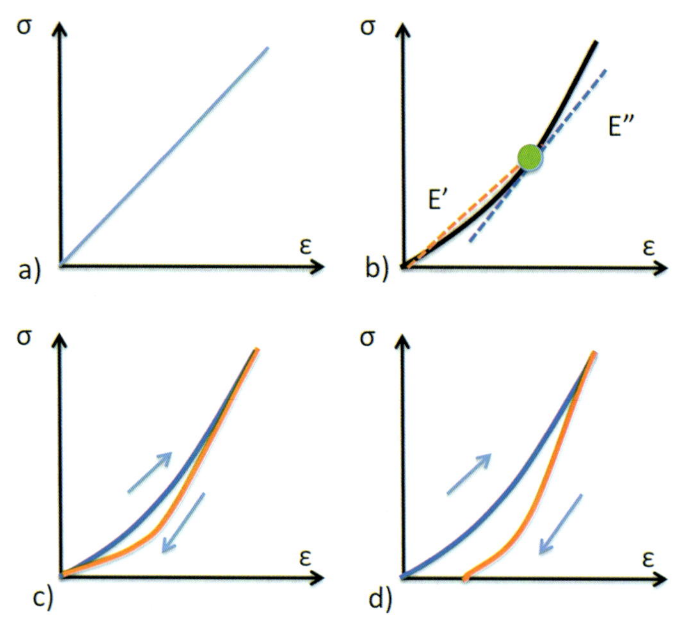

Figure 3.10: Stress/strain for; a) linear elastic material b) perfectly elastic material c) elastic material with hysteresis and d) material undergoing permanent deformation. Modified from Fjær et al. (2008).

the best starting point for obtaining an estimate of the static moduli, when lab data is not available. The conversion from dynamic moduli to static moduli is then done using established relations for the actual lithology groups. The difference between static and dynamic moduli is very dependent on the stress magnitude, Fjær (2009).

Geomechanic practitioners uses some additional moduli, which are convenient for analysis of their data. Bulk modulus K, describes changes in volume as a function of change in mean stress. G or μ, is the shear modulus or the modulus of rigidity and it describes the objects tendency to shear when acted on by shear forces. Young's modulus, E, has been introduced earlier. In addition, we have Poisson's ratio, ν , the Lame parameter and the longitudinal modulus M also referred to as H (uniaxial compaction modulus), The latter describes the axial stress to axial strain when the lateral strain is 0. This modulus is related to the stress path.. The Lame first parameter combines bulk and shear modulus in a convenient manner and are used to simplify the stiffness matrix. Stronger materials also tend to be stiffer, that is, having higher elasticity moduli.

For a linear elastic material, stress/strain relations are constant, figure 3.10a. Up to a certain limit, metal may behave more or less like this. Figure 3.10b shows a perfect elastic material, we just need to be consistent when we define E, while figure 3.10c shows a situation with hysteresis, which is quite commonly observed in rocks. The stress/strain relations vary between the loadings and the unloading follows a different path and we have to use loading moduli and unloading moduli. If the deformation vanishes when the stress is released, the material is elastic. If permanent deformation occurs, as in figure 3.10d, the material is said to be ductile. That is, the rock is still able to resist loading but has passed its yield point.

When rocks are stressed beyond their elastic limit, they will deform permanently, failure does not need to occur until significant higher stresses. In a simplistic model we can divide the failure into three main groups: tensile failure, shear failure (fracturing) and yielding (pore collapse). In tensile failure the rock is simply pulled apart when the pulling forces exceed the tensile strength of the rock. The source for the tensile forces might be tectonic activity, like folding of the rocks.

When the difference between the principal stresses becomes large enough, they will exceed the shear strength of the rock and shear fractures will be formed. The orientation of these failures is controlled by the material properties of the rock. Often, we can see that conjugate sets of fractures have developed, with a specific angle between them.

The third type of deformation occurs when the average stress becomes too big. The failure is no longer isolated to a plane, but a lower part of the rock is impacted. A "band" of rock is experiencing pore collapse. This will happen when the average stress is larger than the yield strength of the rock. When the rock passes the yield line it enters the plastic region. Now we can forget about getting the rock back into the shape it had in the elastic region.

While rock failure relieves the stresses, pore collapse failure does not and the rock continues to be unstable unless the external stresses are relieved. Many rocks enter a zone of unstable behavior when there are sufficiently high stresses applied. The rock may still be able to resist additional loading and is referred to as ductile. The point where the transition from elastic to ductile behavior appears is called the yield point.

The presence of cracks in a rock will also impact its stress/strain behavior and in practice it is difficult to find a rock that shows perfect linear relations between stress and strain.

The effect of applying stress to a rock does not result in an instantaneous complete strain change. The deformation very often continues without any changes in external stress. These time dependencies are normally grouped as creep or consolidation. The latter is linked to how long it takes for the pore pressure to stabilize, while creep is related to the visco-elastic behavior of the rock. As we might expect, it could sometimes be difficult to separate the two. But creep may also occur in unsaturated rock.

3.4.2 Mohr-Coulomb

In a triaxial test, the direction of the load is in the maximum principal direction and the direction of the confining pressure applied is in the minimum principal direction. The normal stress σ, and shear stress τ, on a given plane with an angle θ from the minimum principle stress, $\sigma 3$ can easily be found as the functions of $\sigma 1$ and $\sigma 3$ mathematically:

$$\sigma = \tfrac{1}{2}(\sigma_1 + \sigma_3) + \tfrac{1}{2}(\sigma_1 - \sigma 3)\cos 2\theta \tag{3.24}$$

and

$$\tau = \tfrac{1}{2}(\sigma_1 - \sigma_3)\sin 2\theta \tag{3.25}$$

These relationships can also be shown graphically by using Mohr's circle, figure 3.11.

The Coulomb criteria state that failure occurs when the maximum principal stress reaches either the uniaxial tension strength or the uniaxial compression strength. The Mohr-Coulomb theory, or internal friction theory, is often used to predict the failure of a brittle material. This theory assumes that a shear failure plane is developed in the rock. When failure occurs, the stresses developed on the failure plane are on the strength envelope. Coulomb suggested that the shear strengths of rock are made up of two parts, a constant cohesion C, and a normal stress-dependent frictional component, φ. Figure 3.11 shows the Mohr-Coulomb criteria and a Mohr's circle that touches the failure line. This straight line is often called the strength envelope. Any stress condition below the strength envelope is safe, and once the stress condition meets the envelope, failure will occur. S_0 represents cohesion; φ is the friction angle and is found by testing the rock samples in the laboratory. The Mohr-Coulomb failure criterion may be used for boreholes in stability analysis, water injection design, production optimization techniques, and compaction and sand production prediction, but also relates to the interpretation of failed rocks or structural geology. For 4D analysis, the concepts may come into play when working with reactivation of faults and fractures or for analysis of micro-seismicity.

3.4.3 Deviatoric stresses

A stress tensor σ_{ij} can be expressed as the sum of two other stress tensors, a mean (normal) hydrostatic or volumetric stress tensor σ' that normally results in uniform volumetric effects and a deviatoric component, which tends to distort the volume of the stressed body. We can find the deviatoric stresses by subtracting the hydrostatic stresses from the total stress This decomposition is very useful, as the deviatoric stresses produce shear stress and can result in plastic failures if the stability limits of the material is exceeded.

The deviatoric stress "volume" can be very useful in pointing out areas of the field where we could expect abnormal behavior, i.e., during the drilling of a new well.

3.4.4 Acoustic emission and micro-seismicity

Acoustic Emission (AE) is a naturally occurring phenomenon, which can take place in rocks that are subject to changes in net stress and the subsequent deformations generate a rapid release of energy in the form of elastic waves. The effects are closely associated with dislocation movement following plastic deformation and the initiation or extension of cracks.

AE is used for non-destructive testing of materials in the ultrasonic regime; it typically takes place between 100 kHz and 1 MHz.

Earthquakes less than 2.5 on the Richter scale are rarely felt by humans and we usually refer to them as micro earthquakes Any activity that can result in a sudden shear failure in the subsurface can create this type of event. In nuclear waste disposal sites and in the mining industry, micro-seismic monitoring has been used for input to safety and hazard assessment for a long time. In the oil industry the technology is coming of age for use in monitoring hydraulic fracturing and there is emerging interest in using the technology for compacting oil fields. We will discuss this further in chapter 8.

3.4.5 Poro-elasticity and effective stress principle

A volume of fluid-filled rock can support the external load by the grain or by its fluid or both. A reduction of pore pressure, as we have when producing a reservoir, leads to changes in the effective stress, σ defined by:

$$\sigma_j = S_j - \alpha P_p \tag{3.26}$$

where

$$\alpha = 1 - K_b / K_s$$

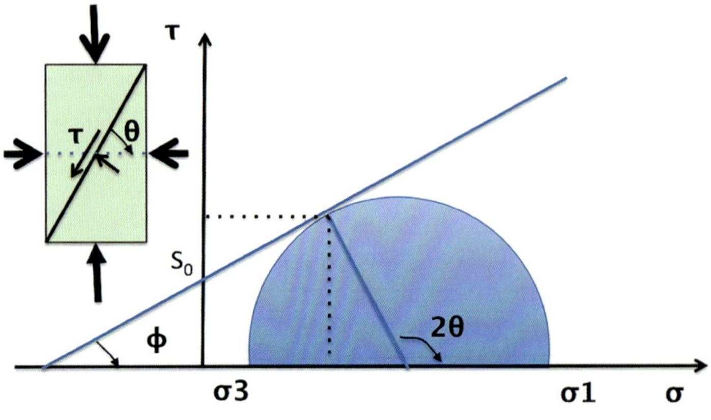

Figure 3.11: Mohr-Coulomb Criteria in the τ-σ pose. The Mohr's circle corresponds to a critical stress state.

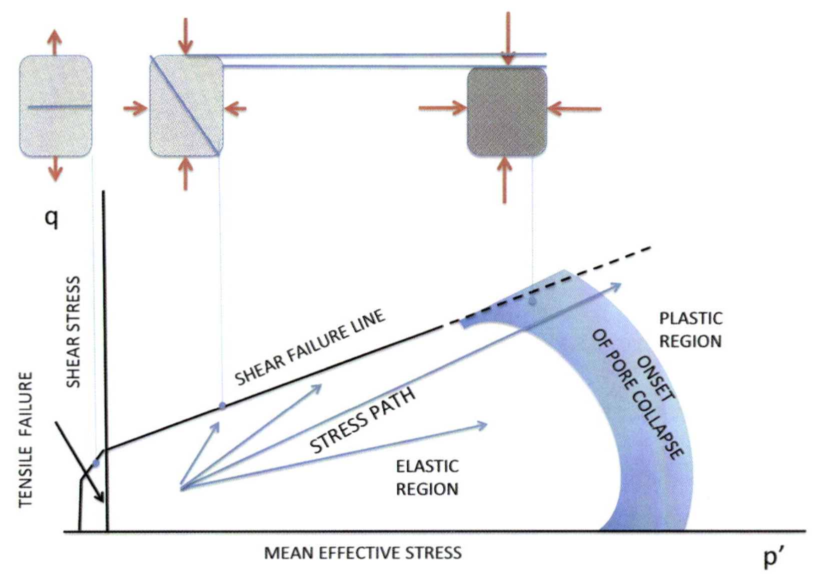

Figure 3.12: p - q' diagram illustrating potential reservoir stress paths during depletion.

S is total stress, $\boldsymbol{P_p}$ is the pore pressure and subscript j can be 1, 2 or 3 and refer to the maximum, intermediate or minimum principal stress. K_b is the bulk modulus of the rock and K_s is the bulk modulus of the individual grains. Normally, $K_b << K_s$ for porous rocks, α also called Biot's constant, is close to 1. Temperature impact is handled via thermal expansion (if heated) or thermal contraction (if cooled), with a resulting change in total stresses, Fjær et al. (1992).

Deformation due to stress state changes is contingent on the stress state "history" or stress path, as this is best described in a p-q diagram, where p' and q are defined as:

$$p' = \frac{(\sigma_1 + \sigma_2 + \sigma_3)}{3} \tag{3.27}$$

and

$$q = \frac{1}{\sqrt{2}}\left[(\sigma_1 - \sigma_2)^2 + (\sigma_2 - \sigma_3)^2 + (\sigma_1 - \sigma_3)^2\right]^{1/2} \tag{3.28}$$

Since we are reducing the pressure, we will move towards the right in the system, figure 3.12. In some cases we will get close to the failure lines, in other cases we could end up in the pore collapse region and experience accelerated compaction. If we assume no lateral deformation (uniaxial strain) and start depleting the reservoir by reducing the pore pressure, $\boldsymbol{P_p}$, the one-dimensional deformation, vertical strain change, ε_{ve} may be described as:

$$\varepsilon_{ve} = C_{me} \Delta P_p \tag{3.29}$$

where $\boldsymbol{C_{me}}$ is the elastic uniaxial compressibility. Sometimes, the effective stress is strong enough for the rock to start to deform plastically. In the p'-q diagram this point is referred to as the yield point, beyond this point some of the deformation will not be reversible and a more complex constituents model will be necessary to model impact. Plastic deformation often increases with time; this process is referred to as creep or consolidation. Fluid interactions, water-weakening and chemical changes may be another important part.

If we summarize the one-dimensional case, the net effect is a reduction, Δh, in initial thickness, h_i, which may be expressed as:

$$\Delta h / h_i = \varepsilon_{vT} = \varepsilon_{ve} + \varepsilon_v(t) + \varepsilon_{vw}(t) \tag{3.30}$$

ε_{vT} is the total vertical strain, $\varepsilon v(t)$ is the creep and ε_{vw} is the changes due to fluid interactions, water weakening or chemical effects. The net impact is a reduction of porosity:

$$\varphi = \frac{\varphi_i - \Delta h/h_i}{1 - \Delta h/h_i} \tag{3.31}$$

The change ranges from virtually no impact at all (low porous fractured granite) to a completely plastic deformation (low permeable, highly unconsolidated sandstone), contingent on the fabric and composition of the rock. In most cases we expect the stress-induced effects to be within the elastic domain and the effect is reversible. Existing micro cracks could be closed or new ones opened, while the load on the contact points could increase or decrease contingent on the pore-pressure acting between the grains.

3.5 Time-Lapse modeling of the reservoir changes

3.5.1 Fluid substitution

As seismic velocities may be expressed as functions of the saturated bulk modulus, shear modulus and densities, we may use the available velocities and densities from well logs or other sources to calculate the initial bulk and shear moduli:

$$K_{sat}^{init} = \rho((V_p^{init})^2 - \tfrac{4}{3}(V_s^{init})^2) \tag{3.32}$$

$$\mu^{init} = \rho(V_s^{init})^2 \tag{3.33}$$

We then use Gassmann's equation to calculate the modified moduli :

$$K^{new} = \frac{K_m C}{1+C} \tag{3.34}$$

where

$$C = \frac{K_{sat}^{init}}{(K_m - K_{sat}^{init})} - \frac{K_{fluid}^{init}}{\varphi(K_m - K_{fluid}^{init})} + \frac{K_{fluid}^{new}}{\varphi(K_m - K_{fluid}^{new})} \tag{3.35}$$

the shear modulus does not change

$$\mu^{new} = \mu^{init} \tag{3.36}$$

but we need to change the bulk density to accommodate the changes in fluid:

$$\rho^{new} = \rho^{init} + \varphi(\rho_{fluid}^{new} - \rho_{fluid}^{init}) \tag{3.37}$$

And finally, we can calculate the new velocities:

$$V_P^{new} = \sqrt{\frac{K_{sat}^{new} + 4/3\mu_{sat}^{new}}{\rho^{new}}} \tag{3.38}$$

and

$$V_S^{new} = \sqrt{\frac{\mu_{sat}^{new}}{\rho^{new}}} \tag{3.39}$$

The basic workflow is illustrated in figure 3.13.

Gassmann's relations were originally derived to describe the changes in rock modulus from one pure state to another. Domenico (1976) proposed that if an effective fluid replaced the mixture of fluids, Gassman's relations could also be used. The effective fluid bulk modulus and density can be calculated by

$$1/\bar{K}_{fluid} = S_{gas}/K_{gas} + S_{oil}/K_{oil} + S_{br}/K_{br} \tag{3.40}$$

and

$$\bar{\rho} = S_{gas}\rho_{gas} + S_{oil}\rho_{oil} + S_{brine}\rho_{brine} \tag{3.41}$$

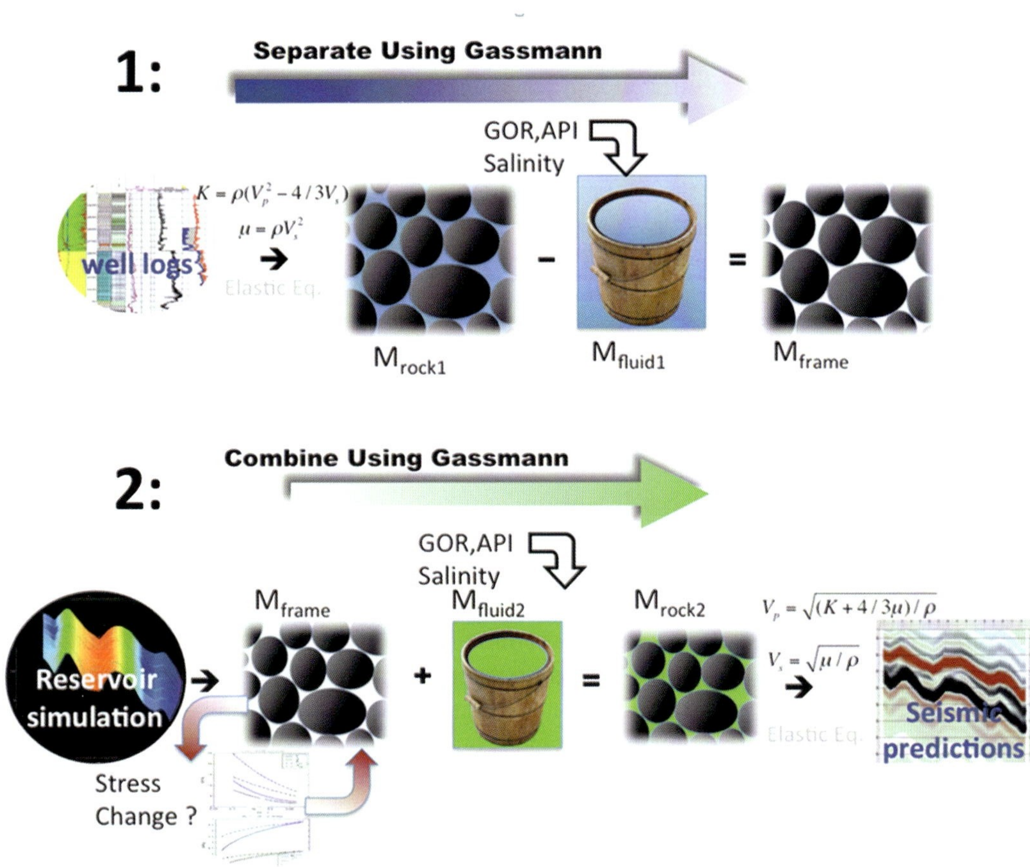

Figure 3.13: The fluid substitution workflow (modified from Askim (2003)).

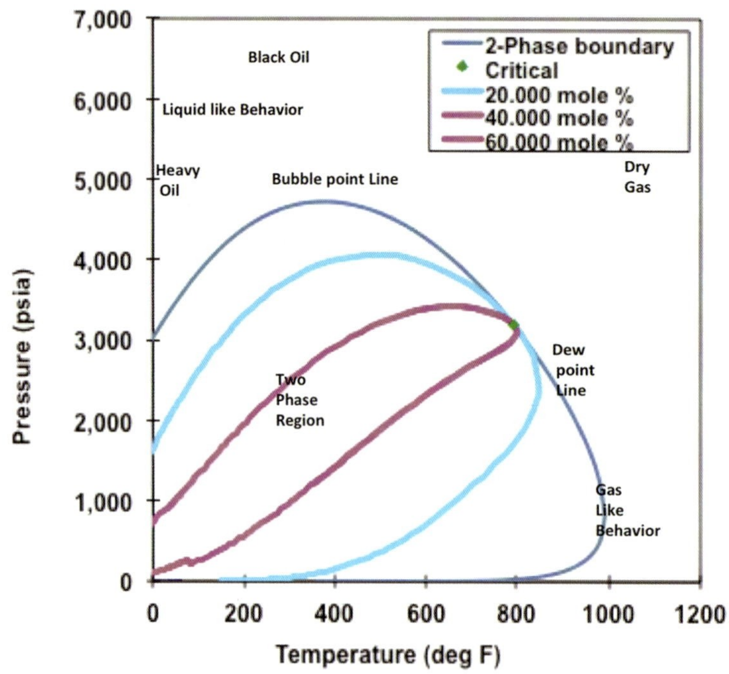

Figure 3.14: PT diagram.

These equations are commonly substituted into Gassmann's relations but it is an issue that the velocities not only depend on the actual saturations, but also how this is distributed. The approach is based on fluids that are mixed uniformly at a very small scale, Avseth et al. (2005). If this is not the case, we refer to the saturations as patchy. We may approximate this case by using the Voigt average of the fluids (Mavko and Mukerji (1998):

$$\bar{K}_{fluid} = S_{gas}K_{gas} + S_{oil}K_{oil} + S_{br}K_{br} \tag{3.42}$$

This appears to give and upper bound solution and an alternative empirical suggestion is, Brie (1995):

$$\bar{K}_{fluid} = (K_{liquid} - K_{gas})(1 - S_{gas})^{e} + K_{gas} \tag{3.43}$$

Where e is an empirical number between 1, which gives the patchy upper bound and ∞, which gives the lower bound, representing the fine-scale mix.

Batzle and Wang (1992) provide a set of empirical relations, which appears to be in general use as a reference for calculating relevant fluid properties. This paper does not include explicit relations for dealing with condensate and the effects of gas saturation in brine may as yet not be well understood. As simple as the relations might read and possibly encoded, there is certainly scope for pitfalls in this trade and consultations with experts when the work is influencing important decisions is strongly advised. The fluid and change in fluid properties in the reservoir is one of the key focuses of the reservoir engineer. The composition of the fluid is controlled by the actual pressure and temperature. The composition could also change over time as the lighter components might flow more easily than the heavier ones. For detailed studies, understanding and keeping track of the fluid states is critical, also for modeling the acoustic time-lapse. Figure 3.14 shows a PT diagram for a specific field, in this case the actual diagram is only valid for a limited region in the reservoir

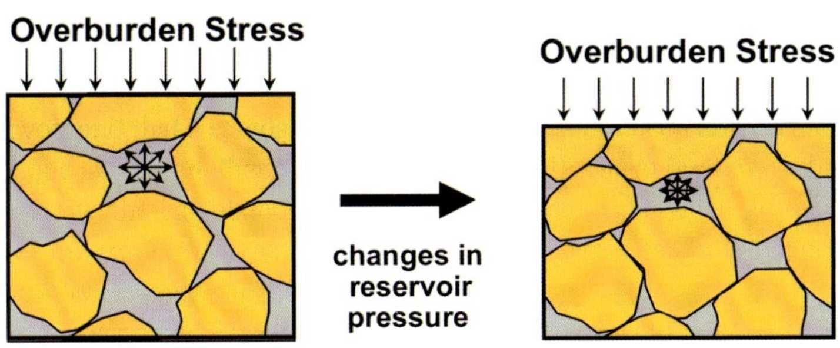

Figure 3.15a: The effective stress concepts. External load minus pore pressure makes up the effective stress the rock grains are subject to.

3.5.2 Pressure effects

The effective stress concept, as discussed earlier, is a effective way to deal with pore pressure changes, figure 3.15a.

Our ability to predict the net stress effect on the minerals from first principles is limited. We might use crack or grain contact models to gain insights into the trends of expected changes and use core measurements of velocity and compaction changes with changes in effective stress for calibrations. The Hertz-Mindlin model, which gives the effective bulk modules for dry random identical sphere packing is given by Dvorkin & Nur (1996):

$$K_{eff} = \sqrt[3]{\frac{C^2(1-\varphi)^2\mu^2}{18\pi^2(1-\upsilon)^2}P_{eff}} \qquad (3.44)$$

And the corresponding effective shear modules are given by:

$$\mu_{eff} = \frac{5-4\upsilon}{5(2-\upsilon)}\sqrt[3]{\frac{3C^2(1-\varphi)^2\mu^2}{2\pi^2(1-\upsilon)^2}P_{eff}} \qquad (3.45)$$

Where υ Poisson's ratio, φ is the porosity, P is the effective pressure and C is the average number of contacts per grain (the coordination number) – which can be approximated by: $n= 20 - 34\varphi + 14\varphi^2$, Murphy (1982)

The stress dependence of wave velocities requires a non-linear stress-strain relationship. Experimentally measured P- and S velocities typically show an exponential behavior, figure 3.15b. Holt et al. (2005) distinguish between three main sources of the non-linearity, which are all related to how the grains in the rock are organized. They suggest that the relations are linked to change in porosity with stress, the presence of sharp grain contacts or generation of cracks/fractures, and argue for that if the porosity is supposed to be the only controlling factor, one will need to include non-linear elasticity to explain significant stress dependency. The Hertz-Mindlin theory addresses effects of the grain contacts, and predicts that the velocities increases with the number of grain to grain contact points, and it may be shown that when the contacts are loaded we increases the stiffness of the rock, and consequently the velocity will increase.

Based on the Hertz Mindlin theory, assuming constant Vp/VS ratio, Landroe and Kvam (2002) shows that the relative changes in P-wave and S-wave velocity can be written as:

$$\frac{\Delta V_P}{V_{P0}} = \left(\frac{P}{P_0}\right)^{1/6} - 1 \qquad (3.46)$$

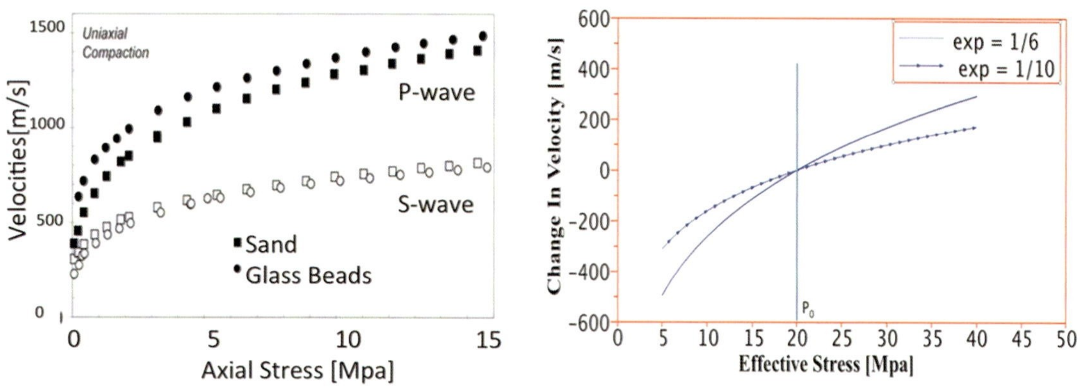

Figure 3.15b: Velocity changes as a function of effective stress. Upper left shows lab measurements, from Holt (2005), upper right the changes in calculated using eq. 3.46 and 3.47.

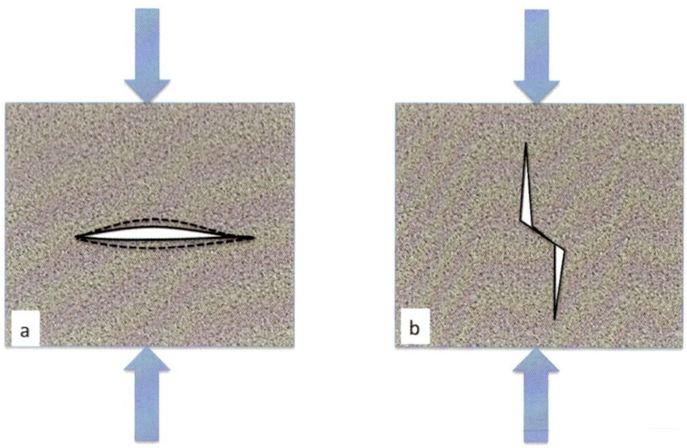

Figure 3.16: Illustration of: a) closure of crack due to normal stress and b) opening of crack due to shear stresses.

and

$$\frac{\Delta V_s}{V_{S0}} = (\frac{P}{P_0})^{1/6} - 1 \tag{3.47}$$

Some lab data indicates that an exponent of 1/10 is better, figure 3.15b. Because potential core damage might introduce cracks, the calibration with lab data is thought to give us the upper bound of the stress sensitivity. Duffault and Landrø (2007) offers additional arguments for that the simple concept of the combining Hertz-Mindlin and Gassmann may actually be sufficient as a first approximation for our 4D modeling when core material is not available, for specific sandstone reservoirs. However, the linkage to number of grain contacts (coordination number) or the stress sensitivity is not well known. The Hertz Mindlin theory is based on perfect spherical grains, but there exist alternative models in the literature that offers this type of models, i.e., Carlson and Ganghi (1985). Presence of cement above certain levels impacts the stress sensitivity of the rocks, Dvorkin et al. (1991).

In practice we may calculate K_{dry} porosity using a polynomial function that has been calibrated to core measurements. A similar treatment may be applied to the shear modulus using an analogous equa-

	Propagation	Polarization	Crack Orientation	Velocity reduction
P-wave	→	↔	⬯	Very strong
P-wave	→	↔	⬭	Weak
P-wave	→	↔	○	Weak
S-wave	→	↕	⬯	Strong
S-wave	→	↕	⬭	Strong
S-wave	→	↕	○	None

Table 3.2 Relations between crack orientation and the directions of propagation and polarization and the impact on velocity reduction, from Fjær, (REGC 10/2006).

tion, Kjelstadli et al. (2005). This approach differs somewhat from methods used by others. See, for example, Vasco et al. (2003), Dong and Oliver (2003), Pannett et al. (2004), Mezghani et al. (2004), Falcone et al. (2004).

Cracks are known to reduce seismic velocity significantly and are likely a major source of stress sensitivity, also the most likely source of anisotropy, as the cracks will first of all impact the velocity propagating (or being polarized) in the orthogonal direction of the cracks. There are a number of effective medium theories that describe the impact of pores and cracks on the velocities, figure 3.16 b illustrated the closing of a crack due to normal stress and the opening of a crack due to shear stress and table 3.2 summarizes the impact on velocity and polarization for various directions of propagations and crack orientations.

3.6 Changes outside the reservoir

If the reservoir deforms during production, this will impact the surrounding rock. Assuming that, the local changes in a contracting sphere with radius r_0 can approximate the reservoir. The radial displacement at a distance R from the sphere can be expressed as:

$$u = \frac{u_0 r_0^2}{R^2} \qquad (3.48)$$

Where $\boldsymbol{u_0}$ is the deformation of the contracting sphere. This simple model assumes constant rock properties: compression and Poisson ratio, through the influenced half space and shows that the influence is reduced with the square of the distance.

Segall (1992), express this poroelastic displacement f in a poroelastic media as a function of the distributions of pressure changes in an axis-symmetric reservoir as

$$u_i(x,t) = \frac{\alpha}{\mu} \int_V \Delta p(\zeta,t)\psi_i(x,\zeta)dV_\zeta \qquad (3.49)$$

Where α is the Biot's coefficient, μ is the shear modulus, Δp is the change in pore pressure, ξ is the integration coordinates over the volume V undergoing pressure changes, t is the time index and ψ is the green's function. The strain may be found by differentiating the equation:

$$\varepsilon_{x_j} = \frac{\partial u_j}{\partial x_j} = \frac{\alpha}{\mu}\sum_{k=1}^{N} \Delta p_k \int_{V_k} \left(\frac{\partial \psi_j(x,\zeta)}{\partial x_j}\right) dV_\zeta \tag{3.50}$$

We can write this as a linear system, which describes the strain at a given point in the subsurface due to a change in pressure in the reservoir sub-volume, Hodgson (2007) illustrates this by a linear system shown in figure 3.17.

The nucleus of strain concept may be implemented in a very effective manner on a computer using the Green's functions. The resulting strain models looks very realistic, figure 3.18.

In reality, the uniaxial model is way too simple. The "side-burden", will try to resist the changes, resulting in stress arching effects, figure 3.19. Stress arching is a result of the rock on the side of the volumes being strained, resisting the movement. In principle, we transfer some of the load from just above the reservoir to the shoulders. This will result in an increase in pressure in this part of the reservoir. If the differential stress induced by these changes becomes too large, the rock will fail and we have an earthquake as a result.

3.7 Will seismic surveillance work in my case?

Not all reservoirs are ideal candidates for seismic surveillance. The best candidates are highly compressible rocks, cases where we have strong contrast in actual fluid compressibility and an area of high seismic quality. Fields of Tertiary ages and high porosity Jurassic sandstones dominated the early 4D seismic implementation in the North Sea. Processing and acquisition technology has continuously been improved and today, 4D seismic are being applied for a much wider range of porosities and reservoir rocks than even the most optimistic practitioner thought possible ten years ago, Foster (2007).

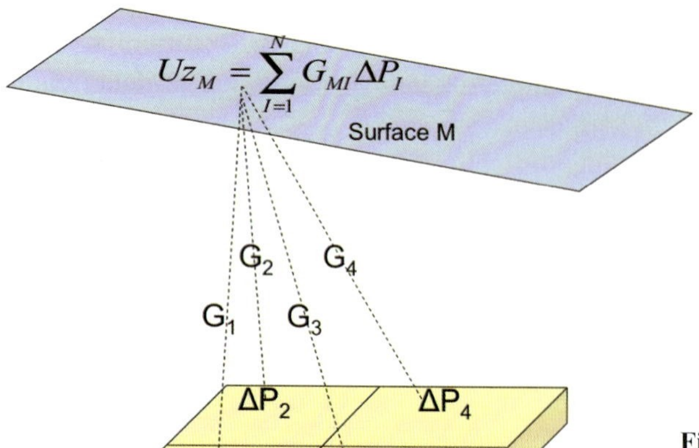

Figure 3.17: the pressure depletion in a rectangular reservoir, split into four cubes contribute to the observed strain at a given point outside the reservoir. Hodgson (2009).

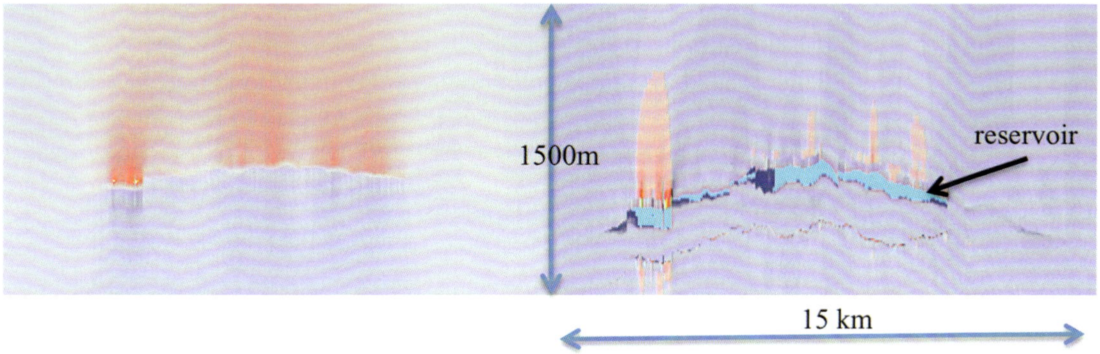

Figure 3.18: Left: modeled reservoir compaction induced strain, red reflects extensions, blue reflects compaction. The strain effect below the reservoir is modeled separately, in the next step and merged into the model. Right: the resulting stress state changes. Red is a decrease in net stress; blue is an increase in net stress.

Figure 3.19: Stress arching.

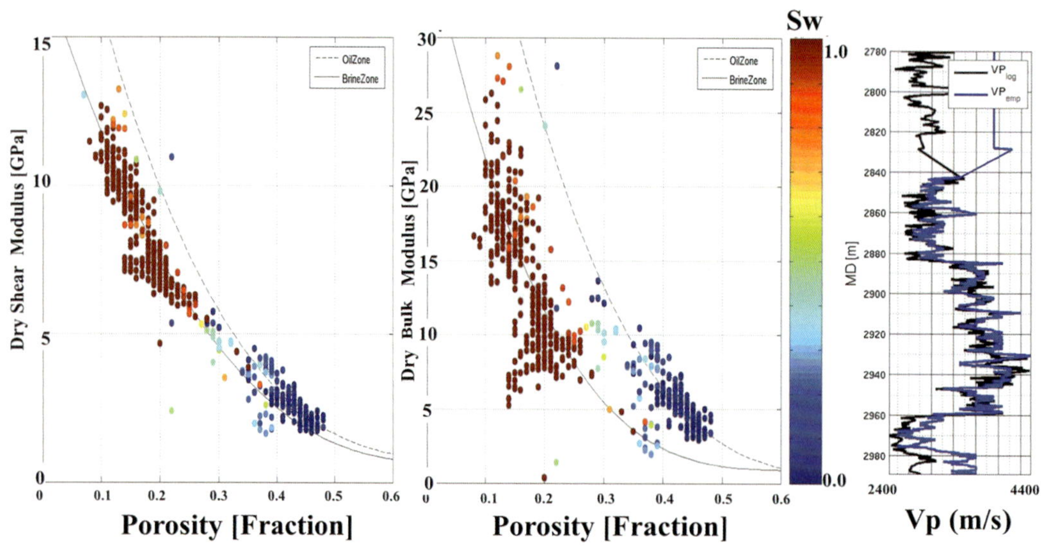

Figure 3.20: Left and middle caption: dry shear and bulk moduli. Rightmost curves show the simulation of the compressional velocities, calculated using the porosity and the saturation logs as input, compared to the velocities from the observed sonic log, from Askim (2003).

3.7.1 Feasibility study

The strength of a reservoir's 4D signal depends on its rock and fluid properties and also the actual depletion strategy. An initial appreciation of the production induced changes can be gained from simplistic fluid substitution modeling or by testing the impact of pressure changes using relevant pressure / velocity ratios, as discussed earlier in Chapter 3. A first step in a seismic monitoring feasibility analysis is to establish the relevant elastic properties of the fluids and the reservoir rocks, from well logs, cores or rock physics models. Figure 3.20 shows the correlation between porosity and dry shear and bulk moduli. The figure also shows the simulation of the compressional velocities, calculated using the porosity and the saturation logs as input. Local pressure sensitivity will typically come from core measurements. We can then test the sensitivity of the reservoir rocks to fluid and effective stress changes.

When the rock physics properties are defined, we might have a look at whether the changes in reservoir properties, pressure and saturation will result in observable changes in the elastic properties of the rock. Figure 3.21 shows the sensitivity of various elastic properties and 4D attributes to changes in gas and pore pressure in a 34% chalk well at 2500 m depth. Modified from Askim (2003).

The actual rock properties were based on available petrophysical data from logs and a comprehensive core database. The initial dry frame (K) and bulk modulus were derived from logs from nearby wells, using the method outlined in figure 3.13. When oil saturated chalks were exposed to brine in the laboratory, it resulted in a significant weakening, Andersen (1995). Due to this we also implemented an option for water weakening. This was based on published data from Guilbot and Smith (2002), and as we can see in figure 3.22, the effects on the moduli are significant. This option could be invoked partly or in full.

The history dependent stress sensitivity was based on a rewrite of the relations published by Zhang and Bentley (2000). As can be seen in the curves in figure 3.22, the loading and unloading curves differ. The loading curves are assumed to include both elastic and plastic deformations. The unloading curves will show leaner behavior as the plastic deformation is not reversible.

The sensitivity analysis of the reservoir rock is based on a homogenous model. The next thing we may want to check is if this holds for a more realistic geological situation, where we may use well data.

In figure 3.23 we show an example were we test the effect of modifying the logs accordingly using the rock physics model we have established.

In figure 3.24a we have tabulated the physical properties for simple block model a number of alternative scenario. The different modeling approach may be useful in developing a good understating of the response, and awareness of possible side effects of simplifying (blocking the model), which will typical be the case when we introduce 3D models. Figure 3.24b shows a seismic angle gather, where the individual traces represent angles from 0-40 degrees.

The next step in the feasibility study is to simulate realistic 3D scenarios, as predicted by the reservoir model and the flow simulator. Each field might have its merits that preclude using the flow model directly, i.e., if the grid-cells are too coarse. However this is seldom the case today. The actual reservoir simulator used in this case was history matched using production until 2002. Eight layers were used to represent the reservoir with an average thickness of 5 meters, Figure 3.25. Cell sizes varied between 100 to 25 m, a grid refinement was used locally to allow for including more details in the model. For the actual time step, the reservoir model provided a full 3D grid with information about pore pressure, saturation (Sw, So and Sg), porosity, and compaction from pressure depletion and from water weakening effects.

Alternatively, we may use the geological model, as a geometrical reference as this may contain more detail and have a finer lateral resolution than the reservoir flow model. In both cases we need to ensure that the physics is consistent with the seismic modeling. We will also have to include at least one layer above and below the reservoir model to properly define the reflectivity at the top and base of the reservoir. The first pass of the model is often used to validate that the geological model is consistent with the seismic. In that case, we will often need to include several layers above and below the reservoir to simulate interference. For difference modeling we might get away with a simpler model. In figure 3.26 we use the reservoir model as the basis for simulating the production induced changes. The cross-section shows a simplified model, where the layers above and below the model is represented by one constant background model.

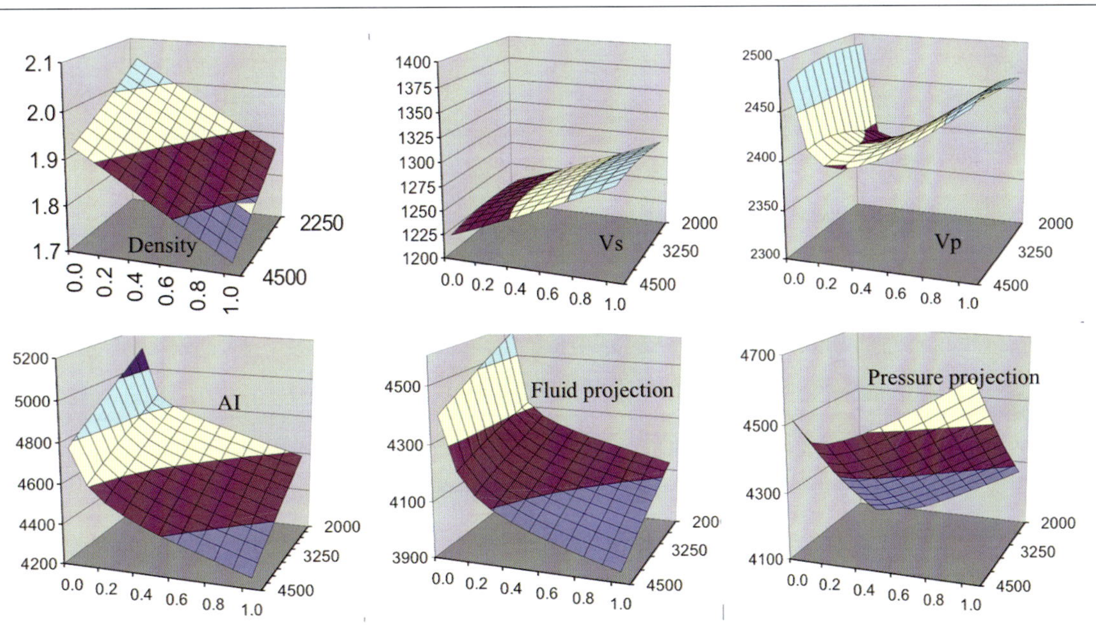

Figure 3.21: Expected changes in various seismic properties from changes in pore pressure and gas saturation. The horizontal axes are fractions of gas saturation and pore pressure. The vertical axis varies and is indicated on the figures, from Askim (2003).

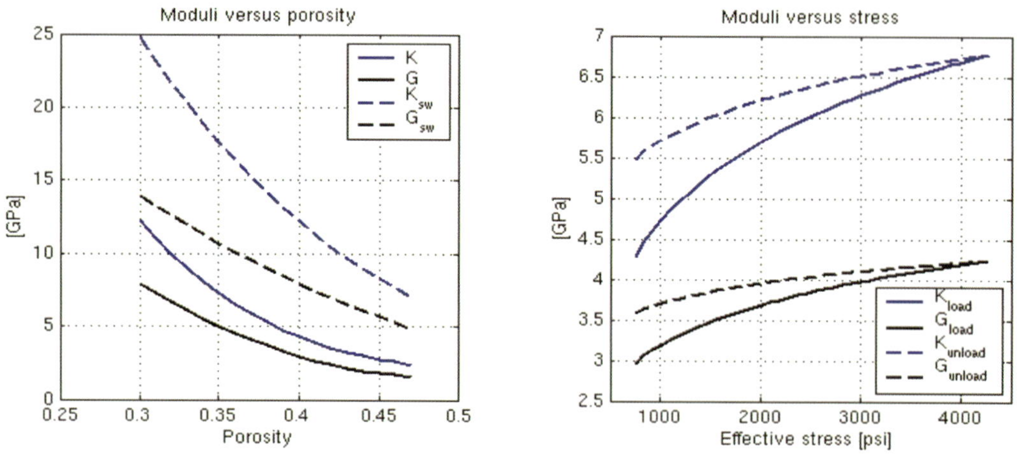

Figure 3.22a: Left: moduli as a function of porosity. The solid lines are the initial frame moduli porosity relations, while the dashed lines are with a full water-weakening impact. Right: moduli versus stress. Solid lines are loading curves, while the dashed lines are unloading for the frame moduli, from Askim (2003).

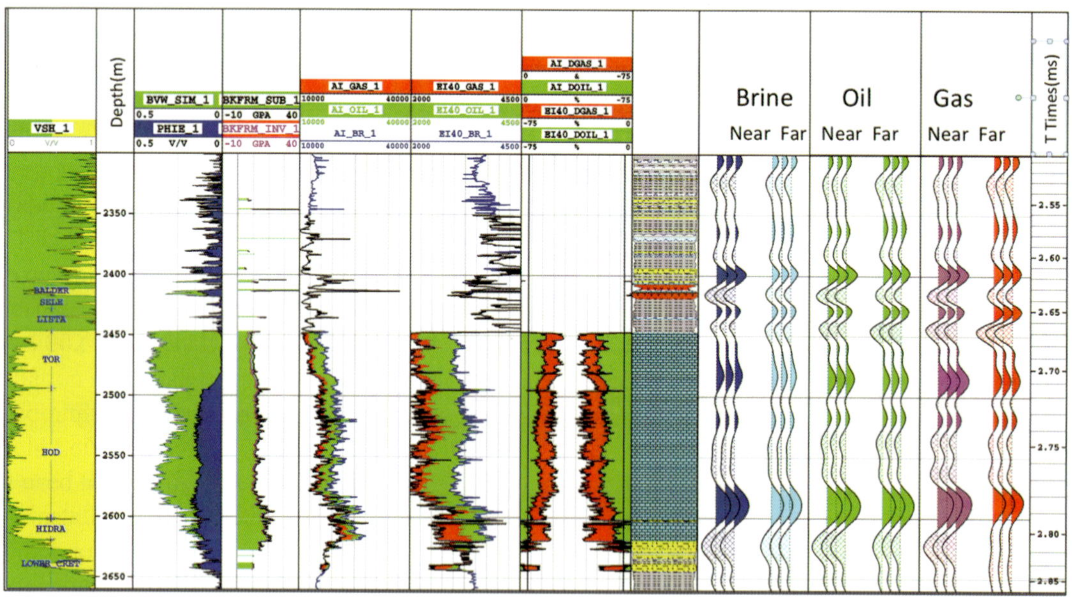

Figure 3.23: Testing the rock physics model at log scale. Courtesy of Ian Stockden.

Fluid properties are then calculated using available pressure, salinity, GOR API and temperature data and published equations and Reuss averaging is used to obtain the effective fluid modulus, given the actual saturations. Alternatively, could we have used Brie's averaging to assess various forms of patchy saturation. The final elastic properties based on combining dry rock and fluid properties are then calculated using Gassmann. The various QC products are essential during this stage, to ensure that the predictions are in line with expectations, and we prefer to qualify the modeled results using actual wells, as shown in 3.23, if possible.

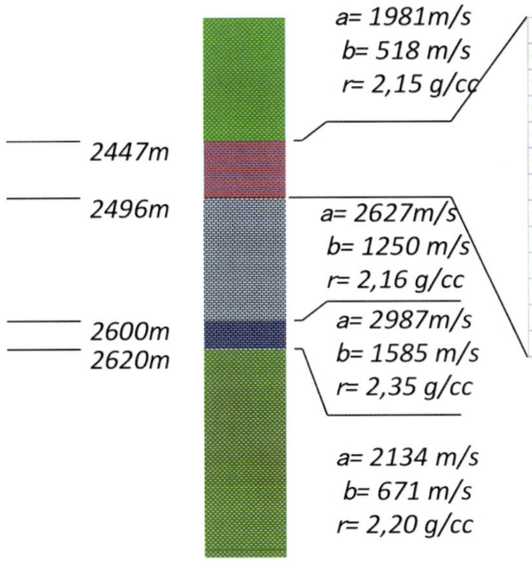

	ϕ	α [m/s]	β [m/s]	ρ [kg/m3]	Sw
Case 1	0,45	2159	998	1855	0,05
Case 2	0,40	2308	1111	1950	0,05
Case 3	0,38	2367	1157	1988	0,05
Case 4	0,36	2427	1202	2026	0,05
Case 5	0,34	2486	1247	2064	0,05
Case 6	0,30	2605	1338	2140	0,05
Case 7	0,45	2467	1008	1941	1,00
Case 8	0,40	2596	1112	2026	1,00
Case 9	0,38	2648	1154	2060	1,00
Case 10	0,36	2700	1195	2094	1,00
Case 11	0,34	2752	1237	2129	1,00
Case 12	0,30	2855	1320	2197	1,00

Figure 3.24a: Simple block model based on well, used for scenario testing and QC of full model. Input data, from Askim (2003).

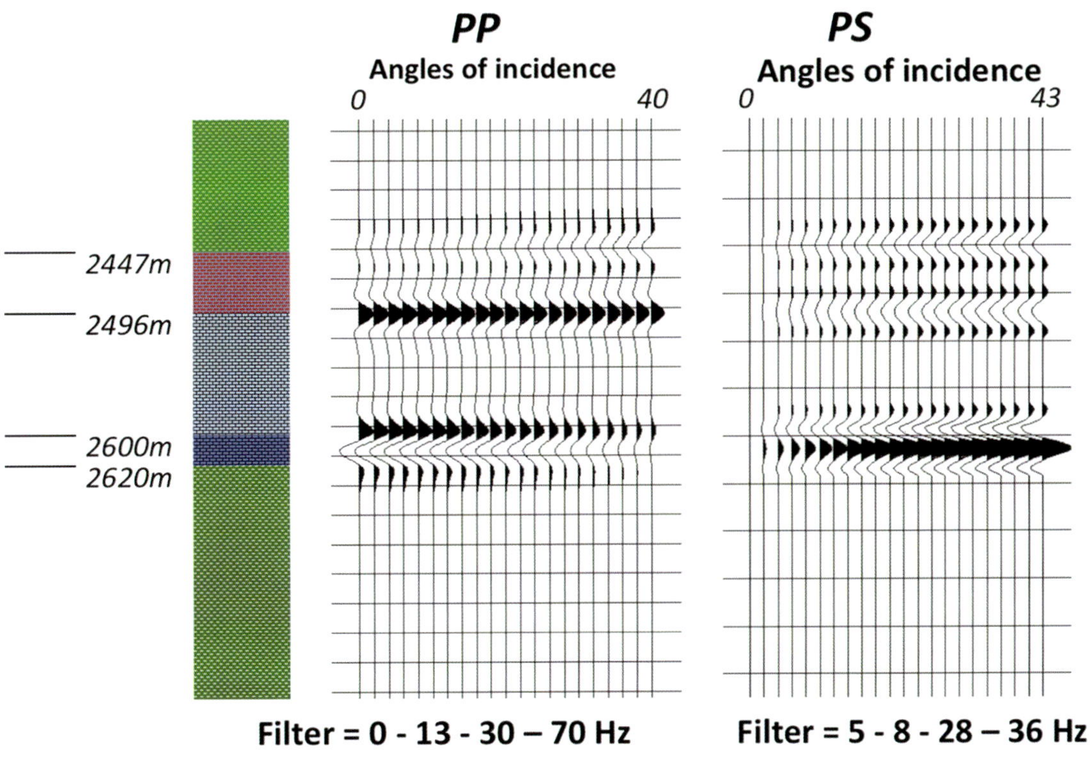

Figure 3.24b: Simple block model based on well 2/8-8, used for scenario testing and QC of full model. Seismic response, from Askim (2003).

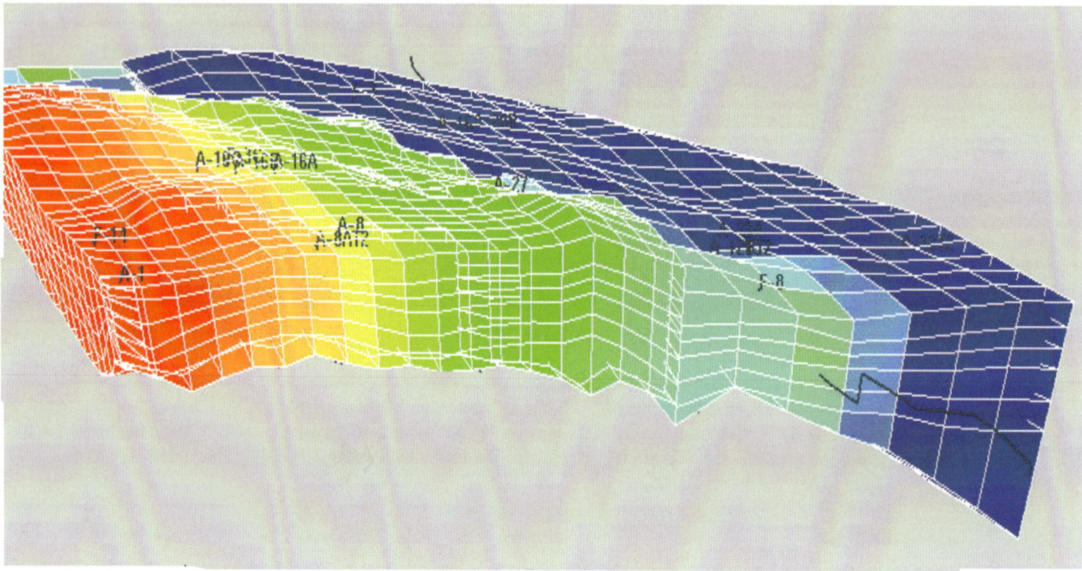

Figure 3.25: The actual reservoir simulator provided information about pore pressure, saturation, initial porosity and compaction at the selected time step, Askim (2003).

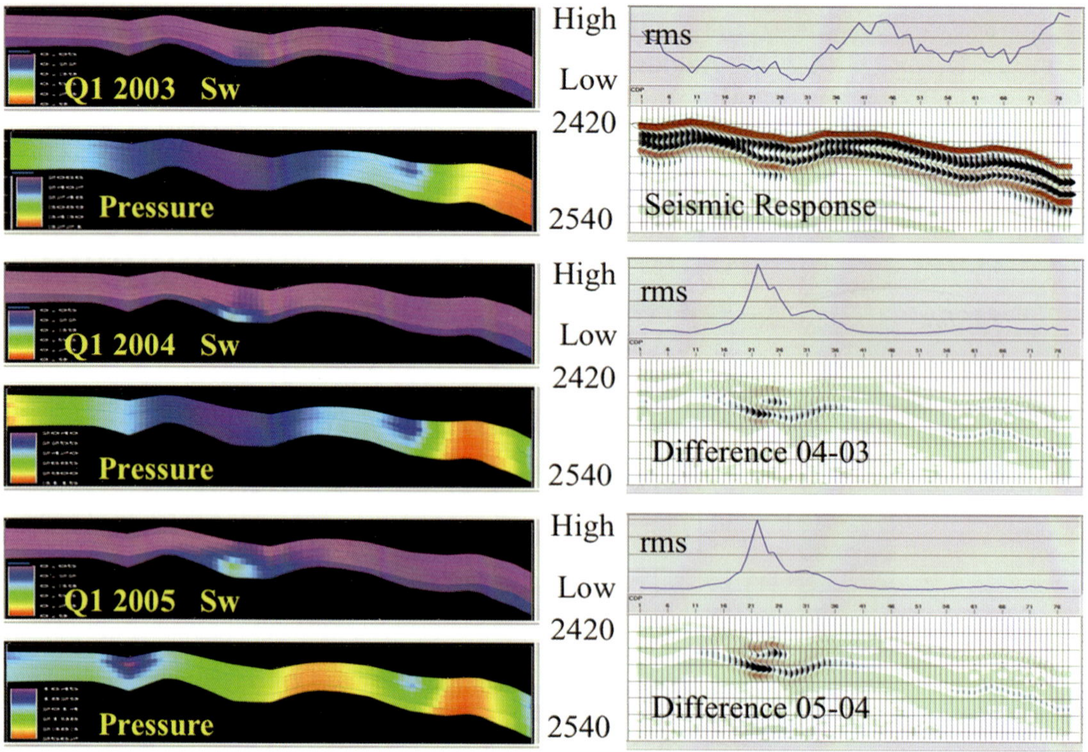

Figure 3.26: Correction showing pressure and saturation changes as predicted by the reservoir model, compared with its seismic response and differences. Modified from Askim (2003).

Due to possible interference effects, the overburden shale model was later refined to allow more detail above the reservoir. In practice this was done based on the stratigraphic layering, using 10-20 m thick layers, and functionality was also included to modify these layer due to the unloading effects from reservoir compaction. Arrival times were determined by using a constant interval velocity for the overburden. The depth of each reflector was adjusted for each time step, as the reservoir thickness changed due to reservoir compaction. Compaction in high porosity chalk causes significant changes in the seismic response, Guilbot and Smith (2002), and Barkved et al. (2003).

As the reservoir simulator grid in part was larger than conventional CDP spacing; all properties were interpolated to 25 m lateral spacing. Angle gathers for both compressional and converted waves were then calculated utilizing the Zoeppritz (1919) equations, figure 3.24b.

Although we are comparing the differences in the AI or EEI space, we were more comfortable analyzing the differences in the reflection space after convolving the acoustic impedance data with a realistic seismic wavelet, as this also included tuning effects. At a later stage, when matching model to real time-lapse data, we refined this approach. Visually it is easier to work with acoustic impedance, as this is a layer property and when working with difference data we applied a color inversion operator to the real data prior to differencing and applied exactly the same procedure to our modeled seismic.

The simulator to seismic conversion is shown in figure 3.26, The RMS and RMS changes were calculated in a 80 msec window and displayed at the top of the seismic sections for quantitative analysis,

We may use a tornado diagram to report the sensitivity of the seismic "attributes" due to the production-induced effects. Figure 3.27 shows links between the rock and fluid properties and the compressional velocity, split between pressure- and saturation-driven effects.

As we move on with our studies and start obtaining data we might have to refine our modeling further to use it actively for re-interpretation, history matching of the model, to guide new well placement and to use for reservoir management. The purpose of the reservoir flow simulator is to predict future production and pressure. If we are to maximize the use of seismic time-lapse data we need to ensure that the reservoir model is fit for modeling the seismic features.

In most cases we will need to add layers above and below the reservoir simulator layers. But to allow for comparing modeled amplitudes on a relevant basis, we need to ensure that the seismic model is comparable with the actual recorded data. Most often when we work on 4D seismic data we will look

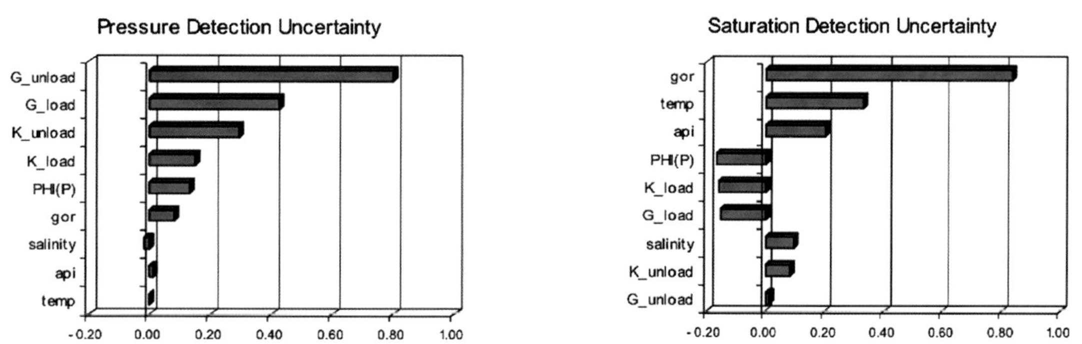

Figure 3.27: Tornado plot, calculated compressional velocity changes due to pressure changes (left) and different fluid phases at constant pressure to the right. From Askim (2003). The properties should be familiar but the subscript load/ unload is related to whether we are increasing or decreasing the effective stress (loading or unloading the rock).

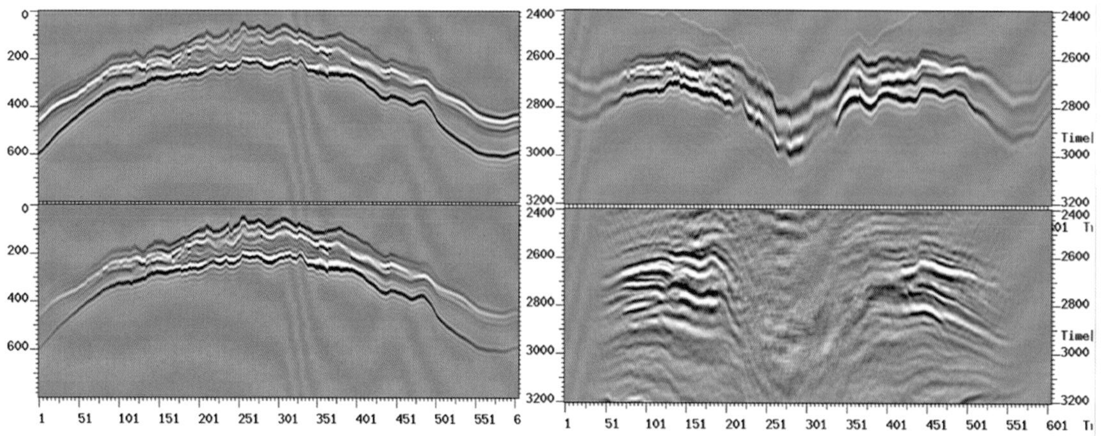

Figure 3.28: Top left picture shows the output from field specific workflow, using just one wavelet. Bottom left: we have applied the source illumination to the 3D representation of the wavelets. Top right: shows the output after applying a wavelet with source illumination and 3D wavelet. The image is now shifted from an idealized time structure to a true time structure. The bottom picture shows the real time data. From Askim (2003).

at differences, which simplifies some issues. However, if we have spatial variation in illumination and frequency, we could also incorporate these effects in the modeled data. In figure 3.28 we see how these effects gradually help in making the comparisons easier. The frequency content now appears similar and one can better observe the similar character of the reservoir section. Overburden propagation and the background model below the reservoir is yet not included at this stage. However, the better match frequency and illumination effects improve the comparisons. Additionally, we might want to scale down amplitude associated with the gas cloud.

3.7.2 Noise

The detectability of a 4D response is a function of 4D signal *versus* 4D seismic noise. Prediction of 4D noise may be as important as predicting the 4D signal in a feasibility study. The modeling we have discussed in the previous section does not include any noise, apart from the bias inherent in the estimated rock properties. Dyce et al. (2004), introduced Calibrated Difference in Reflectivity (CDR) as a measure for comparing 4D noise levels directly with the reflectivity. The CDR is calculated from seismic difference sections where the seismic amplitudes have been calibrated to log data prior to the subtraction. The resulting noise measure, which is estimated from the reservoir where no changes are taking place, can then be directly compared to the modeled changes in reflectivity due to production effects. The CDR represents the noise floor that the modeled 4D changes may be matched against to check if they are detectable or not.

4D noise may be ambient or geologically related. Due to acquisition and processing differences, we might introduce an error between subsequent time-lapse surveys that there may be a systematic error linked to the actual geological settings, inside and outside the reservoir. Dyce et al. (2004) concluded that for the reservoir they analyzed there was a strong correlation between the 4D Noise (CDR) and the "3D signal", see figure 3.29. The 3D signal is the RMS amplitude calculated from band-limited acoustic logs across the same geological intervals as the CDR, or 4D noise, is calculated based on actual seismic data. Such a correlation may be expected as the impact of time-shifts and amplitude errors will be stronger the larger the reflectivity of an interface. Consequently, Dyce et al. suggest using band-limited log data as an initial indication of the expected noise threshold.

Normalized Root Mean Square, NRMS, is the standard reference for seismic repeatability, the lower the number the better the repeatability.
NRMS is defined as:

$$NRMS = 2*(RMS(A-B))/(RMS(A)+RMS(B)) \tag{3.51}$$

where A and B represent two different time-lapse data sets.

$$CDR = RMS(A^{'} - B^{'}) \tag{3.51-1}$$

Where A' and B' are versions of A and B calibrated to the log reflectivity.
The relations between NRMS and CDR can then be expressed as:

$$NRMS = \frac{CDR}{RMS(3Dsignal+3Dnoise)} \tag{3.51-2}$$

NRMS suggests better repeatability when the "3D seismic" signal is strong, which could obviously be misleading.

Will there be an observable signal production induced signal in our data? In the end this is contingent on the properties of the rocks, the repeatability of our data and the amount of 4D noise.

There are many cases of successful 4D seismic in more consolidated or cemented reservoir rocks, but these might need careful attention to ensure optimal repeatability. Steerable marine sources and streamers, retrievable ocean bottom cables and nodes or land geophones, have all demonstrated that very good positional repeatability may be achieved, given that the conditions are reasonable. If positional accuracy and repeatability beyond this is required, a permanent system may be the solution. In marine settings, placing the receivers on the seabed might provide shearwave recordings, which may provide additional insights into changes inside and outside the reservoir.

"Difficult" reservoirs may also require special attention during the processing and analysis. Figure 3.30 shows examples of 4D seismic potential ranking of most of BP's North Sea assets. The magnitude of the hydrocarbon contact reflection coefficient can be used as an indicator of the magnitude of the potential 4D signal resulting from fluid saturation change.

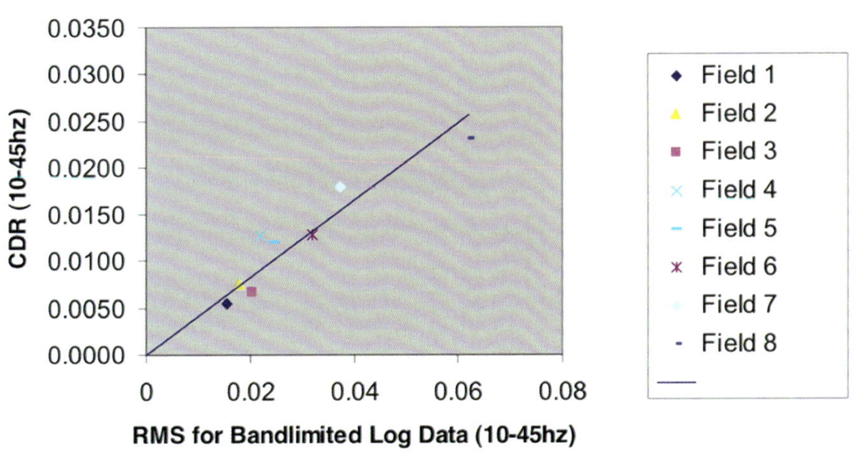

Figure 3.29: 4D Noise versus 3D Signal – measured data, Dyce et al. (2004).

The rightmost field represents the most challenging field. Data for a relevant field in the area is used to illustrate the detection threshold using different processing strategies but the same processing.

As with all attempts of standardizations, there are risks of oversimplification. In the case of using the CDR as a base-lining tool, the model does not honor residual saturation. The detection levels as defined in figure 3.30 is also base on inherent assumption, and the level as indicated are true for a single field, and for a specific point in time. The key message is that improvement in acquisition and processing has dramatically reduced the 4D noise threshold over time, making 4D feasible for even tougher rock properties.

3.7.3 Modeling changes outside the reservoir

When the first 3D seismic data acquired across Valhall were interpreted, we ran into problems tying wells to seismic using check shot and VSP data. VSP has been part of the strategy for imaging below gas and even if the toe of the well was outside the gas-affected areas, the mismatch could be 20–30 msec. Old wells simply did not tie. Interestingly enough, a new well drilled after the 3D seismic data was acquired matched very well and there was only a difference of 2 msec between the two data sets. The geomechanics working the field might have had his suspicions and two years after acquiring the 3D data we ran a densely sampled check-shot survey in a new, close to vertical well in the center of the field. This well was a sidetrack of the first producers drilled at Valhall and was targeting the reservoir only 160 m away from the initial well. Still the difference in travel time between surface and receivers at the same depth were off up to 40 msec, see figure 3.31. Internal and external peers were consulted and the most likely explanation was thought to be lateral variation in gas charges across a fault, which appears to be a common explanation also shared by other chalk fields that had gas charges in the overburden.

It took another 5–10 years before the geophysicists reinvented the geomechanics of a compacting field.

One of the very first published 4D seismic examples of reservoir induced changes in the overburden is by Guilbot et al. (NPF 2002), who presented a detailed interpretation of the first towed streamer survey across the Ekofisk field. The 4D seismic maps showed clear time-lapse effects associated with reservoir compaction and subsidence. Time-shift at the top of the reservoir up to 20 msec, corresponding to a factor 3–4 of the modeled compaction, could be read out from the disclosed figures, figure 3.32. Hall et al. (SEG 2002) published similar results from the Valhall field, matching 1992 towed streamer 3D data with

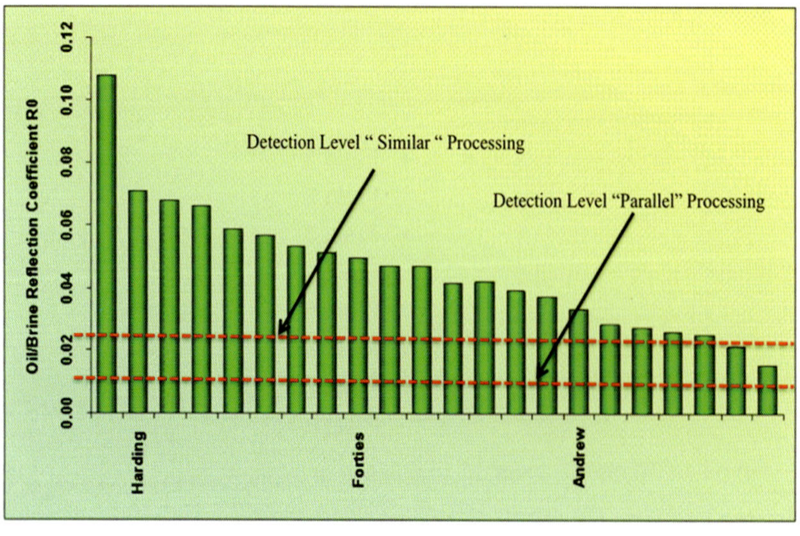

Figure 3.30: Initial ranking of the 4D potentials of BP's North Sea fields, Dyce et al. 2004.

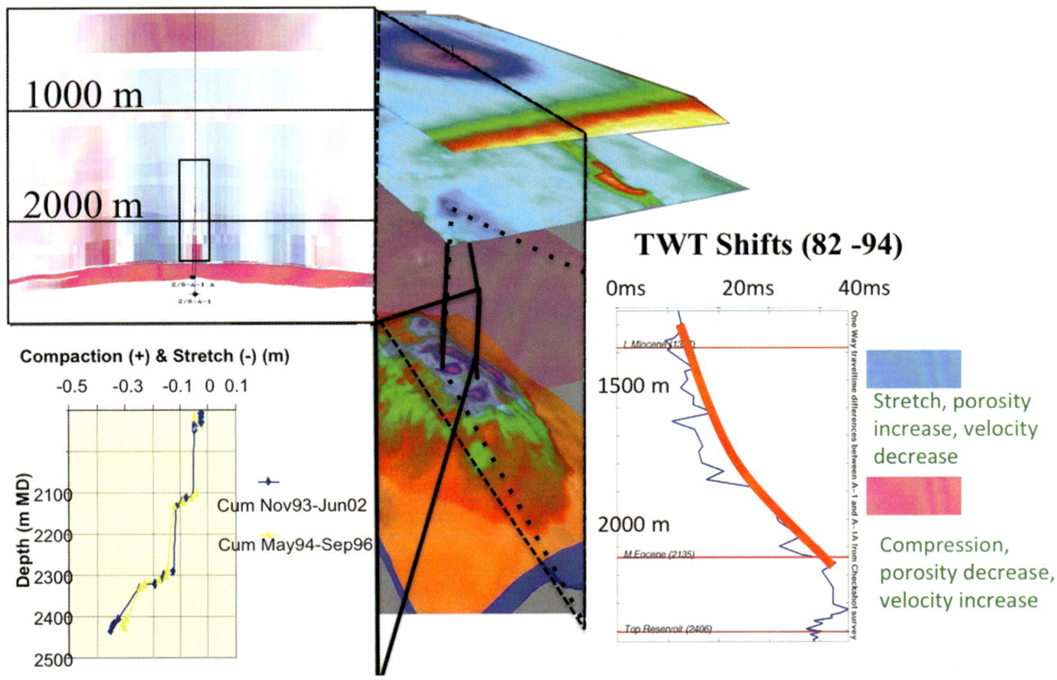

Figure 3.31: Upper left: X-section showing the modeled changes in volumetric strain in the field in 1996, red indicates a reduction in pore volume, blue an increase. **Lower left:** strain measurements in two wells from 1993–2002 and 1994–1996. Center 3D model showing compaction at reservoir level (0–10m) and subsidence at sea-level (0–4 m) in 2002. Lower left: Difference in traveltime between VSP acquired in 1994 and 1982, virtually in the same location, modified from Barkved et al (2005).

1997/98 3D/4C ocean bottom cable data. They observed a time-shift of up to 23 msec at the top reservoir level and suggested that a reduction in velocity of 0.8% could be enough to generate these changes.

Landrø and Jansen (EAGE 2002) and later Landrø and Stammeijer (2004), suggested that traveltime time-shift and near and far traces could be used to discriminate between layer thickness and velocity changes. They proposed a simple 1D formula relating the fractional changes in traveltime Δt/t to fractional changes in thickness Δz/z and velocity Δv/v:

$$\Delta t / t \approx \Delta z / z + \Delta v / v \tag{3.52}$$

The reservoir travel-path is less because of compaction and the velocity will be increased due to increased effective stress (loading), both effects will reduce the traveltime. Outside the reservoir we will see the opposite effects. Above (and below) the reservoir the rock will be unloaded (stretched). The path length will increase and we also have a reduction in effective stress (unloading), which reduces the velocity. The net effect is an increase in seismic traveltimes. However, in the reservoir there may be other effects causing time changes, like temperature or fluid changes. These could often mask the compaction effects. A possible way to separate the fluid effect could to study the amplitude *versus* offset effects, AVO. However, the compaction often introduces local stress-state changes that could result in azimuthal variation or AVOA effects. Figure 3.33, shows an example using the matching surveys 1 and 6 from the Valhall LoFS array. To the left we see how the pressure decline in the reservoir results in an increase in velocity, while the shale above (and tighter chalk below) is experiencing a net stress reduction and

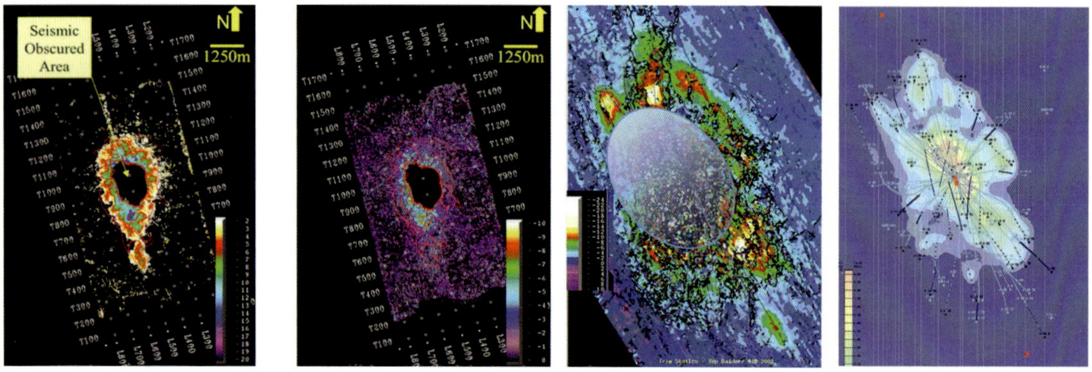

Figure 3.32: Classical examples of early time-shift observations from the Valhall and Ekofisk fields. a) and b) time-shifts and compaction and c) and d) – same from Valhall. In the Ekofisk case the time-shift is 3–4 times larger than the compaction, while in the Valhall case we observed time-shift 5–6 times larger than the modeled compaction. Modified from Guilbot and Smith (2002), and Barkved et al. (2003b)

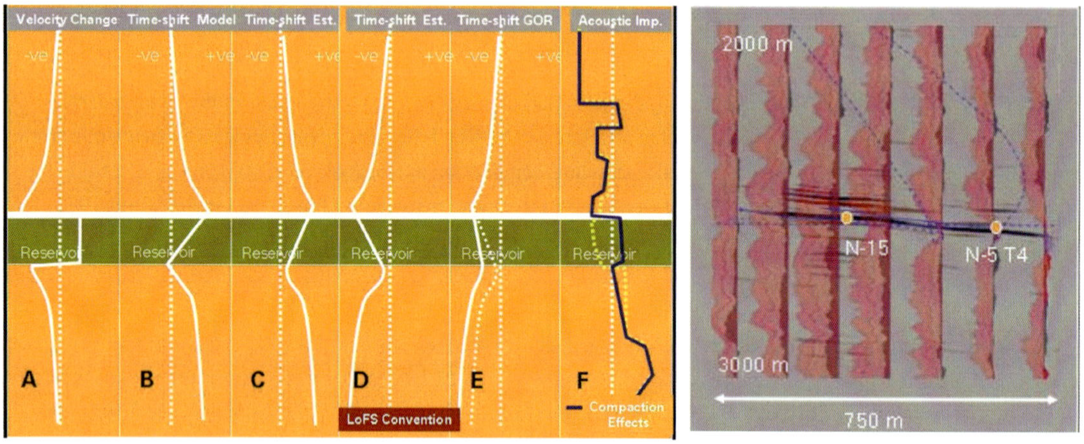

Figure 3.33: Left caption; A) increase in velocity in the reservoir due to production (pressure reduction and compaction) and decrease in velocity above and below. B) Changes in traveltime. C) Using a sliding window operator to do the traveltime calculation. D) Just a sign convention issue E) increases the Gas Oil Ratio result in reduced traveltime effects across the reservoir. F) The velocity changes outside need to be incorporated in the seismic forward modeling. The right-hand figure shows successive time-shifts for six surveys. To the far right we see production from a well at an early stage, the velocity increase in the reservoir is of the same magnitude as the reduction in velocity above. In the center (N-15), where the pressure regimes are lower, we see the gas coming out of solution, impacting the (velocity) time-shift calculations.

reduced velocities. The time-shifts are calculated using a sliding window, which introduces some edge effects. Early in depletion stage we notice a strong "kick-back" in the time-shift, due to increased velocities in the reservoir. As production continues we will eventually pull the pressure below the "bubble" point and obtain an increase in GOR and subsequently, we lose the "kick-back" effect in the reservoir.

Hatchell and Bourne (2005) introduced the R factor as a simple means of relating time-shift changes to compaction induced dilation of the overburden.

$$\Delta t / t = (1+R)\tfrac{\Delta z}{z} = (1+R)E_{zz} \qquad (3.53)$$

Furthermore, based on empirical data Hatchell and Bourne also suggested that a relevant value for R in the overburden was between 4 and 8. E_{zz} denotes vertical strain. Røste et al. (2005) proposed a similar approach. They suggested that the changes in velocity and the changes in dilation are proportional to a given factor α.

$$\Delta v / v = \alpha \tfrac{\Delta z(x_0)}{z(x_0)} \qquad (3.54)$$

The assumption is again a uniaxial stress/strain relation and implicit in their linear strain/ velocity ratio is that the velocity v, is a linear function of porosity, x_0 refers to zero offsets. An initial guess is made of a constant alpha and they developed a method using all offsets to find a constant but laterally varying alpha. The test applied to the Valhall data gave an α of 5, which compares well with Hatchell and Bourne's observations from other fields and internal relations used by the Valhall asset, Røste (1986).

The R factor, due to its simplicity and apparently "narrow" ranges, has gained lot of popularity in the industry. However, there is still a factor of 2 between the low and high estimates. The uni-axial assumptions are definitively valid above a infinite reservoir with infinite permeability figure 3.34, in which case we would not expect to see any time-shift, as the effect is linked to stress arching. Intuitively, we would expect the R-factor concepts to break down on the edges of the reservoir and in some cases we have lateral heterogeneities (fault present). Assuming that the velocity primarily depends on net mean stress, using Hooke's law, Holt et al. (2005) find this expression for R:

$$R = K_{fr}\left(\tfrac{1}{V_z}\tfrac{\Delta V_z}{\Delta \sigma'}\right)\tfrac{\varepsilon_v}{\varepsilon_z} \qquad (3.55)$$

Here, $\boldsymbol{K_{fr}}$ is the bulk modulus of the dry rock (frame) and V is the volumetric strain while εz is the vertical strain, and σ is the net stress. $\varepsilon_V / \varepsilon_z$ will vary around a locally depleted reservoir, see figure 3.34 and consequently, R will vary with distance and position with regard to the depletion zone. Apparently, vertical heterogeneity has less of an impact, according to Fuck et al. (2011).

If we look at a real data example, we will get an idea of possible spatial variation in the R factor. At Valhall, a strong vertical element in the deformation was seen by the seismic waves, figure 3.35, conse-

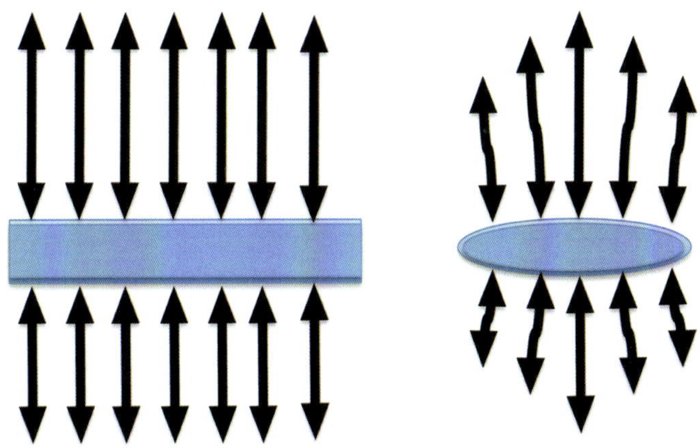

Figure 3.34: Strain/stress behavior around two different types of reservoir. Left reflects a true uniaxial deformation, while in the two on the right we see another geometry.

quently, we are reasonably comfortable using a simplistic approximation, being aware that there could be areas of the field where these relations could break down. In practice, we will need a minimum of geomechanical modeling/assessment before even using the observations for qualitative analysis.

Bauer et al. (2008) introduced an alternative parameter S, describing the stress sensitivity of the P-wave velocity:

$$S = \Delta V_{Pz} / (V_{Pz} \Delta \sigma_z^{'}) \tag{3.56}$$

Where the effective stress

$$\Delta \sigma_z^{'} = \Delta \sigma_z - n_{Pz} P_f \tag{3.57}$$

is controlling the velocity $\boldsymbol{V}_{Pz}$, $\boldsymbol{n}_{P_z}$ is a an effective stress coefficient and $\boldsymbol{P}_f$ is the pore pressure.

Holt and Fjær (2009) compared the R and S parameters using experimental data with a constant mean stress path and found that the contribution is small. This stress path may be valid just above the center of the reservoir. Towards the flanks or the edges of the reservoir, the vertical stress will be constant and the pore pressure response will need to be acknowledged. Holt and Fjær (2005) show that R, as a rock physics parameter measured in the laboratory, depends strongly on stress path. They also found that a wide variety between large *R-values* measured in unsaturated granular media with Hertzian contacts and small *R-values* measured in cemented rocks in virgin conditions. *R* is in general stress sensitive and depends on stress history. However, it turned out that apart from the lateral effects pointed out by Bauer

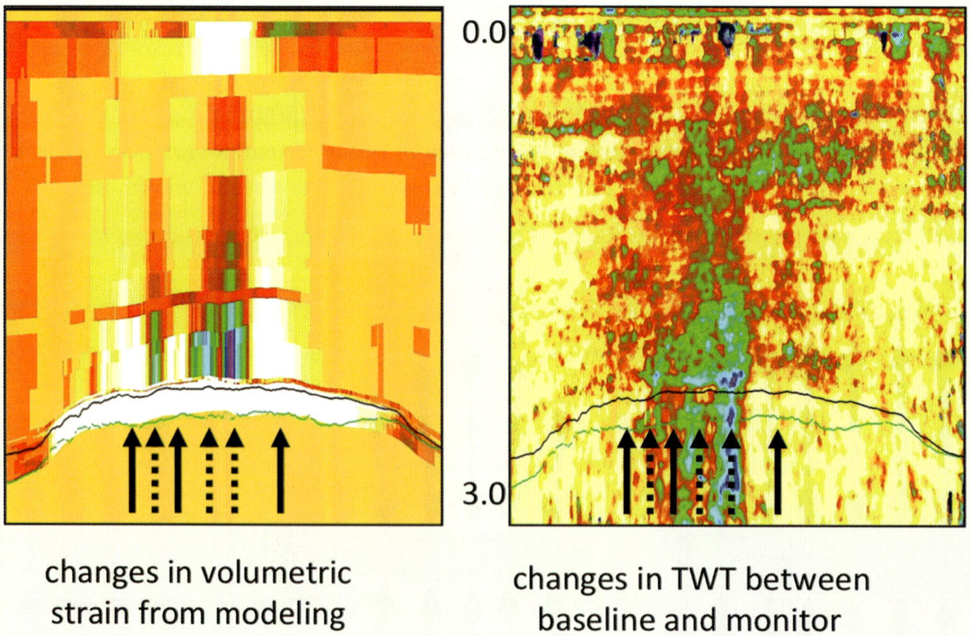

Figure 3.35: Valhall-modeled changes in volumetric strain with observed time-shifts. Barkved and Kristiansen (2005).

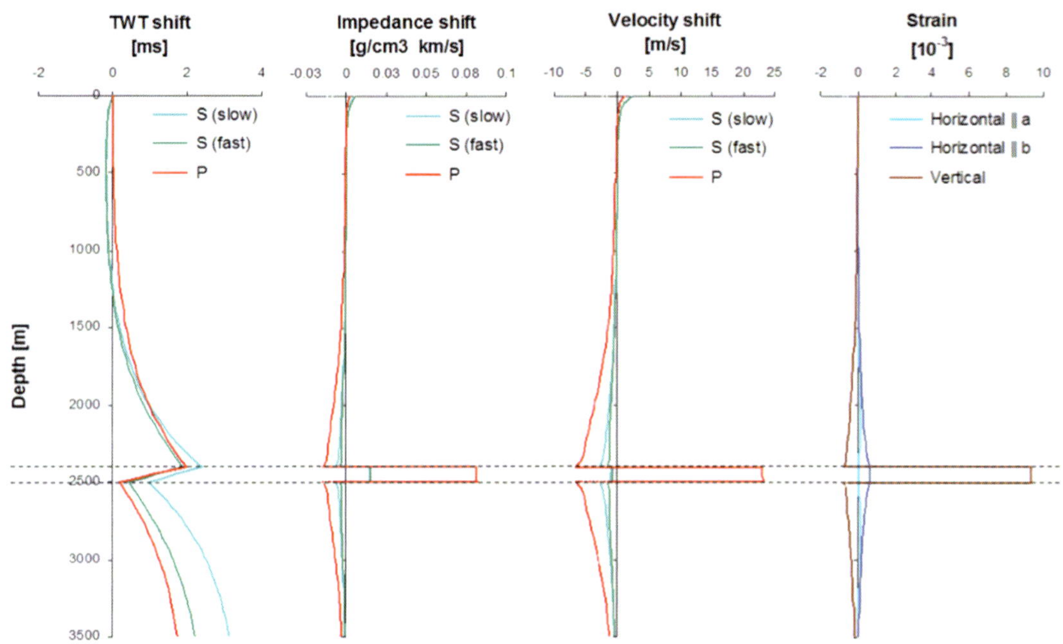

Figure 3.36: Shift in traveltimes, impedances and velocities for zero-offset waves and strain at the center of the drained region. The dashed horizontal lines indicate the location of the reservoir, from Fjær and Kristiansen (2009).

et al. (2008), using a constant R gives a similar predicted TWT shift obtained with a depth dependent *R* based on a rock physics model. Theory and experiments, Sayers (2010), shows that the seismic velocities are dependent on the complete triaxial stress state and Herwanger (2008) shows that by neglecting the influence of horizontal strain, the third-order elasticity reduces to the R factor model as presented by Hatchell and Bourne (2005) and conversely, also the α as introduced by Røste et al. (2005).

A sometimes overlooked phenomenon is that changes in pore pressure in the reservoir changes not only the stress state in the reservoir but will have to couple with and possibly impact the stress state of the surrounding rock. There is no need for a strain change for this to occur. This will of course impact the properties of the surrounding rock and also impact a potential 4D seismic signal.

Not only do we need the properties for the rocks outside the reservoir, we also need a better framework, a geomechanical model, to predict how the stresses will develop.

Fjær and Kristiansen (2009) describe a screening tool to help simulate expected 4D seismic effects due to saturation changes and production induced geomechanical changes. The tool, named mech2seis, consists of a geomechanical module, a rock physics module and a seismic module. The tool can be used to check out conceptual models. One of the examples they describe considers traveltime changes for zero-offset waves at the center of a depleted reservoir section. Figure 3.36 shows the response of a compacting reservoir. The depleted region is assumed to be elliptical (long axis a = 1000 m, short axis b = 700 m). The characteristic increase in P-wave traveltime through the overburden followed by a reduction through the reservoir is exactly as expected for a compacting field; notice the impedance changes in the surrounding formations. For quick look modeling, the core of the model is based on a nucleus of strain concept, Geertsma (1973). The depleted region consists of small, spherical elements that shrink when pore pressure is reduced and the surrounding rocks will be slightly deformed as a result. This model is based on modeling rock deformation in a linear elastic manner and the total displacement effect at points outside the reservoir is given by summing over a relevant assembly of these spheres, similar to the approach shown in figure 3.17. Strain is calculated from the displacement gradients and Hooke's law then estimates stress state changes. For a more elaborate analysis, a FEM module that includes a variety of constitutive models or alternatively, data from an external FEM model can also be imported.

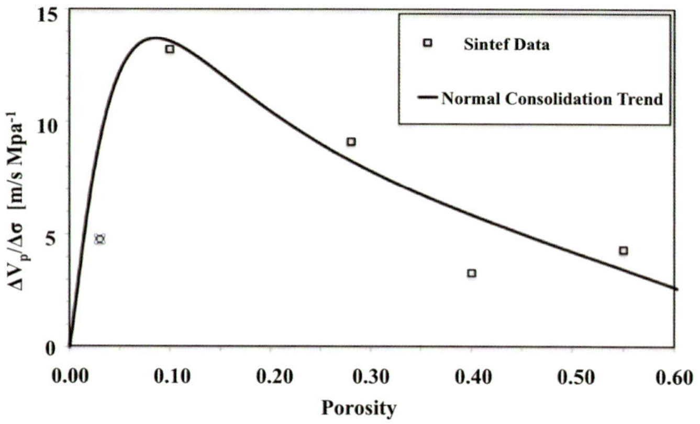

Figure 3.37: Stress sensitivity of shale. From Holt et al. (2005).

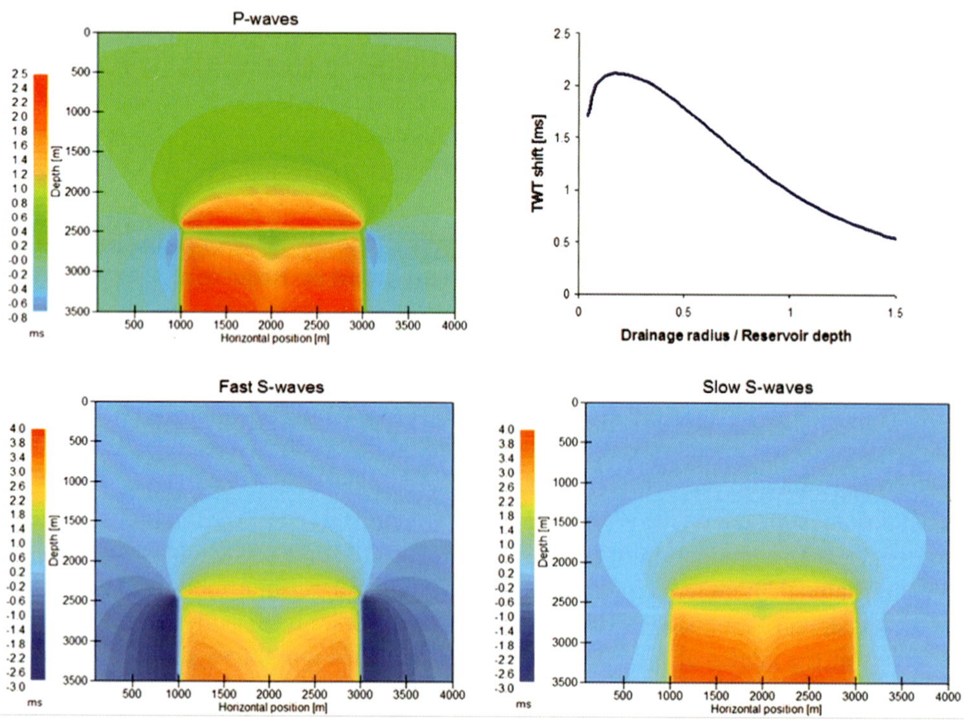

Figure 3.38: Change in two-way traveltime for zero offset waves at a depleted reservoir section. The upper right-hand diagram shows traveltime for P-waves reflected at the top center of the reservoir, *versus* the radius of the depleted region. The depletion is assumed to be constant within the depleted region, from Fjær and Kristiansen (2009).

The rock physics model for the reservoir is based on a model initially designed for weak sandstone, Fjær (2006), and refined to take into account fluid effects. Using fairly general stress geometry, the relevant density and anisotropic compressional and shear velocities are calculated. The surrounding rock utilizes a generalized version of the shale model of Holt and Fjær (2003). The resulting elastic properties can be exported to external software for modeling, or the seismic responses may be modeled using in the seismic module based on simple ray tracing and a convolution model. The results may be viewed as CMP gathers, x-section or maps of the changes in traveltime, impedance, shearwave splitting, etc. The impact of the individual factor depends on the actual rock, depositional history, stress path history and pore pressure. Figure 3.37 shows the estimated stress sensitivity of shale *versus* porosity. Data from an external FEM model can also be imported.

Figure 3.38 shows a cross-section along the long axis of the reservoir and we notice the increase in two-way traveltime above the drained region at 2500 m (under the stress arch), while we see a reduction in traveltime at the sides of the drained region, where the stress pillars are located. Figure 3.39 shows a real example from the Valhall LoFS array for this scenario. We are in the early stages of "stress arch" development. The drainage radius (~125 m) *versus* depth (~2500 m), is 0.05 and as can be seen from the upper right caption in figure 3.38, we are still in a region where the TWT shifts will increase. Notice that the light green/blue colors are in the order of 0.25–0.5 msec. Only when we have a highly repeatable system we will be able to see these details. The consistency of subsequent surveys adds to the confidence.

The lower two captions in figure 3.38, show the fast and slow s-wave. We can calculate the shearwave splitting at varies interfaces through the overburden. Our starting model was isotropic but we now see that compaction induced unloading effects in the overburden has an anisotropic response. Figure 3.40 shows the shear wave splitting at various levels and in figure 8.17 (chapter 8) we see the shearwave splitting response from about 600 m, see also Olofson et al. (2003). This information is then very useful for our calibration of the rock physics model.

The same analysis and product as shown above using the simple screening module can be produced based on input from advanced coupled models, coupling fluid flow models (finite difference) with

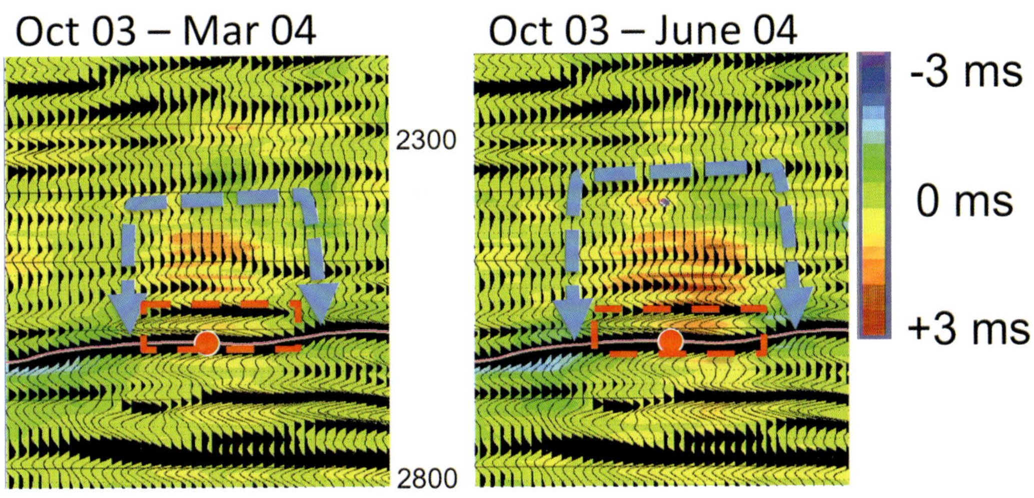

Figure 3.39: Stress arching effects over a producing horizontal well. The well is 1500 m long and the x-section is orthogonal to the well trajectory. The vertical extent of the seismic caption is 600 m. Barkved and Kristiansen (2005).

geomechanics models (finite elements) for both the reservoirs and the surrounding formations, see Herwanger (2010). The modeling will then be used in a history matching process using the seismic response in the reservoir and overburden. Kristiansen and Flatebø (2009) illustrate how this approach is used to verify detailed well-bore stability modeling at Valhall, taking into account both spatial and temporal changes of the stresses in the overburden.

3.7.4 Modeling changes inside the reservoir

Intra-reservoir shale beds are often present in clastic reservoirs, and are typically addressed by using a net to gross ratio (N/G) for volume calculation and flow simulation when they are less than 1-10 m thick, and limited in lateral extent. The same approach is used in the fluid substitutions and seismic time-lapse modeling. MacBeth et al. (2011) found that shales with such thicknesses can also strongly affect the effective elastic wave behavior of the composite reservoir, particularly when observed with frequent time-lapse surveys. We do expect that the reservoir harden due to depletion, however MacBeth et al. shows that the presence of inter-bedded shale can reduce this effects due to a combination of geomechanical and pressure diffusion phenomena that are in turn controlled by the shale thickness, permeability and the mechanical properties. They modeling suggests that a sub-seismic shale layers of approximately 1 m thickness take less than three months to pressure equilibrate whilst thicker shale layers of 10 m can take over 20 years. The research in this area are still in its early stages and the critical set of parameters required to carry out accurate calibration of these predictions is not yet fully available from published

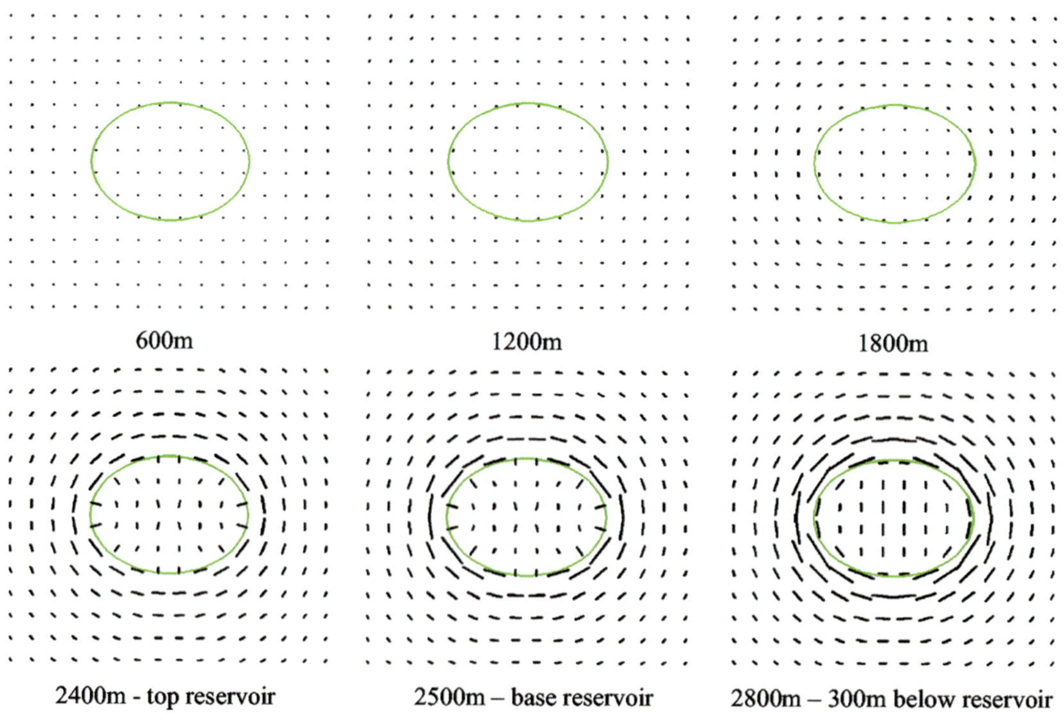

Figure 3.40: Shearwave splitting for waves reflected at various depths from the surface down to 300 m below a depleted reservoir. The length of the black line shows the magnitude of the splitting at this location, while the orientation of the line illustrates the polarization of the fastest component. The green line shows the location and extent of the depleted region, from Fjær and Kristiansen (2009).

literature. They suggest that seismic surveys repeated over 5–10 years are likely to be influenced by 5–10 m thick shales such as those found in low energy depositional systems, whilst thinner shales will not present a problem for these surveys as they will pressure equilibrate over that time period. However, frequent seismic monitoring on a 3–12 monthly basis will be influenced by sub- seismic shales over the entire thickness range of 1–10 m and this is likely to be apparent for a wide range of depositional environments, see MacBeth et al. (2011) for further details.

3.8 Conclusion

In this section have we briefly discussed the fundamental rock physics and geomechanics principles, which may be linked to production-induced changes in the subsurface. Few reservoirs might experience the magnitude of deformation that can be found in some of the North Sea chalk fields. However, unfortunately there are new reports about unexpected consequences associated with production for less consolidated or high-temperature/high-pressure reservoirs.

We have discussed in some detail how simplified concepts like the R-factor can be successfully used in the interpretation of seismic time-lapse data.

Forward modeling remains fundamental in defining what changes should be expected. Including effects outside the reservoir, it can "easily" complement the more conventional fluid substitution/reservoir pressure changes.

Stress state changes will easily introduce an isotropic effects, like shearwave splitting. Few examples have yet been reported using this technique at reservoir level but this effect should also be kept in mind when doing conventional AVO analysis.

4 Acquisition, Processing, 4D analysis and Permanent Systems

"An important scientific innovation rarely makes its way by gradually winning over and converting its opponents: What does happen is that the opponents gradually die out."
- *Max Planck*

The quality of a 4D image is driven by the strength of the 4D signal and the quality of the acquisition and quality of the processing. The purpose of the processing is to reduce 4D noise, to improve the 4D repeatability, compensate for non-repeatability in acquisition and to accentuate the 4D signal due production-induced changes in fluid, pressure and stress. Processing efforts and strategy may vary depending on the acquisition methodology. Permanently installed receiver arrays may reduce the need for cross matching between surveys, and processing can more easily be "automated" and delivered faster. Amplitude versus offsets (AVO) analysis is used to separate fluid and pressure effects. Stress state changes outside the reservoir are addressed by time-shift analysis. For those reservoirs where this applies, it is important to maintain this signal through the processing, and to use relevant algorithms to estimate the time-shift response.

4.1 Introduction

Production induced changes in acoustic impedance in a reservoir depend on the rock properties and amount of depletion are usually very small, only a few percent. The 4D noise due to scattering, non-repeatability in sources and receivers and environmental changes is often of the same magnitude. The smaller the production induced seismic responses we can see, the earlier we may reaffirm our plans or take corrective actions to mitigate unexpected behavior. We need highly repeatable seismic data for our 4D analysis. The acquisition and processing strategies need to be linked, and selected to delivery 4D images of sufficient quality to address the business problem.

In the processing, we use global cross matching of the time-lapse surveys, or matching using large spatial operators is often used to correct for differences in survey characteristics and certain deviations in geometrical repeatability. Workflows are developed to reduce 4D noise. i.e. multiples and scattering effects. The acceptable geometrical deviations between time-lapse surveys are decided based on local experiences. We commonly refer to the deviation between actual and planned source and receiver position as ΔSrc + ΔRec. Acceptable ΔSrc + ΔRec may vary between areas, and are apparently influenced by the presence of shallow inhomogeneity and diffractions, in addition to the expected size of the 4D signal.

In this section we will discuss options for acquiring 4D seismic data and factors we need to take into account to achieve acceptable repeatability. We will also discuss key steps in the time-lapse processing, 4D seismic data analysis, and share a case study on the processing and analysis of data from the Valhall permanent receiver system.

4.2 Maximizing the signal and minimizing the noise

The convolutional model is implied in most seismic interpretation and processing. This model describes the filtering of a seismic signal by the rock layers in the Earth, as illustrated in figure 4.1. We typically state this as:

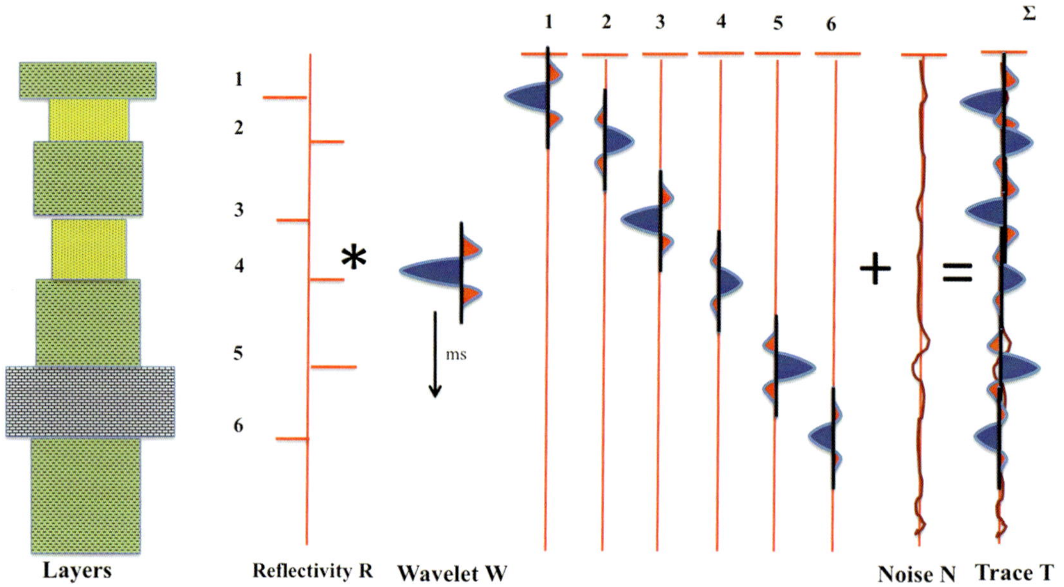

Figure 4.1: The output t is the convolution of r * w + n, which is the sum of all the wavelets shifted and scaled by reflectivity with the noise added to the result.

$$T(t) = r(t) * w(t) + n(t) \tag{4.1}$$

In reality, we use a number of source and receiver pairs at different angles of incidence, θ and positions x, y, z to image a point in the subsurface as shown in figure 4.2. We might describe this as:

$$S(t) = W(t) * R_{x,y,z}(t,\theta) + n(t) \tag{4.2}$$

Where W is the seismic signal, R is the Earths reflectivity', which is linked to locations x, y and z and incidence angle θ. This works well for conventional analysis but is too simplistic for 4D, as we need to take into account the noise that is intimately linked with the acquisition. Calvert (2005) suggested a model that included a term D, which he named distortion, representing the residual Earth transmission effects left in the data after processing. D for a point is linked to the spatial position x, y and z, but also to the source positions (i) and receiver position (j).

$$S(t) = W(t) * R_{x,y,z}(t,\theta) * D_{i,j,x,y,z}(t) + n(t) \tag{4.3}$$

This conceptual model can be used to explain the strength of the 4D seismic concept based on differencing. If we repeat the same errors in the repeat survey as we did in the baseline survey, the distortions should be the same and we should be able to separate out the effects. This will require that we are able to repeat our acquisition efforts. In practice, we will need to repeat the source and receiver responses and they will have to occupy the "same" position. How well this needs to be repeated may vary from area to area, linked to local heterogeneities.

From this conceptual model, what we have called noise in the normal 3D seismic case may also be subtracted or divided out in the 4D response. A simple example could be nasty diffractions from a platform leg or some cumbersome multiples that interfere with the reservoir signal. If the acquisition is repeated, we may be able to see the reservoir better defined in the 4D image than in the 3D image, as the diffractions and the multiples will disappear in the subtraction. Based on the same argument the geology would disappear in the subtraction, but with a sufficiently strong signal the 4D response itself can help to delineate the reservoir. We still need to be wary that the 4D image can be mis-positioned in the same way that a 3D image can be though, (e.g., if both vintages are poorly migrated). Background noise represents more of a challenge, as this will change over time. It is difficult to handle in differencing operations and in the human brain. Systematic noise normally occurs as a pattern, and we might have a better chance of handling this. Ambient noise can change during the day, during the week and during the year. Unfortunately, the infrastructures across an oilfield tend to change over time; new buildings or platforms may make certain areas inaccessible. This is true in the marine setting, where the 500 m safety zone around a platform can generate a significant gap.

Seasonal systematic changes are another challenge, while the primary effects of changes in water velocities between the summer and winter might easily be compensated for, the multiple period will also be affected, which could be more difficult to mitigate. Onshore, changes in the depth of the water table could be even more devastating, Clymer et al. (1986). Sometimes, will we also experience topographic changes due to deposition of sediment or simply movement of sand due to waves or wind.

The characteristics of the source, receivers and recording systems may often change between subsequent surveys. Some of these changes can easily be compensated for or equalized, by using a common signature or system response for individual parts or the whole of the system, others might be more difficult. Compensation for these changes and the associated quality assurance is done in the early stages of the processing and can be very time consuming.

4.3 Quantifying repeatability

Significant efforts have gone into trying to quantify 4D repeatability during the last 20 years. Feasibility studies are run to understand and quantify the seismic sensitivity of the production induced signal, the 4D signal. Part of this analysis will be to assess whether the expected 4D noise level will allow the 4D signal to be detected. In this context, 4D noise is the sum of all nonrepeatable effects not linked to the

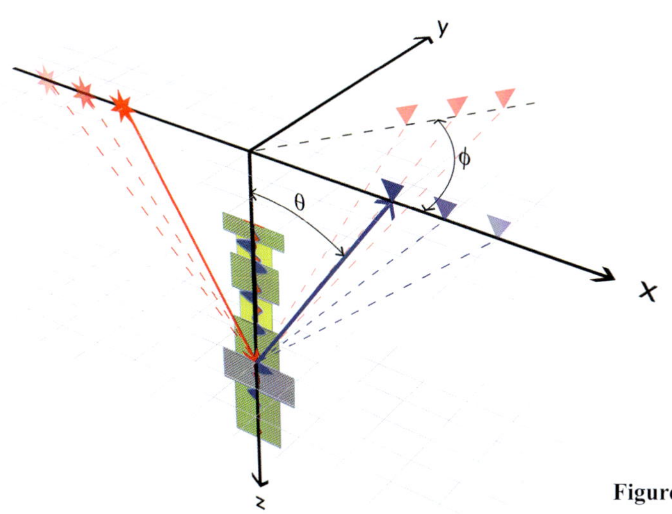

Figure 4.2: A point in the subsurface is illuminated from all source/receiver positions and from all angles.

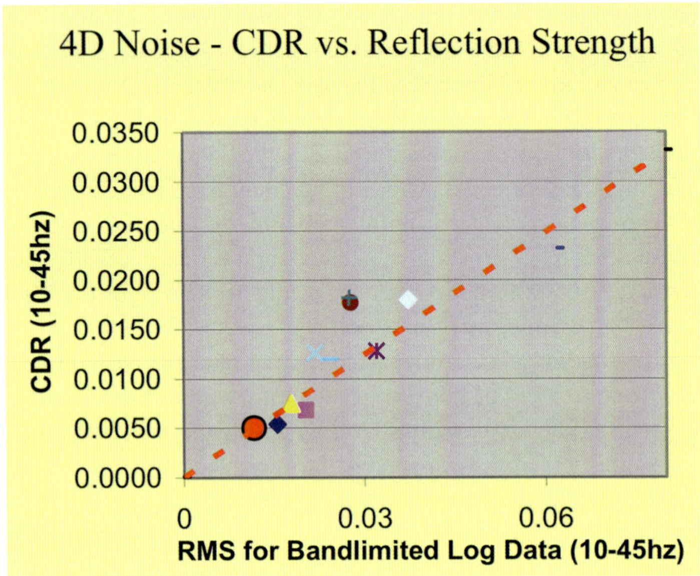

Figure 4.3: 4D noise *versus* the RMS calculated from band-limited log data. This example is from the North Sea and implies that 4D noise (CDR) is directly linked to the strength of the reflectivity. Dyce et al. (2004).

production-induced changes. Normalized Root Mean Square (NRMS) differences are the standard measurements often used to characterize the quality of a 4D seismic survey, see Ronen et al. (1999), Grion (2000) and Kragh and Christie (2002).

NRMS is defined as:

$$NRMS = 2 * \frac{RMS(A-B)}{RMS(A)+RMS(B)} \tag{4.4}$$

Where A and B represent two different time-lapse seismic surveys. RMS is typically calculated is short time windows, trace by trace.

The NRMS measure is controlled by the reflection strength within the window it is calculated in, and as such it is not an objective measure. NRMS gives an indication of the quality of a 4D seismic data set, but is best conditioned for measuring improvement of processing or acquisition within a specific survey area or geological province.

To allow for more direct comparison between the 4D signal and the 4D noise, Dyce et al. (EAGE 2004) introduced the Calibrated Difference Reflectivity, CDR. CDR is based on calibrating the seismic data to reflectivity from well data prior to calculating the difference section and is simply expressed as the RMS of the difference between two surveys. In an area at reservoir level where no 4D seismic response is present, this difference will be an absolute measurement of 4D noise. By cross-plotting the CDR response with the 3D signal strength, figure 4.3, a strong correlation was observed. The implication of this is that the 4D detection level is closely linked to the local geology. That is, it is easier to detect a production change in a low reflectivity reservoir than in a high reflectivity reservoir

All of the datasets in the CDR plot were similar in terms of acquisition, all were narrow azimuth marine surveys, and in processing. We have since demonstrated that improvements in the acquisition repeatability and new developments in 4D processing help to reduce 4D noise and thereby reduce the gradient on the CDR plot.

A fairly rigorous overview on how 4D noise behaves and how it influences our recordings is found in Calvert (2005).

4.4 Acquisition

4.4.1 The importance of source and receiver position

For repeated 3-D marine surveys that have not been designed with acquisition repeatability in mind, the RMS differences could approach 60% between repeated stack sections, Landrø et al. (1999). This is several times higher than what is achieved on dedicated 4D seismic projects and one of the reasons for this is that we try to re-occupy the shot and receiver position.

Landrø's work on Repeatability issues of 3D VSP data, Landrø (1999), is an excellent study on the realization of geometrical differences and repeatability. A fixed downhole receiver (VSP), recorded while the vessel towed an air gun around in planned concentric circles of 40 m radius, and fired the source every 25 m. Ten thousand shots were fired and it turned out to be a data set with shot-point that ended up being spatially very close. To monitor the source signal, a sensor was recording the near field signal. By organizing the data in terms of RMS amplitude difference *versus* nominal shot point distance, Landrø noticed that the RMS difference was fairly stable, 30%, for distances less than 20 m, see figure 4.4. He concluded that for the work at this specific site, a deviation from planned shot-point position of 20 m was acceptable.

Campbell et al. (2005) did a study of the west of Shetland surveys on factors influencing repeatability. He concluded that the single most important variable for improved repeatability was to minimize the distance between initial and repeated source and receiver positions.

In a marine towed seismic survey, feathering effects are a common issue. Due to waves and currents, the cable might be pushed laterally and since these effects increase with increasing distance from the vessel, the cable obtains the shape of a feather. The only way keep track of this and mitigate the impact on repeatability is by detailed planning and close follow-up during data acquisition.

Campbell advocates that keeping within a combined source and receiver "mis-positioning" of 100 m in this area is necessary for any significant uplift in "positional" repeatability. This is about five times larger than Landrø's data showed for the VSP recordings. The reason why we may get away with a higher number is due to the fact that some of the mis-positioning effects are compensated for in the processing by applying residual statics, de-multiple, etc. Calvert et al. (2002), make a point that in a marine towed streamer situation we might have relevant redundancy in the data to compensate for these phenomena in the shooting direction, but we might be worse off in the cross-line direction. We commonly refer to the

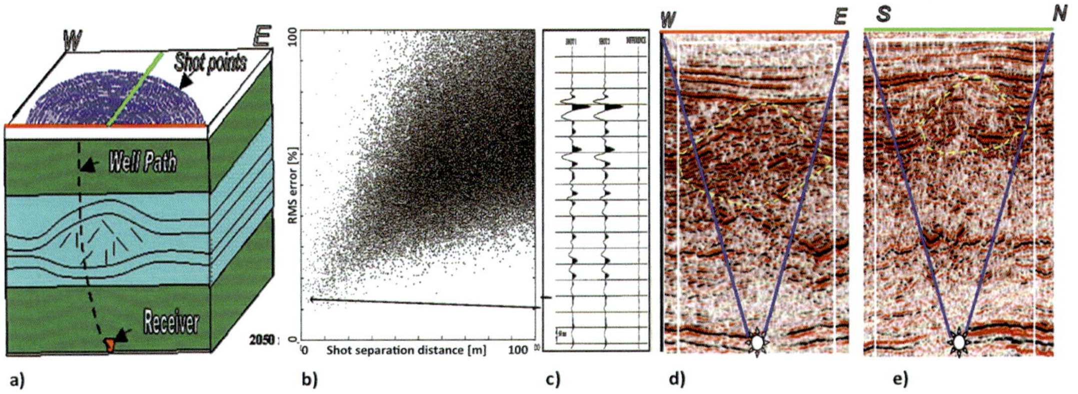

Figure 4.4: a) Ten thousand seismic shots (blue points) were fired following a radial pattern (b), the difference between various shots was calculated and the RMS values were plotted *versus* the distance between the shot c) two races 5 meters apart, the complex overburden is the cause of the large difference d) E-W section and e) N-S section. Courtesy of M. Landrø, NTNU.

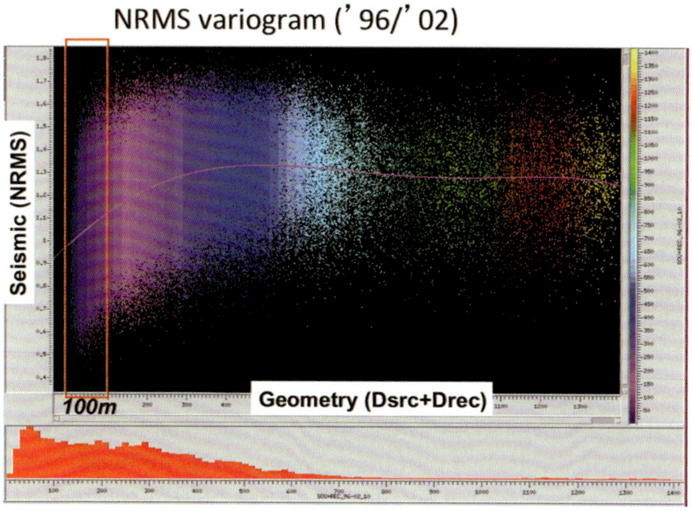

Figure 4.5: Cross-plot of NRMS *versus* geometrical repeatability. Campbell et al. (SEG2005).

deviation between actual and planned source and receiver position as ΔSrc + ΔRec. Acceptable ΔSrc + ΔRec may vary from area to area. The presence of shallow inhomogeneities and diffractions demands good spatial repeatability of source and receiver positions to achieve good 4D repeatability.

An extensive "test of seismic repeatability" is the result from a 13-day seismic monitoring experiment, which was conducted onshore, Schissele et al. (2009). They used two permanent piezoelectric sources and 14 permanent sensors below and above the "weathering" zone. One of the seismic sources was cemented below the weathering zone, figure 4.6. The study revealed that the surface sensors were more sensitive to daily variations than surface sources. The best repeatability was obtained when both the source and the sensors were buried, in that case the variations remained below 6 μs and amplitude variation less than 0.5 %. Forgues et al. (2011), discuss the use of a similar system used for simultaneous recording of both active and passive seismic signal for monitoring of a steam-assisted heavy-oil producers

4.4.2 Acquisition factors

The following parameters are typically addressed during the planning of a seismic time-lapse survey:

- Orientation of survey, shooting direction, receiver array orientation and pattern

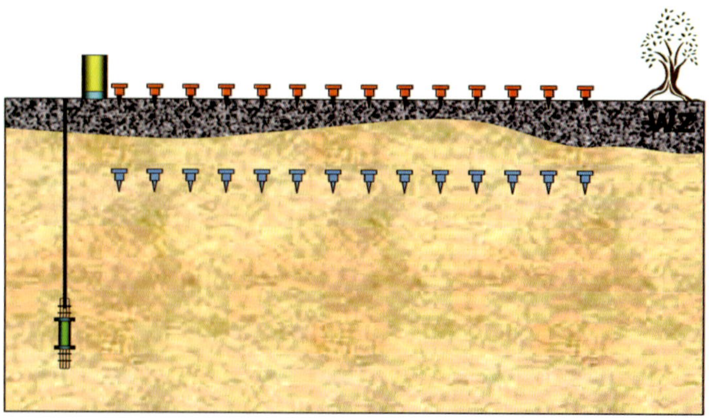

Figure 4.6: Experimental setup: Yellow and green elements are the surface and buried sources; the orange and blue elements are surface and buried sources, modified from Schissele et al. (2009).

- Source:
  - Positions and acceptable deviations (shot and shot line spacing)
  - Source array configuration (marine), volumes, strength
  - Source depths (marine and onshore if dynamite sources are used)
  - Onshore: vibroseis parameters, charge size if explosives are used
- Receiver
  - Positions and acceptable deviations
  - Marine towed streamer: number of streamers and distance between them
  - Receiver array configuration
  - Onshore & OBC: Receiver Coupling
- Fold coverage, total, per offset and within azimuth ranges
- Sample rate and recoding filter settings

Lumley et al. (2005) split 4D projects into three categories: "legacy", "re-shoot" and "4D-design". The "legacy" 4D, is a product of matching several 3D datasets acquired across a reservoir at various points in time, designed without the 4D application in mind. The "re-shoot" is used to define a 4D seismic project where the monitor survey is designed to match a previously acquired 3D seismic dataset. This leaves the 4D design project to the categories where the initial baseline survey was acquired with the objective to optimize a future 4D-monitoring program. Until the turn of the century, the latter covered a very exclusive group of reservoirs.

4.4.3 Seismic sensors

Good understanding of the expected 4D signal and 4D noise is an essential starting point for specifying the acquisition requirement. In the following we will look at the available options.

Onshore, the main challenges are source repeatability, receiver coupling and changes in the weathering zone. In a marine setting, coupling is less of an issue when the receiver and the source are in the water, but positioning might be more of an issue, especially when using a long streamer system. Since 1990, most 3D seismic surveys have been delivered with quality navigation information. Since then, there has been impressive technology development, allowing the towing of very long and wide seismic receiver arrays. There might be 10–18 cables, each up to 6–8 km long. The spread could be 500 m wide and control systems have been constructed that can steer individual cables and thereby mitigate for unfavorable wind or current situations. Other systems have been developed to steer the seismic sources. Most systems are programmed to fire the seismic source at pre-programmed positions in the sailing direction, which might favor less mis-positioning in the in-line directions, while the cross-line position would require that the source be towed into the correct position.

In general, the types of sensor to record seismic waves are based on measuring motion or pressure. We use a motion sensor to record movement, acceleration or velocity of a motion in a specific direction or axis. Three orthogonal sensors will be needed for uniquely recording an arbitrary movement. Hydrophones or pressure sensors based on piezoelectric transducers are the most common. The piezo electrical material changes shape when inserted into an electrical field, this property is utilized in the opposite way by gluing a plate of the material to a disc or a cylinder that changes shape when exposed to pressure changes. Julien Meunier presents a rigorous overview of pressure and movement sensors in his Distinguished Instructor Short Course, Meunier (2011).

Motion sensors have typically been used to measure ground movement onshore, on the seafloor or in a well. Recently, they have also been introduced in marine streamers, Tengham et al. (SEG 2007). The electromagnetic sensor, or geophone, is the most common in the market. The sensor is based on a moving spring-mounted coil moving in a magnetic field. The coil will resist movement and when a passing movement moves the housing and the magnetic field, an electronic current is induced; this is proportional to the velocity of the movement. Figure 4.7 shows that the hydrophone and four geophones could be arranged in a modern 4-component system.

In the 1990s a micro-electrical sensor, MEMS, was introduced, Maxwell (1999). These are the type of sensors we find in cars to control the air-bag system. These MEMS accelerometer sensors are tiny silicon chips; displacement is in the order of nanometers. They are known for their broadband capabilities but have a higher noise floor than in the intermediate frequency ranges, Meunier (2011). It is common to group several sensors to reduce the noise and geophones are generally better suited for this.

Recently sensors based on fiber-optical technology have become available. The geophones consist of an optical fiber wound around a moving mass coil, and uses interferometer techniques to measure the changes in the path length of the light, which will be caused, by pressure or movement impacting the coil. The geophone is elongated in one direction, proportional to the body's acceleration in the same direction. When the size of the body changes, the length of the fiber changes. The fiber has inherent half silvered surfaces or partial reflectors that reflect a particular wavelength, but transmit all others. The latter is called Bragg gratings (FBGs). FBGs are implemented by adding periodic variations to the refractive index of the fiber. This means that one fiber can serve several sensors, utilizing a different bandwidth. The wavelength of the specific sensor is measured before and after it has travelled through the sensor and the acceleration is measured by analyzing the phase of this signal. A similar concept works for optical pressure sensors, the volume of the body will change as a function of pressure. The noise characteristics are probably similar to that of the MEMS system. The size of a sensor is about one cm, connecting the sensor into a grating adds to this. One advantage of fiber optic sensors over electric sensors is that they do not require any electronic components or electrical power at the sensing point. Figure 4.8 shows a drawing of a 4-component optical sensor as used by CGGVeritas / Optoplan. There are currently at least two other vendors in the market, PGS and TGS Stingray. They use slightly different technologies depending on where the reference signals are measured. The interesting aspect of optical sensors is that they have very few elements at the recording point, at the seabed.

In principle, there is no limit to the number of sensor channels that can be multiplexed. Increasing the number of cable modules only requires an increased number of fibers in the lead-in cable. With the small dimensions of optical fibers, a cable with a diameter of a few centimeters can contain on the order of 1,000 fibers.

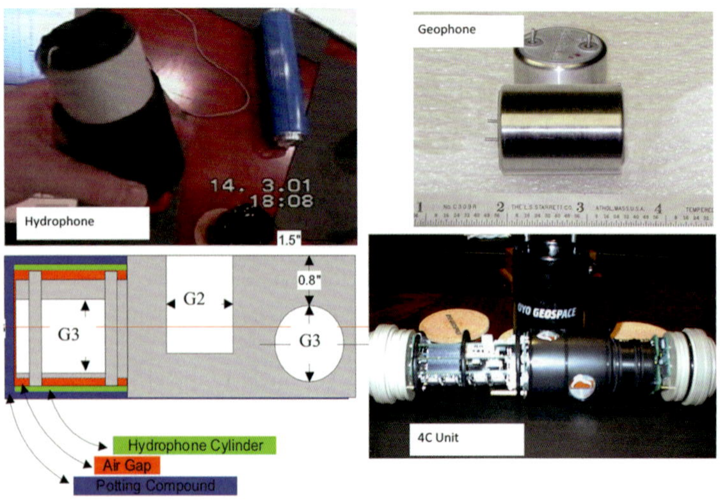

Figure 4.7: upper left: hydrophone cylinder, upper right: geophone, lower left: scetch of a 4 component sensors, 3 geophone and one hydrophone, lower right picture of same, including electronics. Courtesy of Geo Space Engineering Resources International.

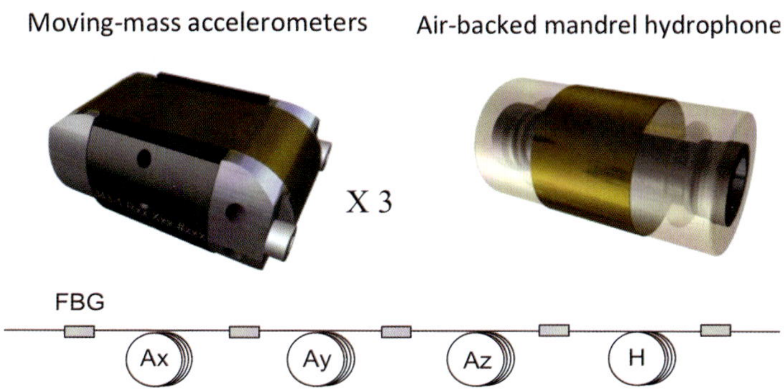

Figure 4.8: Fiber optic sensor configuration. Ax, Ay, Az are geophones oriented in x, y, and z direction accelerometer coils, H is the hydrophone mandrel and ref is a reference coil. All FBGs (in between each coil and mandrel) in a sensor station are adapted to respond to the same wavelength. Figure: courtesy of Optoplan.

4.4.4 Marine seismic sources

Explosives were the first and only available seismic source for a long period of time. Due to safety and environmental impact, this has been almost completely banned in marine settings since the 1970s. The air gun is by far the dominant seismic source used offshore today. Alternatives like marine vibrators and water guns have been tested but did not make it into commercial use. The principles of an operating airgun are fairly simple but the precision of the mechanics is quite impressive. Compressed air at a pressure ranging from 2000–3000 psi is delivered from a compressor to a pneumatic chamber in the air-gun, which is submerged in water. When the air-gun is fired, the system first triggers a solenoid that releases the air into a firing chamber, this in turn causes a piston to move and allows the air to escape from the main chamber, which produces a pulse of acoustic energy.

It is usual to combine several air guns to increase the strength of the pulse. Specific combinations of geometrical arrangement and source volumes may be combined to make a stronger signal, shape the seismic signal and reduce the impact from air bubble effects. The "cluster" of air guns in an array may add up to several tens of guns of various volumes. The total volume of air for a conventional marine tow survey ranges from 2000–5000 cu. inches. At Valhall we used the same seismic air gun array for all LoFS surveys. This is a BOLT annular port gun array, a relatively new design that provides improvements in both the operating efficiency and acoustic output of marine seismic source arrays. The Annular Port Air Gun is a compact design, based on an annulus containing the air chamber and a shuttle valve surrounding a hollow passage for the umbilical cords that support the air supply hoses and electrical control lines, Saunders, (2004). Figure 4.9 shows the Bolt APG source and how it is configured, compared to a conventional air gun arrangement.

In order for a source array to have uniform directivity it must be designed to resemble a point source as closely as possible. The dimensions of the source array must be small relative to the shortest seismic wavelengths of interest. For a typical 4D/4C ocean bottom seismic survey this implies a minimum seismic wavelength of approximately 20 meters, consequently a source array with less than 6 dB of directivity in the bandwidth 10 Hz to 70 Hz must be no larger than 10 meters in any dimension, Saunders et al. (2004). The source also needs to be as symmetrical as possible with respect to the array's center of gravity. Placing the largest element of the array in its center and then placing the smaller elements radially outwards achieves this. This configuration is illustrated in figure 4.9b. The system generates approximately 70 Bar-m of peak-to-peak output measured through a standard DFS-V Out-128 Hz filter. The shown Far-Field signature was recorded in a fjord in Norway using a moored hydrophone deployed at a depth of approximately 200 meters.

The design of the marine seismic source also has important issues for safe operations. The sources are mobilized and retrieved from the back deck of a boat and it is important that the design of this system

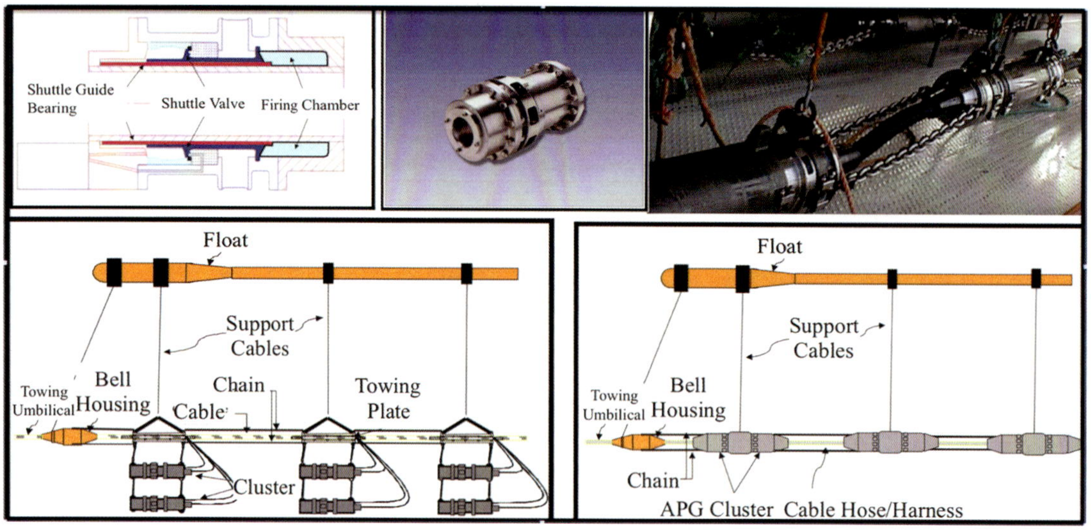

Figure 4.9a: Upper left: cross-section of Annular Port Gun. Upper middle: 150 cu. inch. Element, upper right: part of the array on deck. Lower left: conventional air-gun array. Lower right: APG configuration. Courtesy of Bolt.

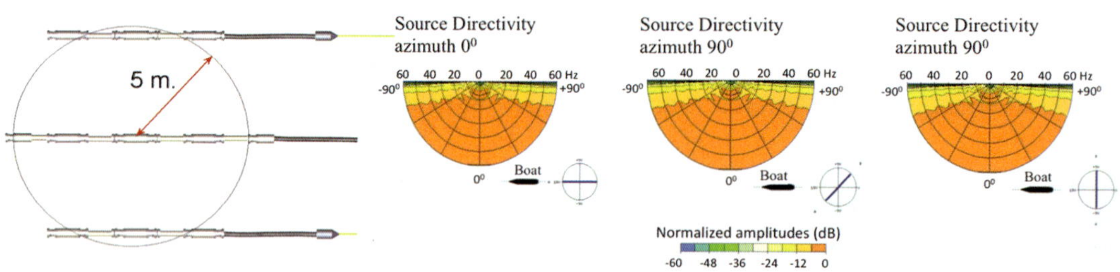

Figure 4.9b: Towing configuration as used at Valhall, the symmetrical compact array is built to provide a symmetrical source response (left). The three other captions show the response along, at an oblique angle and orthogonal to the towing direction. Saunders et al. (2004).

limits the exposure of the seismic crew to a minimum. Figure 4.10. Illustrates the "docked source strings" and the source as it is configured in the water.

4.4.5 Land seismic

The challenges for onshore 4D projects are somewhat different from marine 4D acquisition. Positioning of source and receiver locations is more easily repeated onshore but the variations and changes in near-surface conditions can be orders of magnitude higher than they are in marine settings. Modern GPS positioning systems have made it much simpler to reoccupy the same shot and receiver positions. The most common source used onshore is vibroseis. For 4D surveys is it most common to use point source and receiver configurations in wide azimuth configurations. This allows the use of 3D algorithms for noise canceling.

Like offshore, cable less systems, like for example Firefly by Ion, have been introduced on land to ease the logistics of installing the seismic sensors in the field.

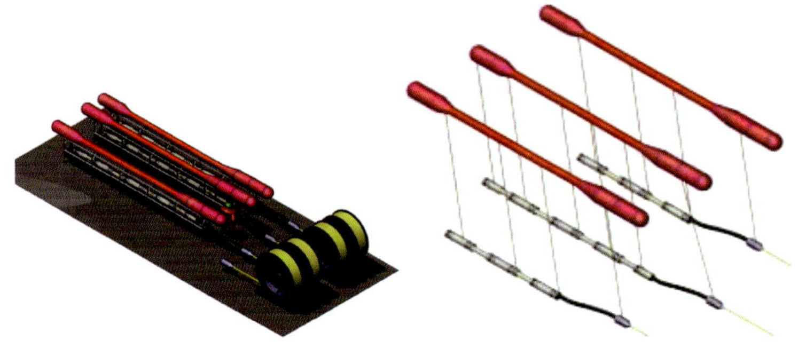

Figure 4.10: Simplified illustration of a 2,000 in^3 Annular Port Air Gun Source Array (on deck and in the water) .The red strings represent the floats that keep the system afloat in the water. In reality, these are split into three separate units for easier handling. Saunders et al. (2004).

SeisMovie is a permanent seismic monitoring solution developed by CGGVeritas in conjunction with Gaz de France and IFP. The system was initially designed for the monitoring of underground gas storage, whereby gas is stored in natural subsurface reservoirs. These storage reservoirs require frequent and accurate observation to monitor gas saturation and the extent of the gas bubbles during usage.

Due to my limited recent exposure to land seismic acquisition, I would recommend the Distinguished Instructor Short Course by Meunier (2011), for those who want an updated overview of options and issues.

4.4.6 Marine towed streamer

Marine Towed Streamer (TS) cases dominate in marine seismic time-lapse projects. Most of the early TS 4D was based on matching new acquisitions to legacy data. The business drivers were often to improve the 3D imaging, which may imply that surveys acquired in orthogonal directions were matched, aiming at detecting time-lapse effects. It was soon realized that matching azimuthal coverage improved repeatability significantly by reducing spatial misalignments. The marine towed streamer technology has evolved significantly and we know more about the factors influencing repeatability.

In order to achieve reasonable data quality, we will normally need to reoccupy the previous positions as best we can to get within acceptable noise margins. As we saw earlier from Landrø's results, in the Oseberg area we needed to be within +/–20 m of the original positions, for a given source receiver pair, to achieve optimal quality. For P-wave surface seismic we can apply the reciprocity theorem and link source and receiver position errors. Since amplitude variation with offset (AVO) might be important parts of the 4D response, we need to make sure that the typical position errors are acceptable for all offsets. As a consequence, planning a 4D seismic project could easily develop into a major survey planning exercise. Modern marine towed streamer survey acquisition for 4D is in many ways a technological masterpiece, including sophisticated steering of the spread (source and receiver layout), real time monitoring of how the new survey matches with the base-line survey and of source performance. Several challenges might have to be addressed, ranging from the impact of a new platform to planning to match tidal variations, requiring significant upfront planning for success. One way these matching needs are handled is to shoot the monitor with receiver redundancy, i.e. by using wider streamer arrays, so that mismatches in feathering can be dealt with by the selection of appropriate source-receiver pairs.

4.4.7 Ocean bottom

The first seismic survey was acquired on the Norwegian Continental Shelf using multi-component sensors placed on the seafloor. Such a system has been used by academia for very long offset recording, used for deep imaging. During the last 20 years the technology has been regularly used by the oil industry and interest has lately gained momentum.

Every receiver station in the OBS survey is often a four-component (4C) sensing system: a three-component geophone and a hydrophone. While the sensing system is stationary on the seabed, a source vessel tows a marine source array and delivers a source signal on a predetermined dense grid on the sea surface. When the shooting is complete, the sensing system may be retrieved and redeployed in the next location and the shooting campaign can continue.

Water bottom cables were introduced in the mid-1980s to provide 3D seismic recording coverage in congested areas and in transition zones, Rigsby (1987). These early systems were equipped with a single geophone and a hydrophone. The next wave of OBS systems came ten years later. The focus was to record converted waves and to improve imaging. To do this we needed three geophones, two recording horizontal movement and one recording vertical movement. The early cable systems used gimbaled orthogonally-oriented geophones; to make sure that the vertical components were oriented downwards. Later on, fixed phones were introduced and these had to be oriented by redundant but known source points, or one could complement the sensor group with a tilt-meter.

Ocean bottom recording systems offer a number of advantages above surface tow. It is generally quieter on the seabed and it is possible to acquire wide azimuth and long offset data more cost "effectively". They allow better coverage in areas around platforms, Koster et al. (2010), and they allow for the recording of shear-waves, Thomsen et al. (1997). They also offer possibilities for multiple attenuation by separating up-coming and down-going wave-fields, Barr and Sanders (1989). Using a seabed system, we can achieve a regular, high-density sampling of the reflected wave-field for all azimuths. Ocean Bottom Seismic (OBS)

Figure 4.11: PGS Ramform towing 18 streamers. Courtesy of PGS.

systems are normally implemented as stand-alone receivers placed at specific locations, (we refer to this as Nodes or OBN), or the receivers are connected via cables, (OBC). OBC may be placed permanently on the seabed or may be retrieved after the survey is acquired, Figure 4.12. OBS systems may offer very good solutions for geometrical repeatability and have the capability to deliver excellent 3D or 4D seismic recordings. Nodes are favorable when the seabed topography is difficult.

Figure 4.12: First generation of 4C retrievable cables. The cable was used at the Valhall field in 1997/98, one of the world's first 3D/4C OBC surveys.

Coupling and vector fidelity has been a major concern for utilizing Ocean Bottom Seismic solutions. Permanent placement may include trenching into the seabed, which appears to give the best coupling.

Multi-component OBS systems can be used down to several thousand meters. In 2005, the PGS Force cable crew recorded data at 1860 m water depth at the Roncador field in the Campos Basin and they claimed to be able to position their cable within 14 meters of the planned target, not using a ROV to steer. Retrievable OBC system acquisition projects may utilize 50–100 km of seismic cables, with 2- or 4-component receivers every 25–50 m.

For 4D applications using the retrievable OBS cable system, repeatability has been a major up-front concern. Because the OBC system is closer to seafloor heterogeneity, we might need better geometrical repeatability than using towed streamers. Local coupling effects may also impact repeatability. In practice, we have seen that the seismic crew is able to deploy the OBC system very close to the baseline survey positions.

Grion et al. (2000) reports a quite comparable number of NRMS comparing permanently installed, re-deployable and marine towed systems in some North Sea locations, Table 4.1, figure 4.13. Grion suggested using noise to signal ratio (NSR), which they showed is equal to:

$$NSR = NRMS / \sqrt{2 - NRMS^2} \tag{4.5}$$

Assuming that the two data sets used for the NRMS calculation have the same variance and the signal has the same strength.

Dataset	NSR	NRMS
3D Permanently installed OBH (Foinaven) Time Migrated	22% ±5%	0.34 ±0.06
3D Consistent surface streamer (Foinaven) Time Migrated	25% ±5%	0.34 ±0.06
2D Permanently installed OBC (Viking Graben) PZ-NMO Stack	27% ± 6%	0.37 ±0.08
2D Permanently installed OBC (Viking Graben) X-NMO Stack	12% ± 3%	0.17 ±0.04
2D Redeployed OBC (Viking Graben) PZ-NMO Stack	27% ± 6%	0.37 ±0.08
2D Redeployed OBC (Viking Graben) X-NMO Stack	36% ± 14%	0.48 ±0.20

Table 4.1: Comparable number of NRMS and NSR for permanently installed, re-deployable and marine towed system in some North Sea location. From Grion et al. (2000).

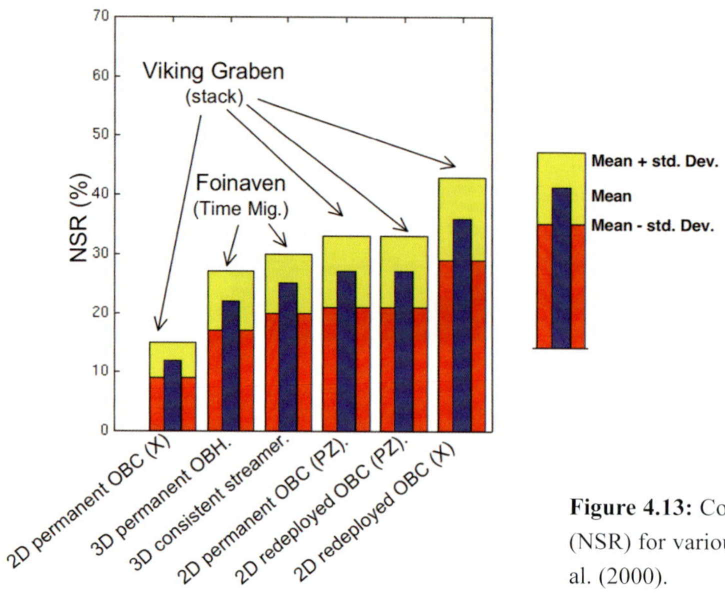

Figure 4.13: Comparing average noise-signal ratios (NSR) for various marine data sets. From Grion et al. (2000).

Ronen et al. (1999) reported on one of these Viking Graben studies and they also looked into the repeatability of the converted shear-waves (C-waves). They observed that the repeatability of C-waves was better than the P-waves if the cable was not moved but the opposite was true if the cable was re-deployed.

4.4.8 Nodes

OBS nodes are autonomous seismic recording units that are typically placed in a grid pattern on the sea floor. The nodes are deployed and recovered from the seafloor at depths up to 3000 m by remotely operated vehicles (ROVs), figure 4.14. Operating the nodes depends on their memory capacity and battery life. A cable system is normally connected to the surface, either to a boat or a buoy.

In 2006 BP acquired one of the largest OBS Node surveys to date across the Atlantis Field in the Gulf of Mexico, Howie et al. (2006). Nine hundred autonomous nodes were used to acquire a survey that covered 240 km^2 at water-depths between 1200–2400 m. The OBS node survey provided a significant improvement in imaging quality compared to the existing towed streamer seismic figure 4.15.

In 2009 the world's first field-wide node on node monitor time lapse was acquired at Atlantis. Five hundred nodes re-occupied a subset of the positions of the base line survey. The ROV managed to deploy 91% of the nodes within 5 meters of the respective baseline location and 98% within 10 meters. The geometric repeatability was quite impressive given the deep water and the challenging seafloor conditions, figure 4.16.

In 2010, Shell acquired a node on node time-lapse monitor survey in the Mars Basin, Gulf of Mexico, Stopin et al. (2011). The Baseline was acquired in 2007. Very high repeatability, and an NRMS of around 6%, has been reported. The shot density resulted in a very rich data set for both surveys. Therefore, they could apply shot receiver pairing and select only the pairs that had acceptable geometric repeatability. Advanced imaging algorithms (VTI reverse time migration) and mirror imaging of the down-going wave field gave a high-quality 4D signal, figure 4.17. The rapid change of sonic velocity in water due to strong loop currents, introduced time-shifts, "cold water static", between the two vintages of data. Dedicated efforts to estimate the time delays and comparisons of the multiple wave-fields that went three times through the water with the ones that traveled only one time through, helped them to remove the majority of these effects. The high repeatability was achieved even within the obstructed facilities area, which would have been difficult using marine-towed streamer acquisition.

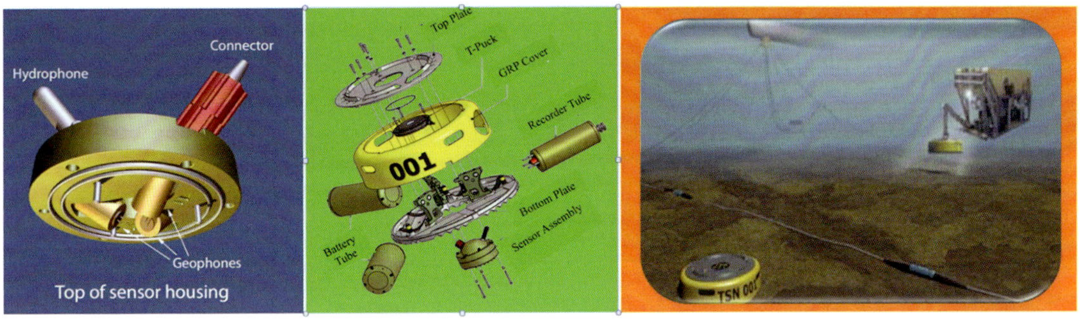

Figure 4.14: CGGVeritas Seabed Equipment, 4-Component Trilobit Node System. Courtesy of CGGVeritas.

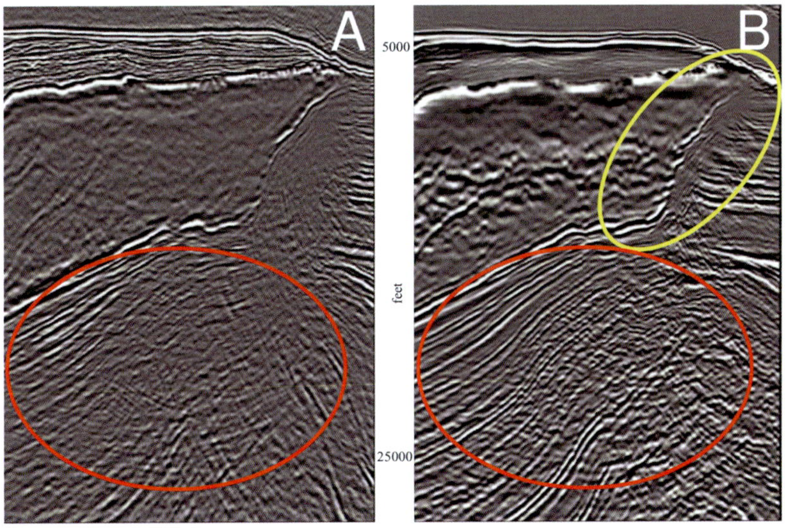

Figure 4.15: Comparison of narrow-azimuth towed streamer (A), shot-migrated OBS node (B). The node images benefit from an improved salt model. Subsalt reflectors (red ellipses) are clearly improved in the node image (B) over the streamer data (A). Howie et al. (2008).

4.5 Processing

Processing for seismic surveillance is driven by the end objective of differencing two or more 3D volumes of seismic that have been acquired at different stages during the production of a reservoir. Lumley et al. (2005) divide the processing efforts into three main categories; these are "post-stack matching", 4D "parallel processing", and "simultaneous" pre-processing. We keep to a similar categorization but allow some minor refinements. An example of post-stack matching is the example of matching 3D/4C OBS to towed streamer, Hall (2003), mentioned in Chapter 2.

There is an additional fourth approach commonly used, and often referred to as "Fast-track". In this case the baseline survey is used as-is or re-processed with the 4D application in mind and the monitor is processed with the same flow, with minimal cross-matching – only the most essential, like compensating for source signature effects and tidal variations. It is not unusual that the 4D results may be available after 3–6 weeks, after the last shot using this approach.

The parallel processing flow is a much more rigorous approach, where the data is matched and quality controlled after each "main" processing step to maximize repeatability in the choice of processing parameters. Included in such a flow is often an update of the velocity model. It is not unusual that a

Figure 4.16: Nodes have been successfully utilized in areas with complex bathymetry, like the Atlantis field in GOM. From Beaudoin and Ross (2007).

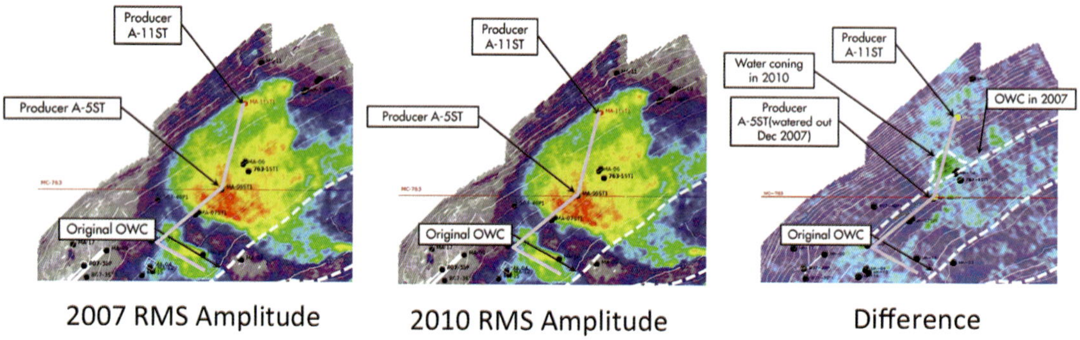

Figure 4.17: Mars 4D. Amplitude map from the H1 reservoir for the 2007 and 2010 data and corresponding 4D difference map between the 2010 and the 2007 survey. A clear 4D signal can be seen near the down dip well where water coning is occurring. Note the clear correspondence between the 4D signal and the depth contour. From Stopin et al. (2011). Copyright Shell, used with permission.

parallel processing flow, including velocity update, can take 6–9 months. Simultaneous 4D processing is used by Lumley et al. (2005) to design 4D data acquisition projects, where "the multi-vintage data sets are merged or intertwined for many of the processing steps, such that processing operators are derived and applied to the multi-vintage data sets jointly or simultaneously.

One of the main tasks in all 4D processing is to match source and receiver pairs as best as possible through 4D binning. Campbell et al. (2011) shows impressive progress in 4D processing, where 4D binning plays an important role. The other key element is to reduce the non-repeatable ambient noise. The sources of non-repeatable noise are quite diverse and could be introduced due to differences in acquisition settings like variable amplitude gain, frequency content, static shifts, waveform phase changes, differences in position or azimuthal effects and event positioning between the multiple vintage data sets. In addition, there might be a difference in ambient noise between two surveys. It would be natural to design matching filters in areas without any expected 4D signal to reduce the mismatch between the surveys. Unfortunately, more than once this has resulted in processing away important signals. Spatial matching filters might also have the undesired effect of smoothing and reducing the real 4D signal in the reservoir.

One obvious mitigation to this is to reduce the need for matching filters by attacking the non-repeatable errors in a systematic manner. This reduces the need for matching filters to where they can mainly be used as a tool for correcting for the difference in the system responses.

The main element of 4D processing includes:

- De-signature and re-datuming
- Statics corrections
    - Small periodic changes in the near surface, in the water column offshore or in the weathering zone on shore, need to be corrected. In marine settings, measured or predicted tidal variations are corrected.
- Regularization
    - Harmonize to a common overlapping data volume
      The acquired seismic data are mapped onto a regular grid. Missing traces are borrowed for other vintages or interpolated from nearby recordings. Rigorous testing might be required to find the approach that gives the least 4D noise. Ignoring missed shots or recordings in one of the dataset is often an unsatisfactory solution.
    - Multidimensional interpolation (x, y, t, offset and azimuth) is often used to homogenize two or more data sets.
    - The purpose of regularization is to pre-condition the data for domain specific pre/processing, e.g. shot receivers, as similar as possible.
- 4D binning
    - The most compatible source/receiver pairs are selected based on their source, receiver, mid-point or azimuth position. Alternative matching schemes and strategies are tested and compared using 4D seismic quality attributes like NRMS.
- Footprint removal
    - The specific acquisition geometry could results in an uneven contribution to the bins or there could be other residual noise linked to the acquisition time. These patterns or footprints hamper detailed analysis. The data are analyzed to find if there is systematic variation that needs to be compensated.
- Matching
    - Not all the differences between time-lapse surveys may be corrected for by deterministic corrections. Rigorous statistical processes are used to compensate for residual differences. At this stage it is important to make sure that the 4D signal is not being removed.

Arguably seismic processing sometimes appears to be more of an art than science, especially in 4D processing. It is without doubt that a team that has worked 4D processing in the same area for a while develops specific insight that sometimes turned out to be critical for the results. The development of standard processing templates with associated testing and quality control requirements helps to ensure consistency and transfer the learning, Campbell et al. (2005). An integral part of these templates is the careful use of NRMS or related measures to monitor the reduction in 4D noise at each step in the processing, figure 4.18.

Another point worth mentioning is that we have seen some significant improvement in repeatability/reduction in 4D noise lately, which will clearly impact the value of 4D, figure 4.19 is an excellent example on this development. For the non- practitioner, the leftmost examples might appear confusing and it requires some experience to delineate the signal from the noise in cases like this. The combined use of map-based analysis is important in this process. In the next session we will discuss 4D analysis in more detail

Some useful references and case histories that can help to develop insight into 4D seismic processing are: Jack (1998), Calvert (2005), Hoeber et al. (2005), Harris et al. (1998), Kommedal et al. (2007), Helgerud et al. (2011) and Campbell et al. (2011).

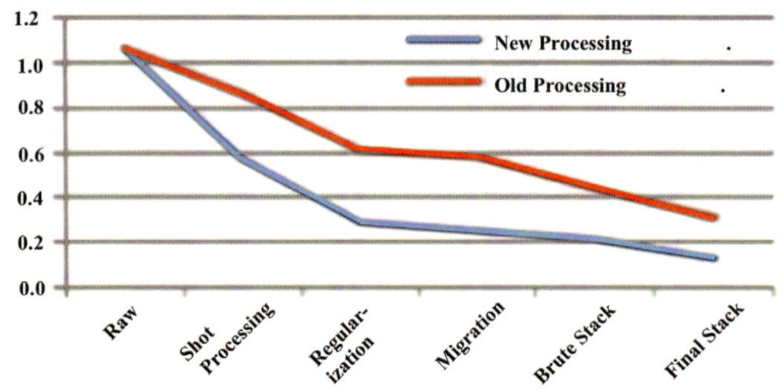

Figure 4.18: The NRMS reduction throughout the processing sequence for the previous (red) and new (blue) processing. From Campbell et al.(2011).

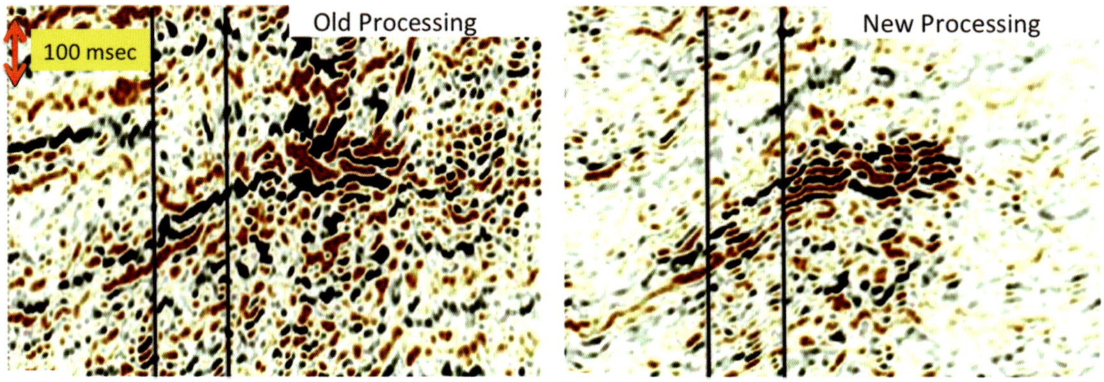

Figure 4.19: The level of 4D noise for new processing (right) has been reduced significantly. The old processing (left) was considered to be of good quality when delivered in 2006, average NRMS of 0.25, the new example is of considerably higher quality. The example is from Campbell et al. (2011).

4.6 Time-lapse analysis

Changes in amplitude or two-way traveltime are the two fundamental 4D seismic attributes, figure 4.20. Detecting amplitude changes was for a long time considered to be the most robust technique and the time-shifts were primarily used as an attribute to delineate effects in the reservoir, by mapping an event at the base or below the reservoir. The use of time shifts due to changes outside the reservoir have gained more popularity lately, as it was recognized that this could be used to delineate changes outside the reservoir. We will discuss time-shift in more detail in the next section.

4.6.1 Time-shifts

The early focus was on the changes across the reservoir but as the awareness of possible production induced stress state changes outside the reservoir become apparent, this became the focus. And simple as it might sound, time-shift estimation has its own inherent pitfalls. In reality, it is the associated velocity change that is of primary interest to us, the time-shifts are just a convenient way to find this. The problem is that it is not only the velocity that changes; we might have thickness changes due to compaction in the reservoir and stretch in the overburden, as seen in figure 4.21, and there might be phase difference in the data itself.

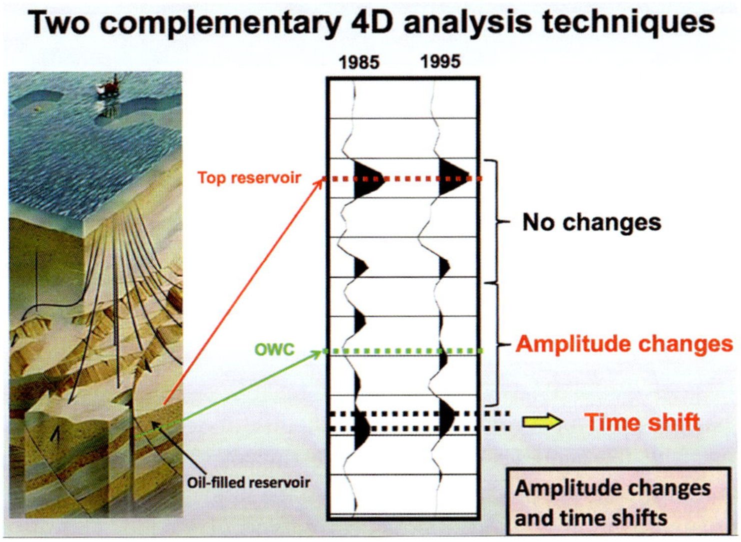

Figure 4.20: Changes in amplitude or two-way traveltime the two fundamental 4D seismic attributes for delineating production-induced effects in the reservoir. After Landrø (2009).

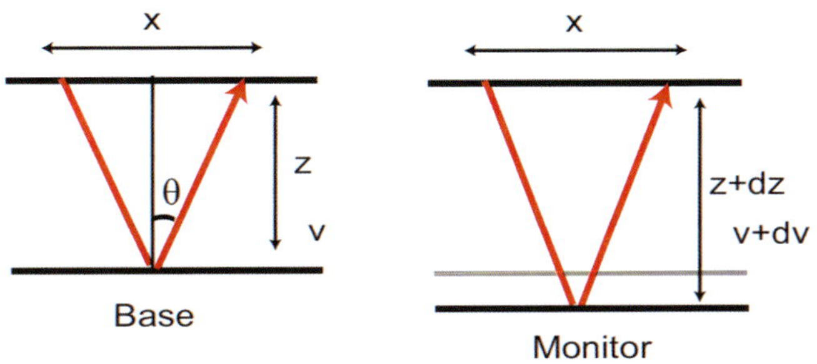

Figure 4.21: Illustration of a reflected seismic ray-path before and after the effects of stretch, or unloading, in the overburden. The stretch is due to compaction of the reservoir. Z is the depth, x indicates offset and v is the velocity. dz and dv are the depth stretch and the velocity changes, respectively.

Guilbot and Smith (2002) observed time-shifts of up to 12–16 ms in their seismic 4D images, resulting from matching surveys from 1989 and 1999, while the subsidence at the top reservoir level was expected to be less than 6 meters, based on geo-mechanical modeling. Guilbot and Smith proposed a 4D constrained tomographic technique, assuming a linear velocity-porosity relation, uniaxial strain and no changes in the depth down to base reservoir. The important observation from this work was that the velocity changes accounted for most of the time-shift effects.

Landrø and Jansen (2002) introduced a method using near and far offset traces to estimate the compaction as well as the velocity changes. This technique is discussed in more detail in Landrø and Stammeijer (2004), who derived the following relationship between time-shifts; Δt and two-way travel time t, thickness z and the velocity V :

$$\Delta t / t \sim \Delta z / z - \Delta V / V \tag{4.6}$$

They also introduce an alternative technique that links the acoustic impedance (AI) change and time-shifts Assuming this we can write

$$\Delta AI / AI \sim \Delta \rho / \rho + \Delta V / V \tag{4.7}$$

Where ρ is the density and, using the relationship between density changes and compaction

$$\Delta\rho / \rho = -a_0 \Delta z / z - a_1 \tag{4.8}$$

Where a_0 and a_1 are parameters related to specific reservoir properties like porosity, fluid and matrix density, Landrø and Stammeijer derived the following relations:

$$\Delta z / z = (\Delta t / t + \Delta AI / AI + a_1) / (1 - a_0) \tag{4.9}$$

And

$$\Delta V / V = (a_0 \Delta t / t + \Delta AI / AI + a_1) / (1 - a_0) \tag{4.10}$$

Historically, the amplitude attribute has been the most robust to use but the use of dedicated 4D surveys has increased the sensitivity and improved the accuracy of time-shift estimates. The link to velocity change and the direct proportional link to the net reservoir thickness have made this a popular attribute to use. The time-shifts are often estimated using a windowed cross-correlation technique. The seismic time-lapse effects might not necessarily be vertical and there might be real and apparent effects associated. In the general case, we will need to address the effects of a vector. For a compacting reservoir, we might experience a sudden transition from a slowdown outside the reservoir to a speed up in the reservoir. If the reservoir is thin, it might be difficult to detect this change using a windowed correlation technique. To further complicate this, the changes in velocity might be so large that they impact the acoustic contrast between the reservoir and its surroundings, which could result in incorrect estimates (Williamson et al. 2007).

Image warping was introduced to account for gross geometrical misalignment between two data sets, such that reasonable amplitude difference volumes could be made (Rickett and Lumley (2001), Hall et al. (2002). If the time-lapse data set was acquired with similar geometry and migrated with the same velocity fields, this gross alignment is not needed and warping could be used to delineate the subtle production distortions, Hall et al. (2002), Hall (2006).

Hodgson (2009) reviewed some of the limitations in common methods proposed for delineating time-shift in the industry. The windowed 1D cross-correlation technique is very fast but also limited in accuracy and resolution, by simply varying the window length we will obtain different results. When using short windows, which are needed to catch the changes from slowdown to speed up across the reservoir, even the tapering used in the window will impact the results, see figure 4.22. The results will also be sensitive to frequency content and the magnitude of the shifts. To overcome these problems it might be necessary to verify the results by comparing estimates from varying window lengths and also expanding the window laterally by including more traces. Special care might be necessary across the reservoir. If these mitigations are needed, we might as well return to a more rigorous warping concept.

Hale (2007) introduced a cross-correlation process that addressed 3D patches of data, which improved the estimates in case lateral variations introduce strong elements of non-uniqueness, i.e., dip *versus* velocity effects.

An alternative to the cross-correlation technique could be to use an inversion approach, i.e., by shifting part of the monitor trace such that the least-squares difference between them are minimized. This will necessarily require that the two trace segments are interpolated and the problem is not necessarily linear. There are several ways of addressing this and to regularize the problem such that the solution adheres to certain characteristics, see Rickett et al. (2007) for further details. Figure 4.23 shows a comparison using cross-correlation and non-linear inversion, NLI.

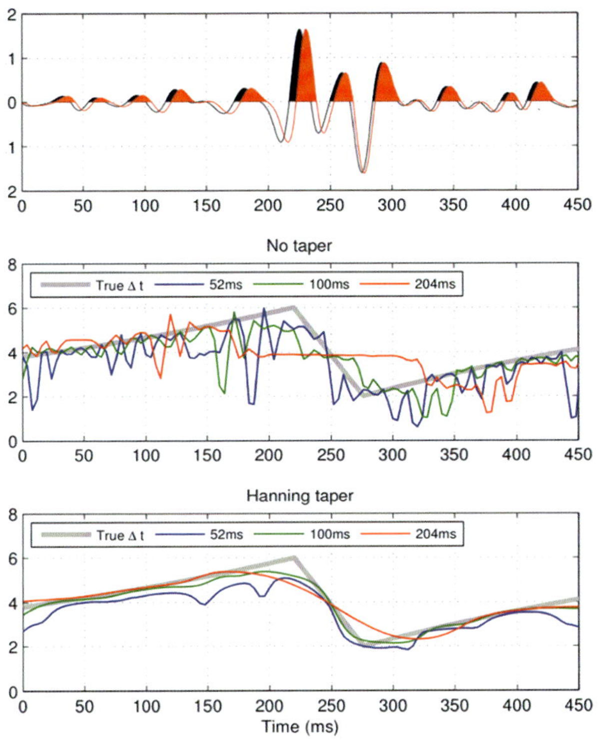

Figure 4.22: Time-shifts measured using variously sized windows to demonstrate some of the limitations with the cross-correlation technique. From Hodgson (2009).

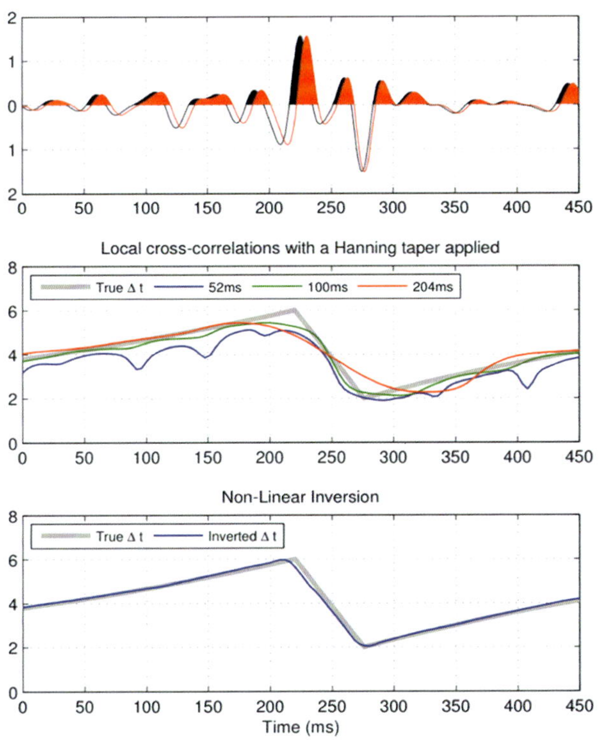

Figure 4.23: Comparisons of using cross-correlation (middle) and non linear inversion (bottom) to calculate production induced changes as reflected by two seismic traces recorded at different stages during production (top). From Hodgson (2009).

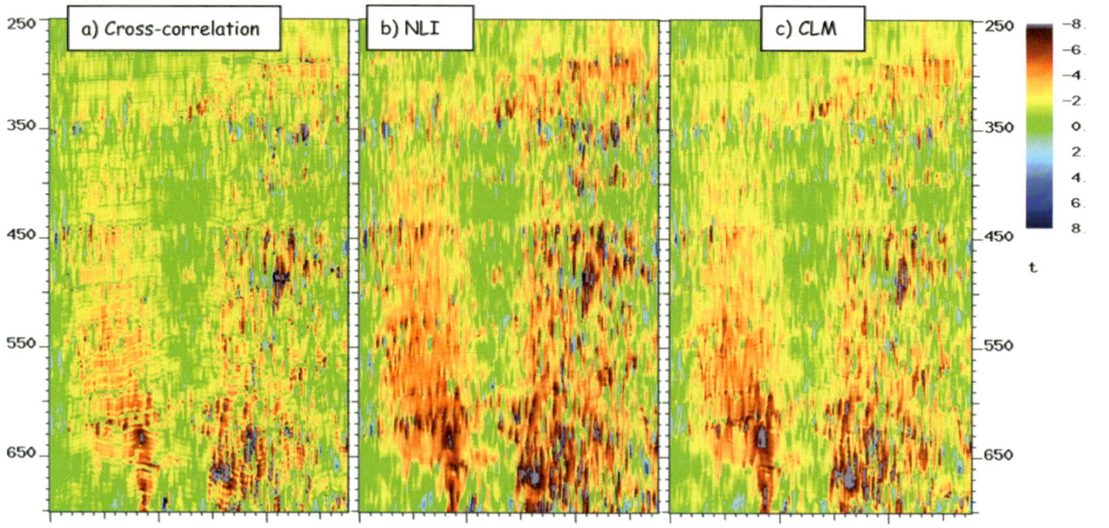

Figure 4.24: Time shifts between Valhall surveys LoFS01 and LoFS08. a) Using the cross-correlation method. b) Using the NLI method. c) Using the CLM technique. From Whitcombe et al. (2010).

The correlated leakage method, CLM, Whitcombe et al (2010), calculates time-shift from the gradient of a line fitted to a cross-plot of the amplitude difference between baseline and monitor (Y axis) against the amplitude difference between the baseline and monitor average and a time-shifted version of this average (X axis). The gradient of this fitted line represents an approximation of the time-shift between a seismic trace from the baseline survey, B=f(t) and a trace from a monitor survey, M=f(t+a), where "a" is the time-shift between both seismic traces.

In order for this method to work accurately we have to assume M is a shifted version of B (no large amplitude or waveform changes between baseline and monitor data). Given these conditions and assuming "a" is small, we can approximate our X and Y-axes through a Taylor series expansion. This method works best in cases where small time-shifts occur between baseline and monitor data. For large time-shift, it is recommended to follow a two-step procedure with either a NLI based approach or a cross-correlation based approach with a large time window as the first step. Figure 4.24 shows a "raw" comparison between time-shifts calculated using cross correlation, non-linear inversion and the correlated leakage method.

Time-shifts are accumulated effects, while understanding local variations between the layers and in which layers the changes take place is sometimes more useful to us. Ricket et al. (2007) introduced the time-strain, which is the fraction change in travel time. This is generally a noisier attribute than time-shifts, but has its merits in that it can be used as a layer property.

An alternative would be to invert for the velocity anomaly directly. Albertin et al. (2006) demonstrated that this could be done, utilizing a wave equation velocity analysis methodology. The method is based on back-projection of image differences using downward and upward continuation to recover slowness perturbations. Figure 4.25 shows the results from applying this technique to a time-lapse data set from the Valhall LoFS array. Here, slowness associated with compaction in the reservoir over time were recovered to a vertical resolution of about 60 m by minimizing the difference between the two migrated sections from data acquired 2 years apart.

In the next chapter, we will discuss how to interpret the time-shift in the overburden and relate this to compaction in the reservoir.

4.6.2 AVO and 4D inversions

Separating the seismic responses of fluid changes and pressure changes in the reservoir is one of the key objectives in 4D seismic analysis. One approach is to utilize the amplitude *versus* offset (AVO) information. There are several tools available for decoupling lithology and fluid effects using AVO cross-plots; (Verm and Hilterman (1995), Castagna and Swan (1997), Ricketts et al (2008), based on Huey's two term approximation:

$$R(\theta) = A + B\sin^2(\theta) \tag{4.11}$$

Traditionally, AVO analysis used to be done in the reflectivity domain. Goodway et al. (1997) used the shear modulus μ and Lame's constant λ as the optimum lithology and fluid indicators. Connolly (1999) introduced Elastic Impedance (EI) as an approach to calibrate inverted long offset data sets. EI is a function of Vp, Vs, and ρ ***(α, β, and ρ)*** and the incidence angle θ. When the incidence angle θ is zero, EI corresponds to the acoustic impedance (AI):

$$EI(\theta) = \alpha^a \beta^b \rho^c \tag{4.12}$$

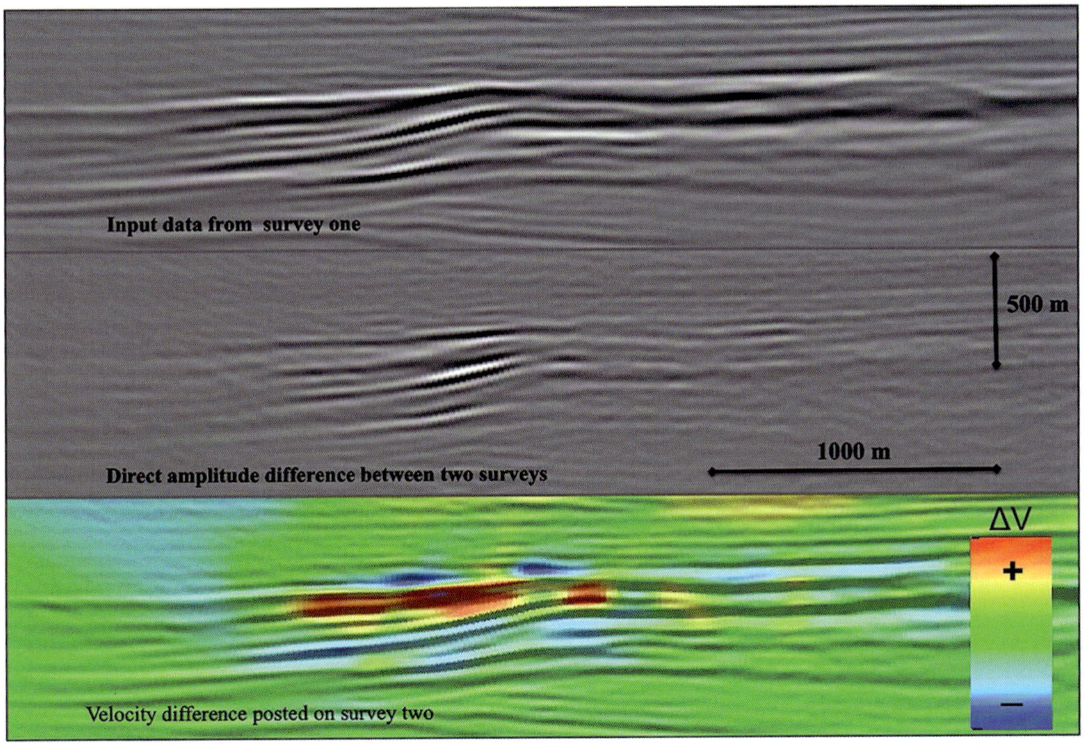

Figure 4.25: 4D wavefield-difference wave-equation velocity back-projection from an image difference. (top) shows a section from two 4D seismic sections (middle) shows the direct image difference obtained by subtracting the two 4D images data sets (bottom) shows an overlay of the velocity back-projection of the difference posted on the same section as shown at the top. This section begins at a depth of about 2300 m and has an extent of about 1100 m. From Albertin et al. (2006).

where

$$\begin{aligned} a &= (1+\sin^2\theta) \\ b &= -8K\sin^2\theta \\ c &= (1-4K\sin^2\theta) \end{aligned} \quad (4.13)$$

and K is a constant, "usually set to the average of ***(β/α)*** over the interval of interest", Connolly (1999).

Whitcombe (2001) redefined the definition of EI to remove the unit dependency on the angle θ, by introducing the reference constant α_0, β_0 ***and*** ρ_0, in order to extend the definition of EI to all angles θ:

$$EI(\theta) = \alpha_0\rho_0\left[(\tfrac{\alpha}{\alpha_0})^a(\tfrac{\beta}{\beta_0})^b(\tfrac{\rho}{\rho_0})^c\right] \quad (4.14)$$

Extended Elastic Impedance (EEI), Whitcombe et al. (2002) was introduced to allow the analysis to be carried out on data that was tuned to lithology or fluid variations

$$EEI(\chi) = \alpha_0\rho_0\left[(\tfrac{\alpha}{\alpha_0})^p(\tfrac{\beta}{\beta_0})^q(\tfrac{\rho}{\rho_0})^r\right] \quad (4.15)$$

where

$$\begin{aligned} \tan(\chi) &= \sin^2(\theta), \\ p &= (\cos\chi + \sin\chi), \\ q &= -8K\sin\chi, \\ r &= (\cos\chi - 4K\sin\chi). \end{aligned} \quad (4.16)$$

The concept Gradient Impedance (GI) was also introduced, which is the impedance equivalent to Shueys AVO reflectivity gradient B;

$$GI = V_p V_s^{-8K}\rho^{-4K} \quad (4.17)$$

Whitcombe et al. (2000) show that bulk moduli and Lame's parameter tend to lie within the an area with values of χ between 10 to 30 degrees around 0.15–0.5, and μ will lie in the range of χ between -31 to -90 degrees, which are good starting points for optimum fluid and lithology in the EEI space, see figure 4.26.

Landrø (2001) introduced a simple method for deducing changes in saturation and pressure from time-lapse amplitude *versus* offset data, and showed that for the Gullfaks field, the changes in saturation ΔS can be estimated by :

$$\Delta S \approx 8(\Delta R_0 + \Delta G) \quad (4.18)$$

which correspond to an optimal saturation angle of $\chi = 45$ degree
while the changes in pressure can be estimated by:

$$\Delta P \approx 23\Delta R_0 - 35\Delta G \quad (4.19)$$

which corresponds to a χ angle of -57 degree.
where R_0 is equal to A in Shuey's equation and G is equal B.
Again this derivation is based on reflectivity and analyzing the reflectivity changes at the interface above

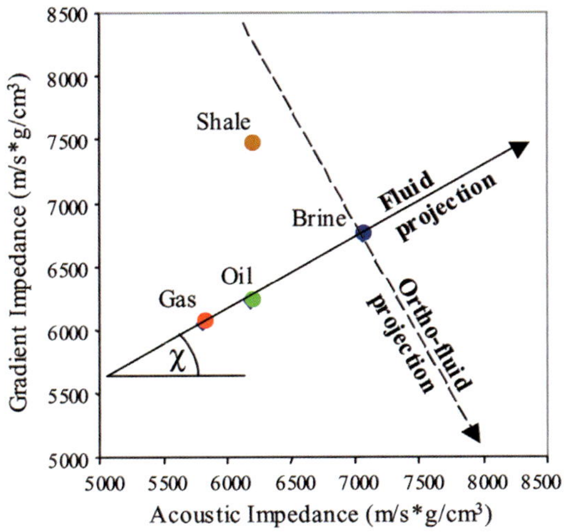

Figure 4.26: The AIGI cross-plot generated using rock properties from a North Sea field. The general relationships shown here are common to all reservoirs, although the exact angle χ and the separation of the points will depend on the details of a particular reservoir.

where the changes took place if we look at the interface below the reservoir, the relation will swap sign.

The early approaches to inverting 4D seismic data for pressure and saturation changes were based on first inverting the AVO data from relative changes in P-wave and S-wave impedance, alternatively including log data to obtain absolute impedance from the relative changes from the seismic. The cross-plotting was used in relating the seismic time-lapse changes to changes in reservoir properties, Tura and Lumley (1999).

Andersen and Wijngarden (2007) propose an alternative combination in their analysis, where they combine 3D and 4D classifications of AVO. Rock physics templates are used to translate from 3D inversion to lithology and 4D data into fluid and pressure changes The 4D observation is classified by projecting into a space defined by the AI ratio (acoustic impedance) and Vp/Vs ratio and include porosity as their third parameter. The resulting classification is interpreted and analyzed using virtual reality visualization.

Figure 4.27 shows various combinations of AVO attributes and how they relate to the physical properties. The AIGI projection has a clear advantage in separating the fluid and pressure effects. By cross-plotting changes in acoustic and gradient impedance, ΔAI and ΔGI, for many reservoirs the pres-

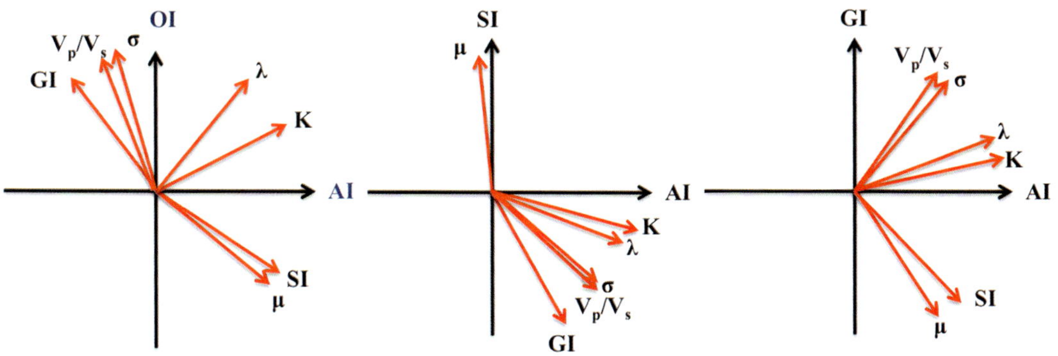

Figure 4.27: Alternative AVO projections (modified from Connolly (2010)). Note that all of these projections but CI and AI are based on the assumptions that Vp/Vs =2.0.

sure changes will often appear in an orientation approximately orthogonal to the saturation changes, figure 4.28. The EEI concept allows for maximum separation of fluid and saturation effects by optimal combination of the AVO attributes.

A common issue in AVO inversion is the presence of noise. The gradient data is typically noisier than the intercept data, Spratt et al. (1993), Cambois (1998), Simm et al. (2000). Ricketts et al. (2008) suggest an approach based on extracting the background noise levels directly from the 4D data. They measure the noise as a function of angle and the 4D response is normalized against this level and the fluid and pres-

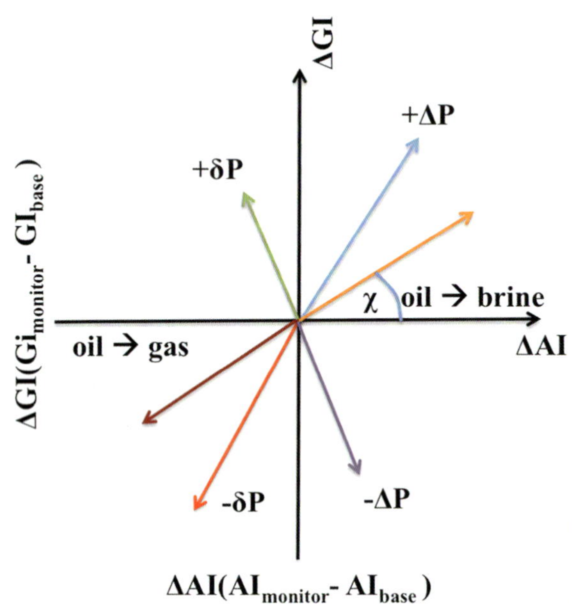

Figure 4.28: Cross-plot of 4D data, ΔAI *versus* ΔGI, shows typical projections for fluid and pressure changes using the EEI concept. Ricketts et al. (2008).

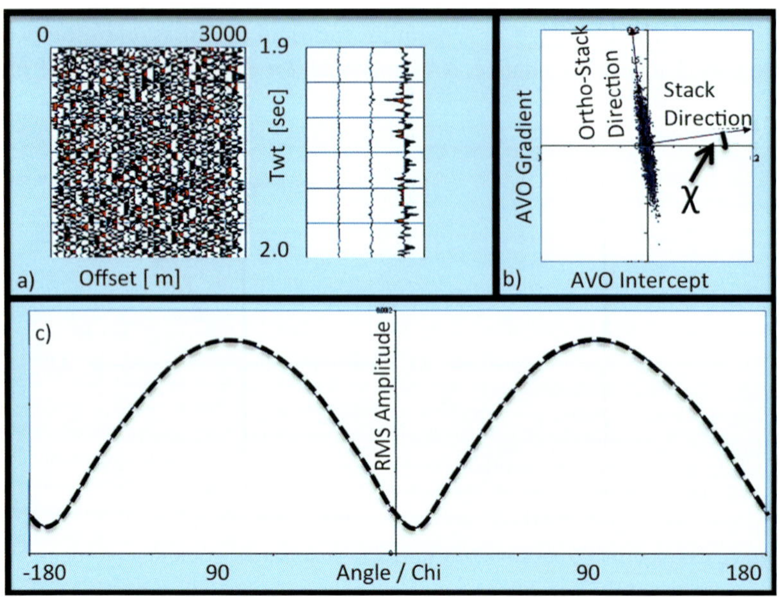

Figure 4.29: Impact of Random Noise on AVO attributes.(a) Stack, Intercept and Gradient traces computed from a gather containing only random noise, (b) intercept- gradient crossplot, highlighting minimum noise in the stack direction and maximum noise in the orthostack direction, (c) cross-plot of RMS amplitude vs chi for the same dataset ; Ricketts et al. 2008

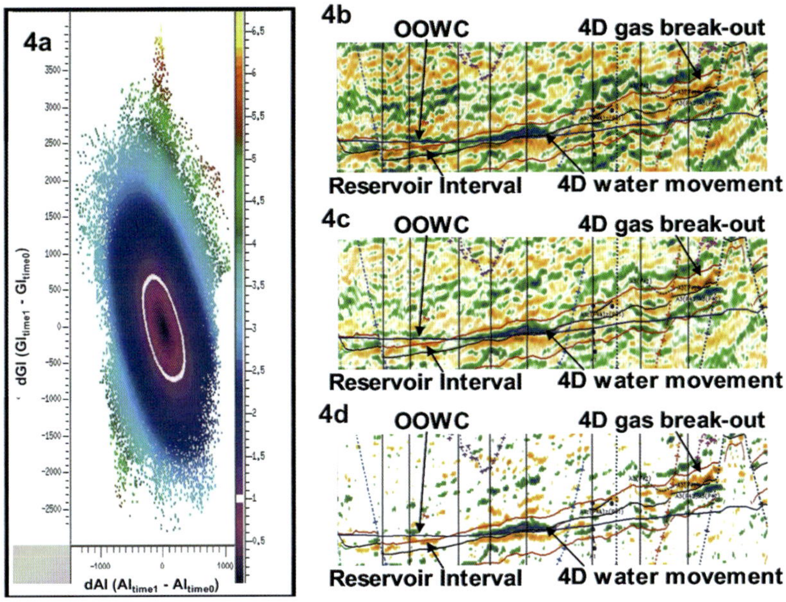

Figure 4.30: a) ΔGI versus ΔAI, colored by Significance b) conventional 4D χ=30 (fluid direction) section c) 4D Fluid Significance section d) Thresholded Fluid Significance section. From Ricketts et al. (2008).

sure projection clipped below the noise trend, are generated. The variation of noise as a function of χ is generated and the magnitude (Saturation or Pressure), volume is normalized by dividing by the appropriate noise value according to the χ directions. Figure 4.29 shows the impact of random noise on AVO attributes, see Ricketts et al. (2008) for further details. Figure 4.30 shows the results from applying the technique to an East Foinaven data example.

The optimal projections are calculated from available measurements; logs or cores and models. Figure 4.31 show the optimal projections as calculated for the Tor formation at Valhall.

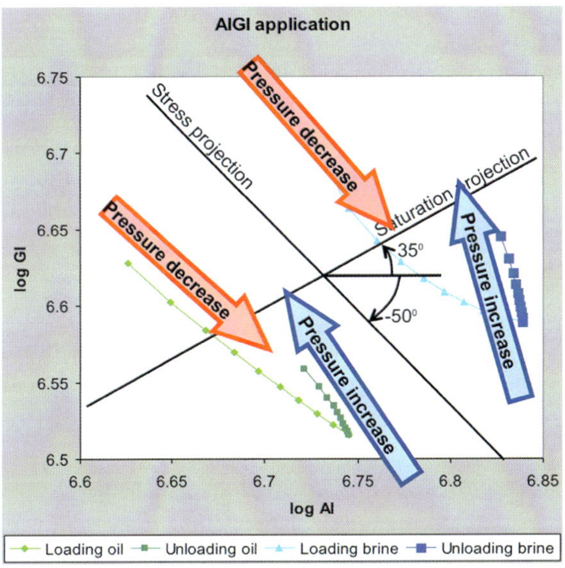

Figure 4.31: Optimal AIGI pressure and saturation projections for Valhall Chalk. From Askim et al. (2003).

4.7 Processing & analysis strategies for permanent reservoir monitoring -A Valhall case study

Seismic surveillance indirectly implies the acquisition of frequent or multiple seismic surveys. Turnaround time is important for being able to impact business decisions. The decision being impacted may be of such a nature that it requires rigorous quality assurance. During the last 20 years, the industry has established a sound workflow and a common understanding of the processing task. Still, noise characteristics, repeatability issues and the nature of the 4D seismic response might be such that local adjustment to the established flow is relevant.

There are several tools on the market with functionality for 4D analysis, often based on interactive workflows. It is important that the analysis steps and parameters used are documented, such that the workflow can be reproduced if necessary. A significant amount of data might be generated in this process and managing these data can easily become a nightmare if not addressed properly.

In the following we will discuss the approach used at Valhall. The frequency of surveying has pushed us to keep a strong focus on the turnaround of processing and analysis. The amount of data makes using interactive tools for the analysis challenging and boring. To deal with this, we have ended up developing script-based tools, which help us to standardize the workflow and naming conventions. The workflow is reproducible and new surveys are easy to add. Testing and including improvements or new products in the workflow are also straightforward. And most importantly, the end user has the seismic surveillance information easily available, in a format they can use in directly in their delivery workflows.

4.7.1 Accuracy of receiver positions

For seismic reservoir surveillance, the objectives of data processing are to maximize 4D resolution and reduce processing turnaround time to a minimum. We are in a unique situation with respect to repeatability, in that acquisition has been designed to keep as many of the acquisition variables as possible constant. We use the same omni-directional, seismic source for every survey and the receivers are permanently entrenched in the seabed, Kommedal et al. (2004). Figure 4.32 shows the accuracy of the shot points for one of the surveys. Each survey has about 50 000 shots on a 50 × 50 m grid, and the average differences between the pre plotted and the actual shot position is very close to zero, figure 4.32, Kommedal et al. (2005).

For a permanently installed system, the receivers are supposed to stay were you put them. However, there might be reasons for retrieving parts of the system, due to the installation of other facilities (rigs, pipelines) or due to maintenance. At Valhall, the initial receiver positions were measured during installation of the array and checked using seismic data from the baseline survey, the absolute position of each receiver is given by a standard deviation of 0.26 m. After the 5th survey was acquired, we did maintenance work on some of the cables and some faulty sensors were replaced. This was done by retrieving the cable at the seafloor, lifting it to the surface, replacing a section of the cable and placing it back into the original trench. Wet mate connectors were used to avoid excessive cable at the seabed. One hundred and seventeen receiver stations were impacted by this intervention. Based on analysis done in subsequent survey 6, in 2006 these stations had moved on average 1.4 m north and 1.8 m east. Later surveys confirmed that these positions have been constant until a second intervention was done in 2010. Given the very small source position variations and the fixed receiver positions, there is no need to put great efforts into careful regularization and binning compared to 4D streamer data, see, i.e., Hoeber et al. (2005).

The orientation of the geophones also includes a vector infidelity correction, which has been well documented in the literature (Dellinger and Clarke, 2001; Dellinger and Clarke, 2002). The vector fidelity of the geophones is excellent and has been consistent for the different surveys, apart from the sensors, which have been physically moved, figure 4.33, (Clarke and Dellinger, 2003).

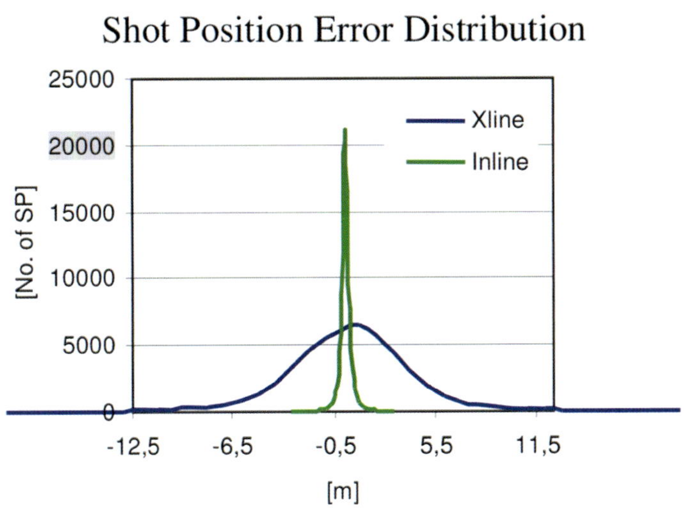

Figure 4.32: Difference in actual *versus* planned shot point position for survey 3, a total of 50 000 shots on a 50×50 m grid, measured in the in-line and cross-line directions. Kommedal et al. (2005).

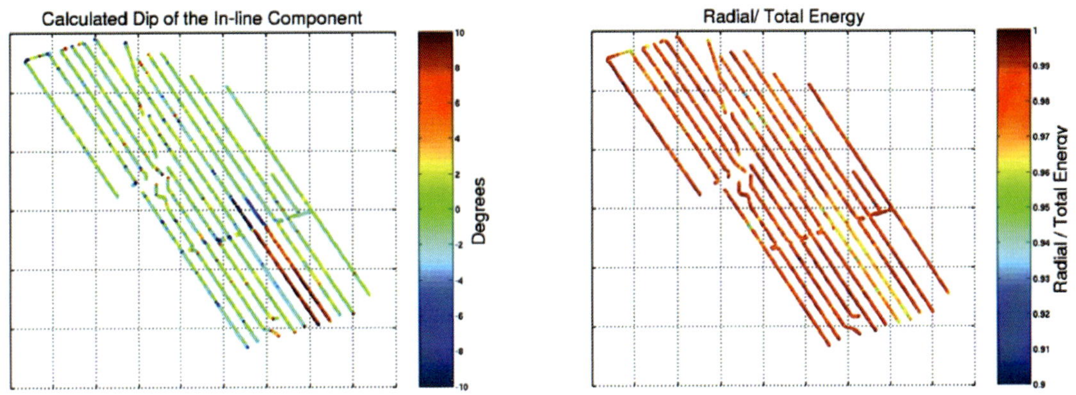

Figure 4.33: (Left) Calculated dip of the in-line component calculated using an initial subset of the data. The scale goes from −10 to +10 degrees. (Right) the ratio of the radial to the total energy (in the direct-arrival window) after vector infidelity correction. The color scale goes from 0.9 to 1.0. From Kommedal et al. (2005).

4.7.2 Processing flow

The high degree of acquisition repeatability allows us to use the same processing flow for each survey with a minimum of change between surveys. The advantage of highly repeatable processing is that it can be automated and turnaround time can be minimized. This approach would not work for non-permanent systems or fields with a weaker 4D signal than Valhall. In most cases parallel processing, involving the matching of seismic vintages at various stages of the processing flow, is required to extract a 4D response. We decided to run two different processing efforts in parallel with slightly different focus. The intention was to minimize non-uniqueness in the interpretation. We refer to these as the company internal processing flow and the in-house contractor flow.

PGS delivered these services and the team from PGS was initially co-located with the Valhall Subsurface team. Two vintages of streamer seismic had already been processed for 4D effects and had given valuable insight into what time-lapse effects could be expected.

The company internal processing flow was tuned towards rapid turnaround. Due to the high degree of repeatability, a very minimal processing flow was selected. All processing was done in the common receiver domain, which makes it trivial to parallelize the processing flow across a supercomputer: each CPU processes a common receiver gather, Clarke et al. (2005).

The P-wave processing flow consists of: sort to common receiver gathers, noise removal, source designature, rotation of the geophone components to the X, Y and Z directions, P-Z summation, deconvolution, and depth migration. Reciprocal wave equation shot migration (i.e., receiver migration) was used. By sorting the data to common receiver gathers, non-repeatable noise will be randomized and can be removed by standard noise reduction methods. Multiple removal steps are also included. We have shown that it is possible to obtain 4D products from a new survey on the interpretation workstation three days after the last shot, using this scheme, in most cases the turnaround time was 2–4 weeks.

The processing team from PGS, which was co-located in BP Norway's offices, performed the alternative flow. This team used a more rigorous pre-processing flow, similar to what is standard for streamer 4D processing (see Calvert 2005). For each survey this flow includes checking of navigation data, orientation of the 4C receivers, matching filters for the PZ summation (for each receiver separately), tidal statics corrections, shot and receiver static corrections and changes to the mute functions due to changes in water velocity. Turnaround time has typically been 6–8 weeks.

Neither of the two processing efforts included classical matching between vintages of data to optimize the repeatability metric from one processing step to the next, which is often done as part of conventional parallel 4D processing. In contrast, our choice was repeated sequential processing but we benefit from using frequent surveying in constraining the analysis using the trends established across the vintages.

In most cases f-k dip filters or frequency filtering has been applied to reduce the non-repeatable noise in the contractor processing flow, the parameters were optimized for each survey. An exception is the removal of seismic interference (SI) , where two entirely different solutions were used for the different surveys, figure 4.35, see also Kommedal et al. (2006), but with no significant difference in repeatability for the final stacked images. Figure 4.34 shows some of the characteristics of the non-repeatable noise.

The velocity model used for the migrations was established during the processing of the first and second surveys. Since then it has been kept unaltered until 2009 when an improved velocity model was implemented based on full waveform inversion (see chapter 2.10). After six years, every processing step was reviewed to include recent best practice. This resulted in a revised workflow, which included more rigorous interpolation and noise canceling schemes, while the overall flow was kept the same. Figure 4.35 shows the main steps of the initial workflow and indicates how often the processing parameters were changed.

Production induced subsidence in the overburden is observed by significant 4D time-shifts between the surveys. The observed 4D time shifts are really an expression of changes in the seismic velocities and Albertin et al. (2006) have shown how it is possible to invert these time shifts for changes to the velocity model. As the maximum time-shift is less than 20 msec at a depth of 2500 m, it is unlikely that the effect will impact focusing, however residual move-out correction prior to stack is checked from survey to survey to avoid artifacts from applications of residual move out correction.

Whenever a new monitor survey has been processed, an automated workflow rapidly analyzes 4D effects and feeds the results into the interpretation database.

At Valhall, the reservoir is compacting during production and this causes subsidence all the way from the sea floor. Hence, a second important 4D attribute is the time-shifts of the different events caused by subsidence. By using a sliding window cross-correlation, these time-shifts can be found and used to monitor the effects of production and water injection, Barkved and Kristiansen (2005). Figures 4.36 and 4.37 show that the in-house and contractor processing produce results that are remarkably similar. The interpreter's preference is the internal processing flow and we attribute the slightly more consistent amplitude to be due to a more advanced imaging algorithm (WEM *versus* Kirchoff), Etgen (2002). On the other hand, when comparing the time-shift results, the preference is for the contractor processing

Noise Type	Amplitude level	Frequency	Comments
Ambient noise	10-15 µBar Peak > 25 µBar	3 – 250 Hz	Remote platforms, undefined
Seismic Interference	Peak > 40µBar Typical 30µBar	10 – 20Hz	Severe < 50km away Significant < 100km
Platform noise	Peak > 30µBar Typical 20µBar	5 - 15Hz	Significant < 500m away from rig
Crossfeed Spikes	Peak > 100uBar Variable	All frequencies	Individual noise burst <1% of stations
Vessels	Peak > 100µBar Typical 40µBar	50 - 80Hz	Significant < 500m away from vessel
Shooting vessel	Typical < 20 µBar	50 - 80Hz	
Multiples		All	Changes with water velocities

Figure 4.34: Noise characteristics. Kommedal et al. (2007).

	Time of establishing parameter							
Processing Step/Parameter Survey:	1	2	3	4	5	6	7	8
Receiver Position / Orientation	1	1	1	1	1	6	6	6
Hydrophone calibration	1	1	1	1	1	6	6	6
PZ Summation matching filter	1	1	1	1	1	6	6	6
First Break Mute	1	2	1	2	2	1	1	1
Seismic Interference Attenuation	1	1	1	-	5	6	6	6
Tidal Statics	1	2	3	4	5	6	7	8
Shot Statics	1	2	3	4	5	6	7	8
Tau-p transform	1	1	1	1	1	1	1	1
Tau-p deconvolution	1	1	1	1	1	1	1	1
Tau-p mute	1	1	1	1	1	1	1	1
Attenuation of platform noise	1	1	1	1	1	1	1	1

Figure 4.35: Key element of contractor's workflow. The numbers indicated in which survey the parameters have been changed. The first break needed to be changed due to water velocity changes, while we did some maintenance work on the array after survey 5, leading to an update of a few sensors locations and vector rotation. Kommedal et al. (2007) used by permission.

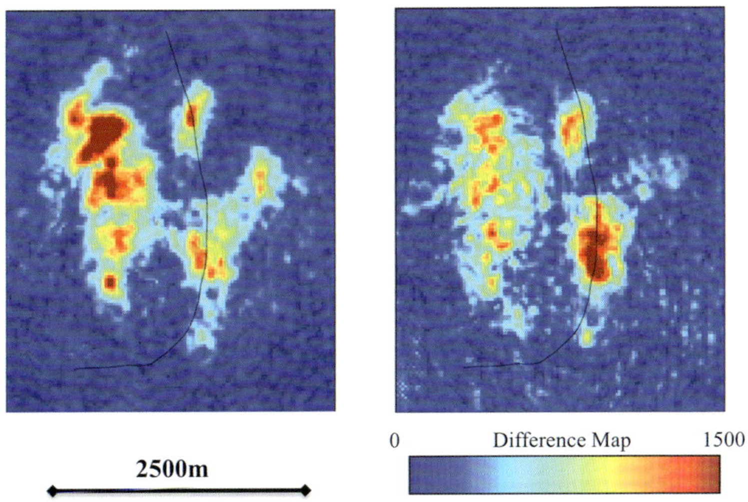

Figure 4.36: a) Amplitude difference map of internal processing at top reservoir, b) Amplitude difference map of contractor processing. Kommedal et al. (2007).

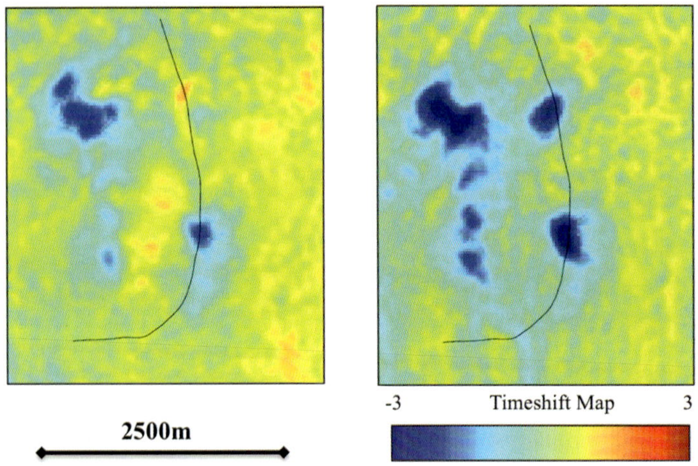

Figure 4.37: a) Time-shift at top reservoir level from internal processing, b) time-shift based contractor processing Kommedal et al. (2007).

flow. We attribute this difference to the fact that the Kirchoff based scheme balances the fold coverage for each offset bin, while the WEM-based scheme migrates all data and the results will be biased toward the higher offset ranges. In principle, the time-shift is based on the vertical incidence ray model. The differences in illumination as a function of azimuth will also affect the stability of estimated time shifts.

So far it seems that the two processing approaches have different strengths and complement each other.

In terms of resolution of 4D attributes, it was established early on that multiples were probably the main challenge. Kommedal et al. (2005) noted that water velocities change with the seasons and varying tides; both would affect the multiple period of the water layer and Pettersen et al. (2006) show the impact this has on the 4D time-shift curves. Tests have been done to further suppress multiples during pre-processing and indications are that this can be achieved either by more aggressive use of the standard decon methods in the cross-spread domain, or by using more sophisticated PZ-summation methods, or both. The current processing flows were designed in 2003. At the time, wave equation pre-SDM was not a viable choice with contractor processing.

We have not taken full advantage of processing 3D gathers or used the most sophisticated PZ- summation methods or multiple removal algorithms (see Boelle at al. 2006). Offset dependent Q compensation is one other possible step of the processing flow that might increase resolution but so far, has this not been fully utilized. Seismic processing technology is a rapidly developing field, a fact suggesting that the processing flow will have to change over time to take advantage of the most recent and best processing tools

A somewhat related issue is the task of ensuring that the processing software does not change over time and we have seen examples of software updates introducing unintentional changes, forcing us to re-do processing steps. The current processing solutions have been successful in showing clear 4D effects between surveys shot only months apart and with relatively small depletion or injection volumes, in the order of 200,000 bbls.

It is possible to make observations of time-shifts down to one quarter of a millisecond, and as these mapped anomalies develop from survey to survey we are able to build confidence in their meaning. So far, we have experience from eight years of data acquisition and processing with the LoFS project, using our initial processing strategy. Currently, we are revisiting the processing strategy based on this experience. One of the new challenges this poses is how to manage the transition to new processing flows, still providing better 4D products in a timely fashion and also such that we can build on the collection of vintage data. How this will work out, remains to be seen.

4.7.3 Processing strategy

Choice of strategy for processing of multiple repeat 4D data will depend on the specific objectives of the seismic surveillance program. Our experience indicates that with good repeatability in the acquisition stage and the limited amounts of non-repeatable noise, sequential repeated processing flows work well and allow for short turnaround times. In terms of 4D amplitude changes, including all offsets appears to give better S/N, but for time-shifts azimuth, restricted data provide the better results.

How often it is necessary to update velocity models to account for changes over time will have to be judged case by case depending on local geology and its response to production at each particular field. In our case, we were able to use the same velocity model for seven surveys, even though there are significant subsidence effects related to the production of the field.

4.7.4 Data management

A field wide permanent seismic system has the potential of generating overwhelming volumes of data. In surveillance strategy based on the 4D seismic concept, the primary entities are 3D volumes of data and maps. We have included graphic images as part of the key entities; we basically have two types, horizontal or map display and vertical cross-section. And of course there are the field recordings, 10,000 channels, 8–18 second recorded at a sampling rate of 4 msec, once for each 53,000 shots. The number of volumes, including all the relevant difference products is large and need a strategy for management.

The data management strategy agreed to upfront has served us well until now. The fundamental idea has been:

- Keep it simple
- Rigorous standardization of products and naming conventions
- Fast and reproducible workflows
- Documented and auditable

The purpose of the Valhall LoFS system is to support the implementations of the field's depletion plan and to provide input to maintaining and refining this plan based on new learning's. The conceptual framework is outlined in figure 4.38. The primary objectives are achieved by using the data to support reservoir management, the delivery of new wells and to optimize well work. In addition, the surveillance data from the array is expected to support and monitor possible production induced or natural changes in the subsurface.

The fundamental basis for all seismic interpretation and analysis is the interpretation system. We decided quite early to continue using the interpretation's database as a hub for the 3D and 4 D seismic volumes. To allow non-geoscientists easy access to the 4D analysis, we decided to implement a web-based information system, figure 4.39. This provided access to general documentation, details on the workflow and links to images (maps or cross-sections) for all wells or logical areas of the field. A set of predefined integration "posters" was also defined.

For every survey there are two data sets or formats that are critical to us: the raw recordings or field data and the final migrated stack 3D volumes for each survey. There are typically 8–24 standard products per survey, including optional offset domain and azimuth products. These two sets of well documented products need to be stored in the official database with full protection. The derived 4D products, which consist of all perturbations of differences of amplitudes, impedance, time-shift of other relevant products, may be reproduced and does not necessarily have the same need for full protection, including backup. In fact, some volumes may be used temporary for the generation of attribute maps, which are stored as part of the documentation package.

The total number of 3D and 4D volumes created has regularly hit limitations in our interpretation system and several tens of thousands of graphic displays have been generated. Due to the strict naming convention and a logical structure for accessing the resulting time-lapse attributes via a HTML browser,

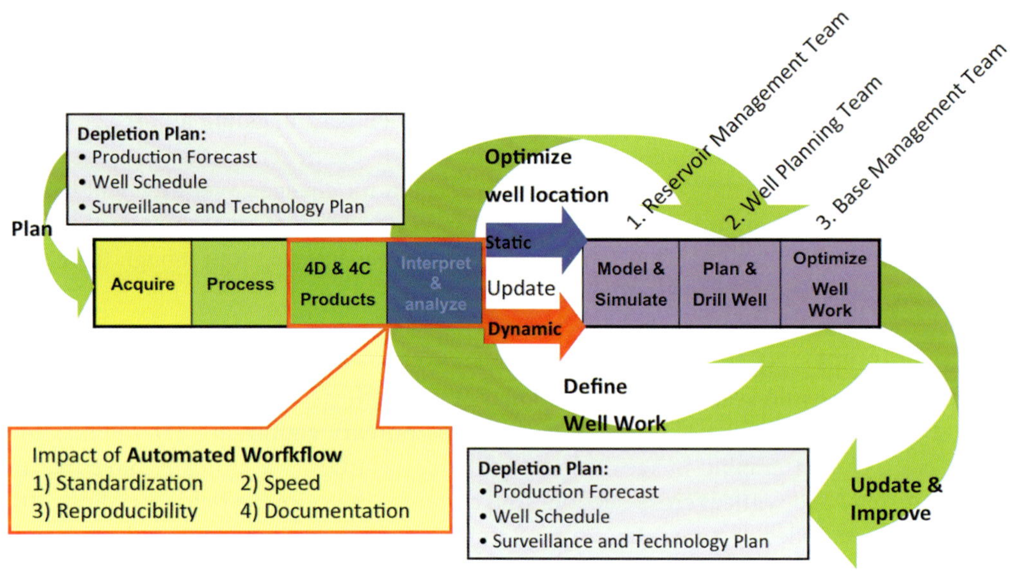

Figure 4.38 Conceptual framework for the Valhall LoFS data flow and business interactions.

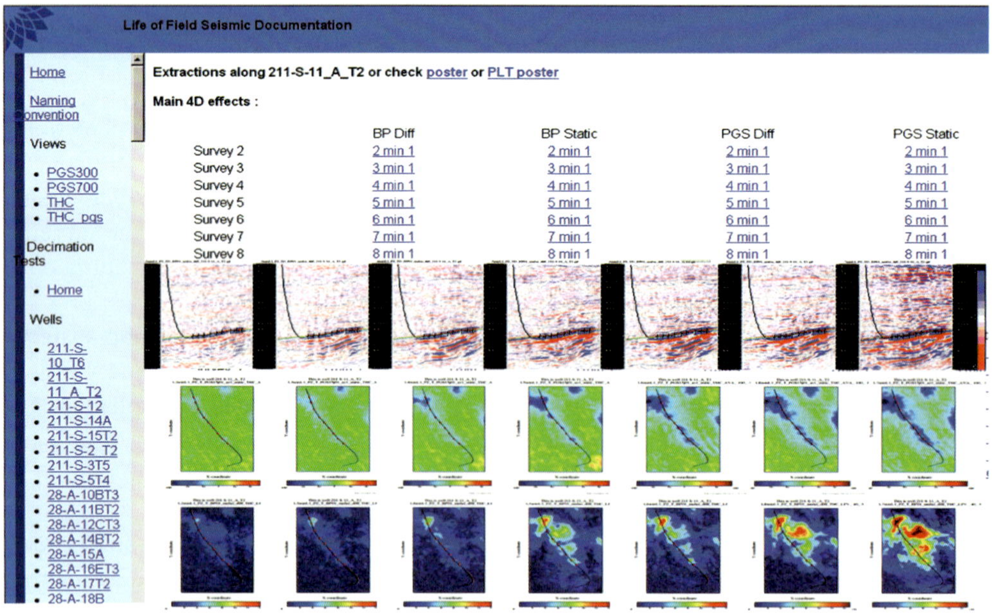

Figure 4.39 An HTML-based website provides documentation of the workflow and links to images from all surveys. van Gestel et al. (2005).

we can quickly focus our time on studying the important time sequences in more detail for any well or local area of the reservoir. Do we need this amount of information? Probably not, but it is difficult to find the optimal set for every location and for every purpose and it gives the user some confidence to be able to easily look at alternative attribute sets.

4.7.5 Automated workflows

The first four surveys acquired were analyzed partly using standard interactive 4D seismic software packages and partly using automated script based processing workflows, to test out alternative parameter settings and algorithms. Valhall was one of the early fields where the importance of geomechanical changes on 4D seismic time-lapse effects was noticed. This required us to develop new algorithms for time-shift analysis, etc. And it was obvious to elaborate on this experience. The tool set and the workflows are continuously being refined but the strategy has been the same, in the following we will use a snapshot of the workflow as presented by van Gestel et al. (2005).

A main motivation behind the automated workflow concept was the experience of going through the same interactive parameter selections for the fourth time, with limited or no help in defining either the name of a sub product or setting the right parameters. The other key motivators are (van Gestel et al. (2005)):

- **Standardization**
  - Changes in the post-processing, time-depth conversions and calculations of time-shift and difference volumes can quickly lead to changes that are larger than the time-lapse changes we are trying to observe (Jack 1998). By standardizing the workflow, we eliminate these changes and significantly reduce the potential for user error in generating the time-lapse volumes.
- **Speed**
  - Most of the modules from the manual workflow were not computer-intensive but required a lot of user input to run the various applications and transfer the volumes between those applications. The implementation of a shell script-based automated workflow significantly reduces this time, especially when repeating an identical workflow for various surveys, azimuths and the like. The ultimate goal is to move the main time of the geophysicist from the generation of time-lapse volumes to the interpretation of these volumes.
- **Reproducible**
  - An automated workflow enables re- running of the entire flow from scratch when necessary. If the workflow is updated, parameters changed and/or new modules added, then the complete workflow can easily be re-run and all volumes re-generated. Automation also enables any parameters to be easily tested throughout the workflow, thereby allowing immediate observations of the effects of changing any parameter in the final extracted map.
- **Documentation**
  - When implementing a workflow using shell scripts, it is simple to add log files that keep track of what has happened to the data by whom and when. This makes it easy to find the history of any dataset. Furthermore, we added scripts that generate graphic files and HTML-based documentation that make files easily accessible for quick first-pass analysis.

Figure 4.40 shows The LoFS time-lapse analysis workflow as it had been established after the fourth survey. The main elements of the flow are Modules 1 through 5. **Module** 1 takes care of loading the migrated volumes into the interpretation system. Residual post-stack corrections are applied to the data and the phase is adjusted to make the data zero-phase. In **Module 2** the data is transferred from depth to time to avoid artifacts in relevant attribute sets. **Module 3** contains the functions for the time-lapse analysis. The time-shift volumes are calculated using a sliding window cross-correlation. The time-shifts are then applied to the monitor survey to align it with the base survey and calculate the difference volumes. In **Module 4** the time-lapse changes are transformed from a volume attribute to a map presentation. This is done using windows around the reservoir, extracting max. or min. or simply calculating the acoustic impedance change within the reservoir. The generation of extractions around the reservoir that best show the time-lapse effects from both the time-shift and the difference volumes. In **Module 5** we convert the data from time to true depth domain by matching the wells and the reservoir model horizons.

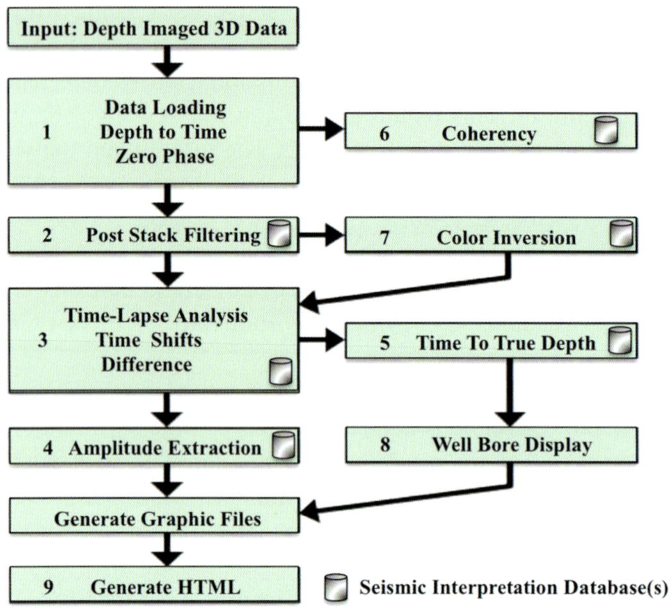

Figure 4.40: The workflow that was automated can be divided up into nine separate modules. van Gestel et al. (2005)

A set of horizons has been interpreted in the time data and a combination of a global velocity mode and residual "stretch and squeeze" to match the horizons of the reservoir model.

During the development of the automated workflow, several other modules have been added. **Module 6** generates the coherency volumes and an extraction at the reservoir level from these volumes. **Module 7** generates CI volumes from the time data, which are subsequently run through Modules 3 through 5 to generate CI difference volumes.

Two dedicated **modules 4 and 8,** include the generation of graphic images from the main attribute extraction (maps along key horizons and cross-sections through wells). And combine this into a separate "movie-set" that show the changes from survey to survey. The image sets are posted in a graphic file directory and **Module 9** contains functions to query the graphic file directory and builds an HTML-based documentation that links all files. Figure 4.38 shows the front page of the website that has links to all generated figures.

The initial automated workflows were implemented using a shell script-based workflow that runs using Unix Seismic Processing (USP), BP's in-house seismic processing package. USP has recently been replaced by DDS (Data Dictionary System), and organized in a more functional style with additional functionality.

A key element of the concept is a rigorous naming convention that is used for both the seismic volumes and the horizon extractions. This has significantly improved the usability of the data and has simplified the comparison of the time-lapse effects for the various surveys.

Having the modular automated workflow available makes it straightforward to now very simply and quickly test alternative parameter settings or include additional post processing functionality, i.e., including new and improved algorithms that become available, such as using frequency dependent structural conformance filtering for stabilizing the low frequency in our data, Whitcombe et al. (2008). Any changes in the workflow parameters can be directly linked to the effect in the final extraction. The workflow is a living process that is constantly fine-tuned and improved. Because the whole workflow is automated, updating all the volumes is a simple task. Currently, the expectation is for the workflow to be updated about once a year.

4.7.6 Reservoir model data

The automated workflows are designed to also handle the output from the reservoir model. The initial model and the relevant time-steps, normally dates that correspond to seismic survey dates, may be transcribed into data-structures that are accessed by the automated workflow. It is then straightforward to extract maps from the simulator model and present them in the same format as the seismic time-lapse attributes. Alternatively, we may convert the modeled changes in saturation, pressure and pore volumes into acoustic impedance changes. By adding additional layers, outside the reservoir we may also look at the reflectivity changes. The standard workflow for using seismic time-lapse data for history matching, includes converting the simulator parameters to reflectivity for each time-step and then calculating the changes in reflectivity or in acoustic impedance using exactly the same algorithms as used for the seismic observations. We will discuss this in more detail in chapter 5

4.8 Conclusion

In this chapter we have shown that improvement in both acquisition design and 4D seismic processing has resulted in significant uplift in the quality of resulting 4D seismic images.We have seen examples of high quality 4D seismic images based on marine towed streamer, ocean bottom cables and from nodes. Processing efforts may vary contingent on acquisition methodology and the size of the 4D response, and Permanent system offers some options for less processing efforts and quicker turn around of data. Results from fast-track processing of surface tow data often have delivered business value. The Valhall case has enabled less processing efforts than on other N Sea fields by virtue of a strong 4D signal and highly repeatable acquisition. That in turn has enabled an automated processing workflow, delivering rapid turn round of the data, appropriate to the reservoir management needs of the data. By utilizing AVO and projecting the 4D signal along the optimal angle in the AIGI impedance or Intercept/Gradient reflectivity or the AVO domains, we may maximize the separation between the fluid and the pressure signal in our time-lapse data. Maintaining the time-shifts, due to production induced stress state changes, during the processing and having appropriate methods for estimations of times shifts, is critical for certain type of reservoirs.

5 Case Study: Seismic Surveillance for Reservoir Management

"It is a very sad thing that nowadays there is so little useless information."
- Oscar Wilde

Integrating seismic surveillance data into reservoir management covers a wide range of workflows and processes. There are times when a single 4D images provides the answer; other times an exhaustive analysis of a wide range of scenarios might have to be tested to verify the best model. In this chapter we will try to illustrate how seismic surveillance can contribute to an improved understanding of dynamic properties of the reservoir. We will share a Valhall example of applying a BP proprietary technology, TDRM™, a workflow that allows for the objective analysis of a wide range of reservoir models, which improves the predictive quality of the models. Most of the applications we will discuss are linked to the frequency of repeated surveillance we may have when operating a permanent monitoring system, but should still be applicable to standard marine tow cases. Finally, we look at a specific study, monitoring of a short injection test in one well, this is only possible in cases where opportunities exist for acquisition of frequent surveillance data, still the study could have been done using a high-end conventional tow streamer 4D acquisition.

5.1 Reservoir management applications

The reservoir model is a fundamental tool for understanding past performance, predicting future performance and to support the drilling of new wells. The structural and stratigraphic description of the reservoir is referred to as the geological model, and provides the framework for the static reservoir description, which also includes the fluid and flow model. This model is typically built with a lateral resolution of 25×25 m and as many layers as deemed necessary, 10–30 layers are not unusual. The static model is based on the integration of seismic, well and production data and is the starting point for defining the dynamics for the reservoir fluid flow model.

The LoFS data has already allowed us to reduce the size of the seismically obscured gas cloud area and allowed better fault interpretation in the crest of the field, resulting in an improved static starting model of the field. It is considered best practice to ensure that the static model is consistent with the seismic data, which may be done by seismic forward modeling. The 4D seismic data offers an additional dimension to this, as we are now also able to assess the validity of the dynamic model. This process is often referred to as "closing the loop", but to utilize the opportunity requires that we honor all aspects of the physics in the modeling, a short-cut in the static model may not influence the flow model but will destroy the options for realistic seismic modeling.

A common outcome of most 4D seismic projects are new insights that reduce the uncertainties in the geological model. In this chapter we will briefly outline "a" reservoir modeling approach, then show how seismic time-lapse data could reduce the uncertainties in the geological modeling before we show a few examples of using 4D seismic in the history matching of the dynamic model. The seismic surveillance data becomes an essential part of our plan-do measure-loop and provide a basis for updating the depletion plans for the field as needed, figure 5.1.

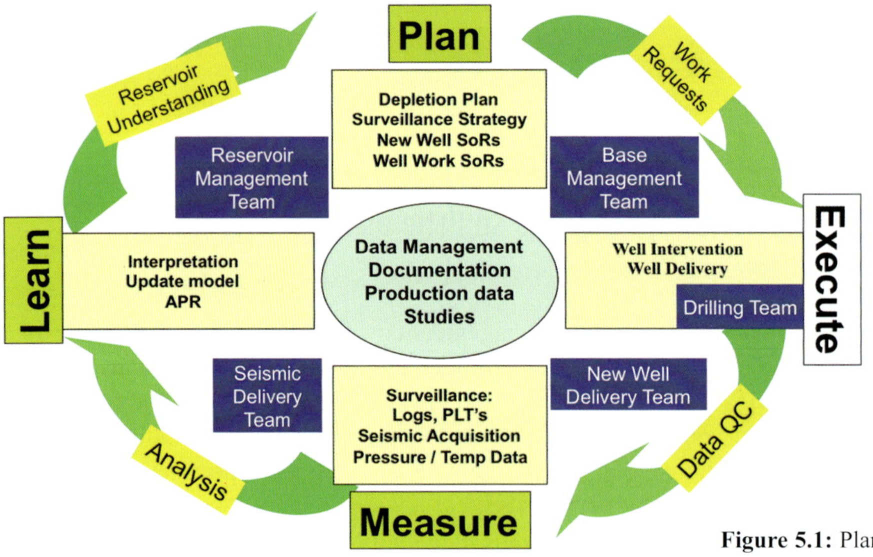

Figure 5.1: Plan-Do-Measure-Learn.

5.1.1 Reservoir modeling

Each reservoir has its own characteristics and challenges that may influence the process of building the model as well as the model itself. However, the model building process may be generalized as shown in figure 5.2. From being a map-based process 20 years ago, today the work is almost entirely done or at least verified, using various 3D modeling tools. Seismic data is used to establish the structural framework, defined by seismic horizons and faults. Using observations in the well and geological conceptual models as a guide, the detailed geometrical model is built. The cell dimensions are a compromise between the minimum features that we wish to resolve and what is practical to handle. Typical resolution could be 25×25 m laterally 1–100 meters vertically. Each cell in the model is assigned a rock type. The distribution of these rock types could be specified in various ways, for example, a layer based polygon or a probability map. Geo-statistical methods are often used to populate the individual cells in the model with appropriate porosity, permeability, level of cementation, clay content, etc., away from the well control.

While the majority of the rocks are saturated by saline water, brine, our primary interest is in modeling the more rare occurrence of hydrocarbons. Distribution of the hydrocarbons could be very complex. Capillary forces and fluid densities control the movement of fluids. Surface tensions and pore throat sizes are important factors. A common method for the calculation of saturation is to establish a saturation height function that incorporates these effects. The saturation in a cell is then given with the height above free surface level. These functions could often vary within the field due to the geological history, geometry, timing of fluid movements, etc.

The next step is to start validating the model by integrating the production history. The reservoir simulator model is a dynamic finite difference model. The geological model is typically transcribed into coarser grids, reducing the number of cells by a factor two to four. The properties are re-calculated, often using rigorous upscaling processes. The validity of the model is then tested against the static properties (volumes) of the geological model and then *versus* the production history of the field.

A key application of 4D seismic data is to support the history matching process. Seismic derived properties or attributes often provide a key source of information to guide the geological modeling. The process of maintaining the geological model of a field also includes incorporating new data as they arrive. If a well is drilled into a depleted area, the geologist may have to re-engineer the initial properties. Time-lapse seismic could be a very useful tool to assist in this.

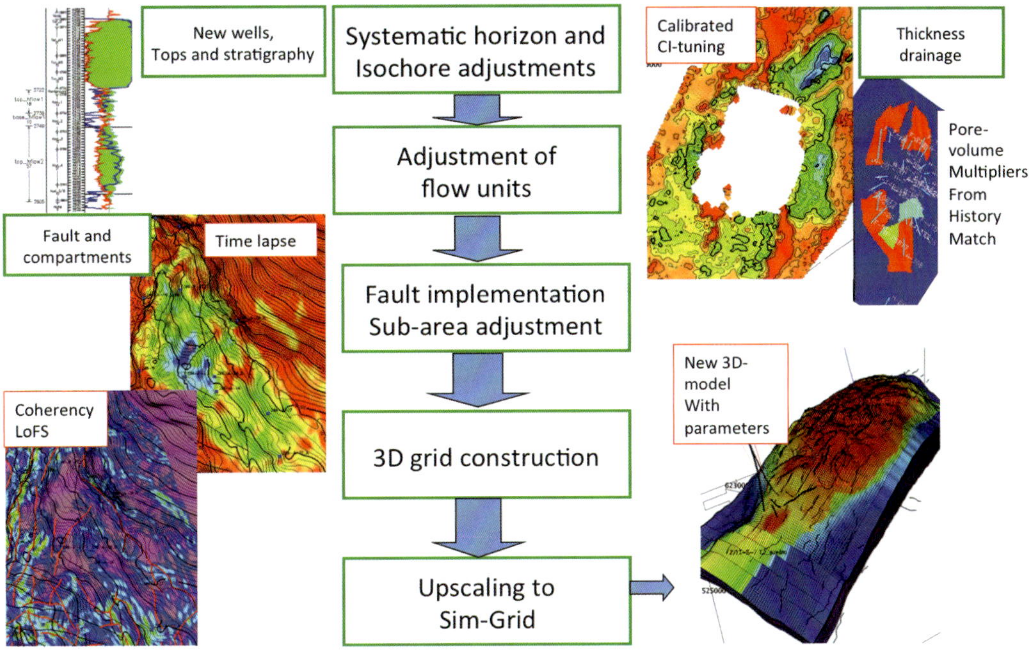

Figure 5.2: Workflow of geo-metrical modeling, from Barkved and Buer (2005).

In the following we will share some examples of how 4D seismic observation can help to reduce the uncertainties during the various modeling and history matching processes.

5.2 Static reservoir modeling is a dynamic process

5.2.1 Reducing geological structural model uncertainties using 4D data

At first glance, the geometrical model of the Valhall field might appear as a simple, well-behaved draping of layers over a mild anticlinal features, figure 5.3. Unfortunately, the details turn out to be much more complex. The layers are fairly thin, with a significant amount of small-scale faults. The structural story is complicated by crestal collapse during the inversion, re-activation of syn-depositional faults and reworking of sediments. A fundamental problem to the geologist is how many faults to include in the model. Reasonable numbers for this size of full field model 12–20 km by 6–10 km and structural complexities is probably in the order of fifty to several hundred, contingent on the purpose of the model. Local models may be built for individual wells, containing similar numbers of structural elements.

Building a geometrical model is complicated. The starting point is typically a detailed structural seismic interpretation. Any interpretation has inherent uncertainties associated with it and there will always a need for judgment of how much detail to carry forward. Figure 5.4 shows a dip/azimuth map, fault lineaments from an extremely comprehensive mapping of "seismic" breaks, used in detailed well planning and the structural elements that used to be in an early 1990 reservoir model. It is close to impossible to carry all the details into individual 2D fault planes that could be carried forward into the geological modeling. All the maps are from top reservoir level. We therefore have to rank the individual faults based on some characteristics. This could be: visible vertical displacement across, tie to observed faults in wells, need to explain production history, etc.

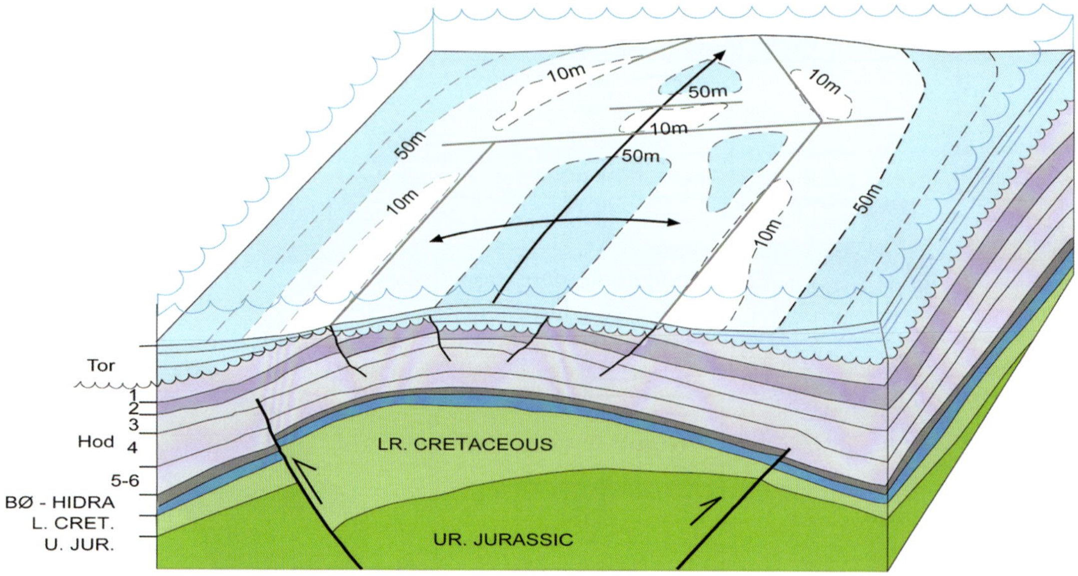

Figure 5.3: The Valhall geometrical model may appear as a simple well-behaved draping of layers across a mild anticlinal feature. Notice that the vertical scale is 10× the horizontal (copyright BP, used with permission).

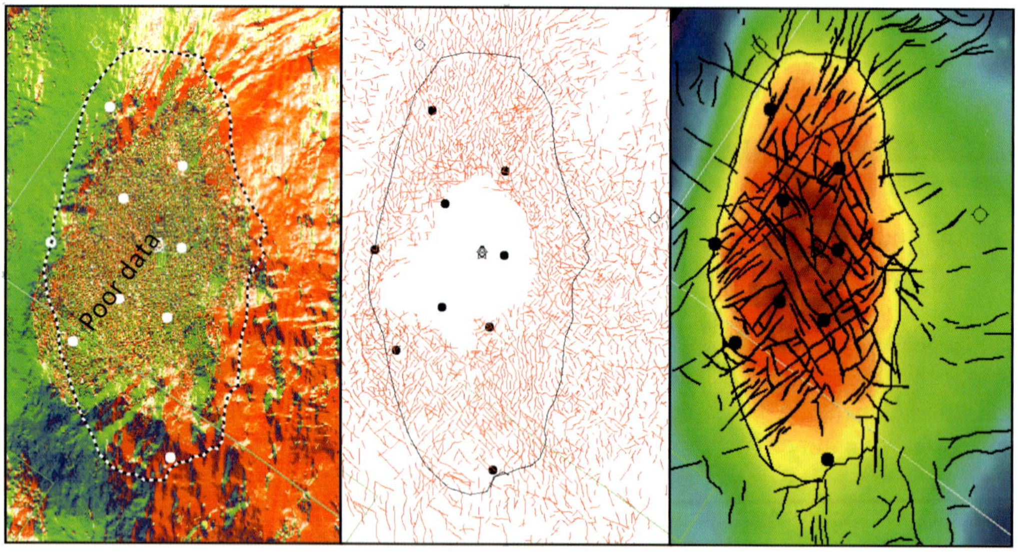

Figure 5.4: Structural attributes and maps of the top reservoir. Left: dip/azimuth map. Middle: structural elements extracted and used as a basis for detailed well planning, Right: structural map used in the 1995 vintage 3D reservoir model, ref Barkved (1998).

Figure 5.5 shows an example of a 2500 m vertical section along a horizontal well. The pre-drilling full field model only contained a single fault in the reservoir. Predicted thicknesses ranged between 20–30 m. As expected, the 4D images, acoustic impedance changes, after 4 years of production, reveal

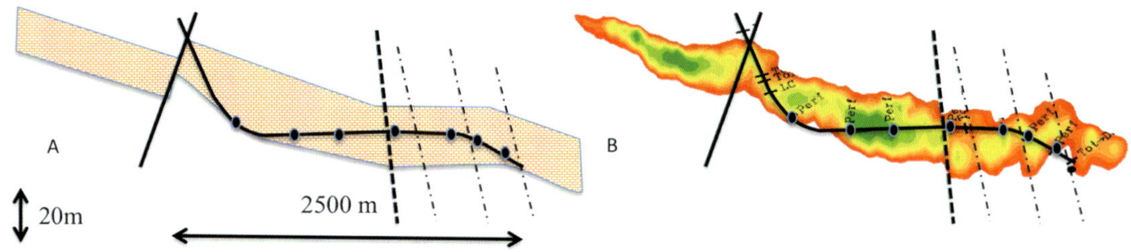

Figure 5.5: A: 2500 m long vertical section along a horizontal well. The pre-drill full field model, left, only contained a single fault in the reservoir. Predicted thicknesses ranged between 20–30 m. B: 4D observation, acoustic impedance changes, reveals significantly more small-scale structural complexities, from Barkved et al. (2003b).

that the reality is somewhat more complex. Some of these faults, the one at the mid-section and the one at the heel of the well, were identified and built into the pre-drilling prognosis; however, the conclusion was that we probably did not have any significant restriction of flow along the well. The fault to the left of where the well enters the reservoir was interpreted to represent a "baffle" zone, with possibly reducing flow. As we see from the results, the reality turns out to be the opposite. The well appears to perfectly drain the block to the left, while there seems to be two main segments along the well, which appear to be less well connected. The segment along the last part of the well has its own perforations, which appear to be less effective. This may be due to restrictions in the well bore and is clearly a candidate for a well intervention to check the conditions.

From a primary depletion point of view, some of these structural details may be irrelevant. However, if we plan to drill another well into this area or for a secondary recovery schemes like water injection, this may be critical information. Can we be more successful in identifying faults that might respond as a baffle zone? Simple analyses of cross-fault juxtaposition are apparently not conclusive. It actually appears that the minor secondary antithetic faults are more important in this analysis, as can be seen in figure 5.6, where the apparently minor stipple fault appears to be "baffling" the flow. Fortunately, this fault only appears to reduce the flow and we finally end up with quite effective depletion. A possible interpretation is that the flow is reduced by fault-smear of some kind. These observations have increased the awareness of the role of minor faults, i.e., while drilling a new well.

The acoustic impedance difference sections provides an excellent image of the geometry of the reservoir. It goes without saying that the model thickness uncertainty or even reservoir presence uncertainty in part of the field is highly reduced when these effects are observed.

4D seismic images have become an important tool for reducing the uncertainty inherent in structural interpretations of seismic data. 4D maps provide the overall structural context, figure 5.7, while details can be further studied by analyzing the 4D image volumes.

5.2.2 Reducing geological properties uncertainty using 4D data

The product of porosity, saturation and thickness gives the total volume of hydrocarbons. In a reservoir that compacts as it is produced, thickness, priority and saturation will change over time. It is reasonable to assume that the velocity of the rock varies linearly as the porosity (and inherently the density) changes. The porosity reduction is built into the reservoir flow simulator and the shapes of these curves are established from laboratory data and matched to field observations. Figure 5.8 shows a realization of the porosity *versus* effective stress. By simply cross-plotting the changes in porosities *versus* the acoustic impedance changes, we may find out if this is a realistic assumption. Figure 5.9 shows that this appears to be the case using the 4D streamer data, covering production from 1992 to 2002.

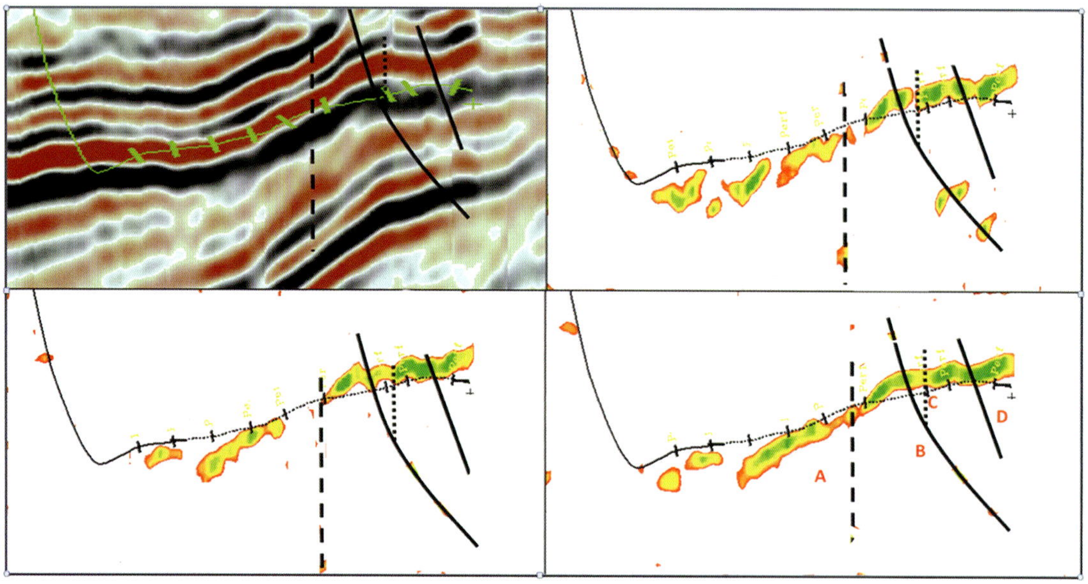

Figure 5.6: Cross-section along a horizontal well showing the amplitude data (top left), difference in amplitude after 6- (upper right), 9- (lower left) and 12-months (lower right) of production. The changes are due to changes in amplitude, traveltime changes are corrected for. The captions are 3000 m horizontally and 500 m vertically. The well is shown in green/black, the vertical sticks are perforations, from Barkved et al. (2005).

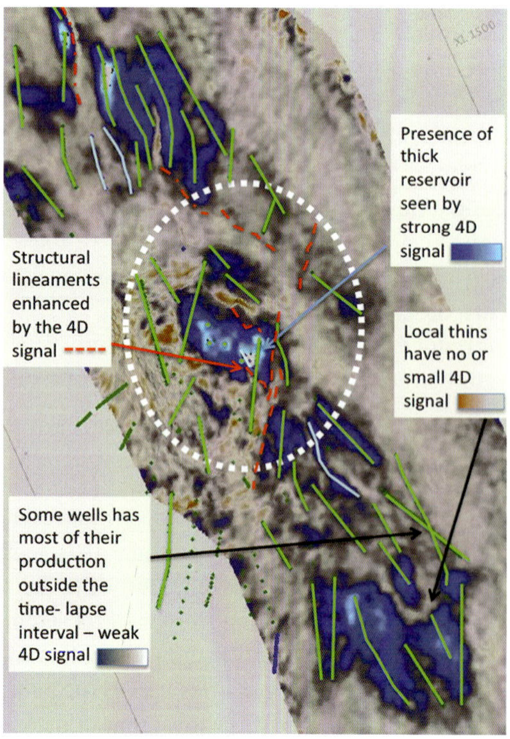

Figure 5.7: 4D images showing the production induced changes between 2003 and 2010. The circle indicates the areas where the presence of gas in the overburden distorts the seismic images such that conventional 3D interpretation on P-wave reflection data is very difficult. The green lines represent wells. The map is a NRMS amplitude map in a 250 ms long window, centered on the reservoir, from Barkved et al. (2010).

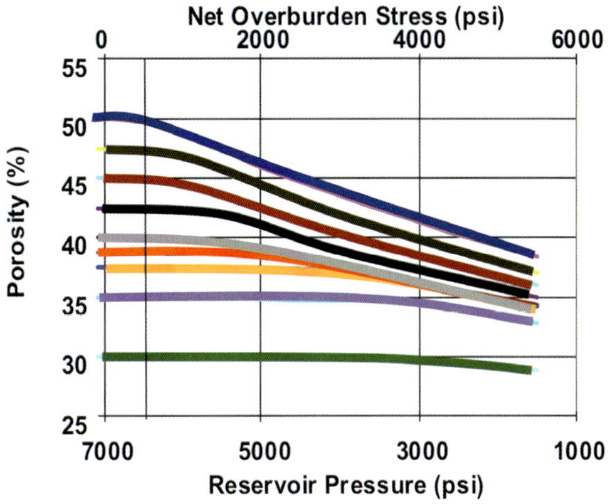

Figure 5.8: In a compacting reservoir the porosity changes as a function of effective stress. Andersen (1995), used by permission.

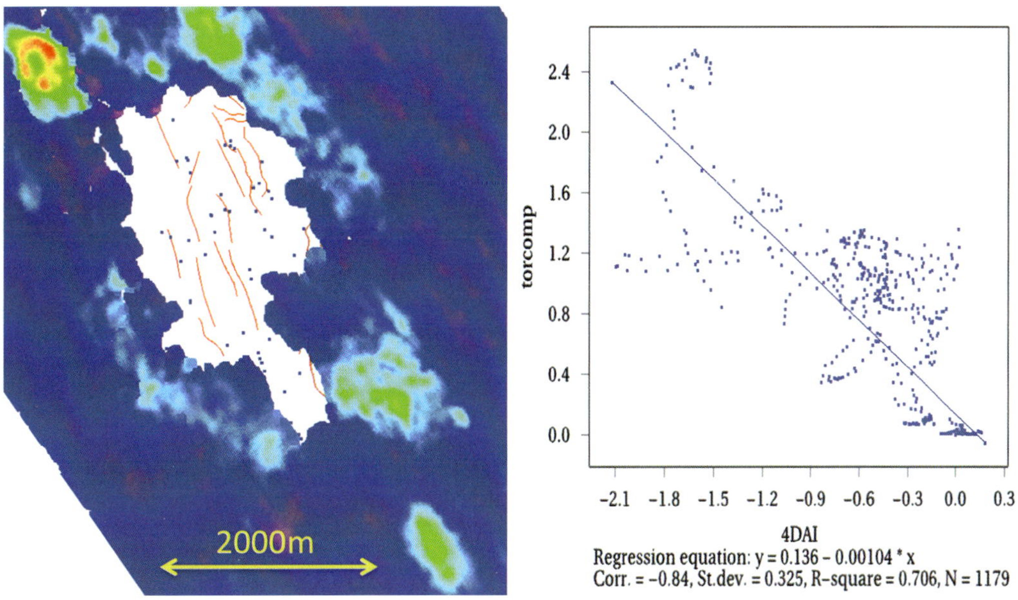

Figure 5.9: Left: changes in relative acoustic impedance during 10 years of production, 1992–2002. Right: cross-plotting the calculated porosity reduction, vertical axis, versus the un-calibrated changes in acoustic impedance, red represent large changes, light blue is small changes (Barkved et al. (2005d).

Figure 5.9 shows the relative acoustic impedance changes, but we have excluded areas where GOR changes, may distort the correlation using a pressure map as a guide. Red indicates large changes while blue and pink colors indicate no changes. As the results are based on un-scaled colored impedance volumes, the horizontal scale is meaningless. However, it is now possible to establish a first-order absolute scale, based on the linear relations to the changes in acoustic impedance. We have two other measurements that can be used to inform us about the thickness of the reservoir, the traveltime changes across the reservoir and the traveltime changes due to the compaction induced changes above the reservoir.

Reduction in the pore pressure results in more of the external load being transferred to the rock grains, which for a weak rock results in deformations and a reduction of the pore volume, Kristiansen (2009). Assuming uniaxial strain conditions (no lateral deformation in the reservoir), we have the following relation for compaction

$$\Delta h = \Delta P C_m h_i \qquad (5.1)$$

Where Δh is compaction, ΔP is pressure depletion; C_m is rock compressibility and h_i is initial reservoir thickness. We might find Δh, assuming that this is linearly related to the unloading effects in the overburden, which is observed as an increase in traveltime.

If we also assume no compaction of the grain, see figure 5.10, the compaction, Δh, will be a result of a reduction in pore volume:

$$\Delta h = h(\varphi_i - \varphi)/(1-\varphi) \qquad (5.2)$$

Where φ is the actual porosity and φ_i the initial porosity.

5.2.3 A workflow for inverting for pressure from time-lapse attributes

Given that we know the pressure and the porosity, we can find first-order estimates of the total thickness. However, while these manipulations are useful we also need to take into account the pore-fluid and the compressibility of the rock. In the next section we will see how these effects can be combined into a functional relationship using an approach as proposed by MacBeth et al. (2006). A scheme of linear relations can be used to link several seismic time-lapse difference attributes that respond independently to the reservoir's saturation and pressure variations. The approach was tested on synthetic data and on data from the Schiehallion Field on the UK continental shelf, with favorable results, MacBeth et al. (2005). In a paper by Corzo et al. (in press 2012), the concept is modified to take into account the change in porosity of a compacting reservoir to changes in pressure ΔP. Corzo et al. start out with a formulation that includes porosity reduction as a function of pressure:

$$\Delta A = (C_1\varphi_i + C_2)\Delta P \qquad (5.3)$$

Where C1 and C2 are specific constants for this reservoir, while **φi** is the initial porosity. It is further assumed initially that the effect of the stress redistribution may be of secondary impact.

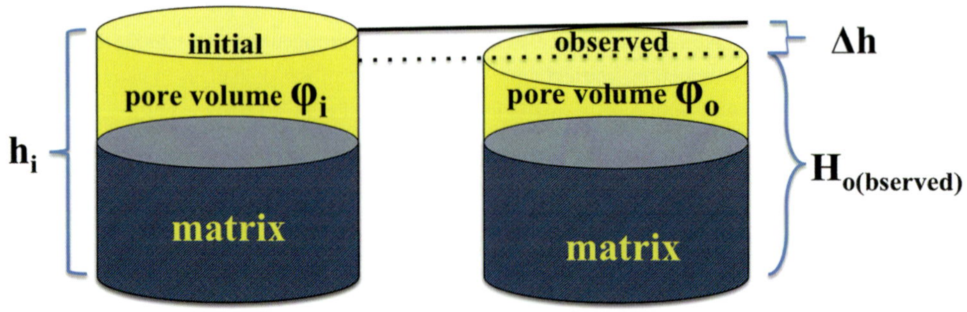

Figure 5.10: Simplified model for compaction. Assuming incompressible grains and uniaxial strain condition, the rock compaction will be assimilated by a reduction in pore volume.

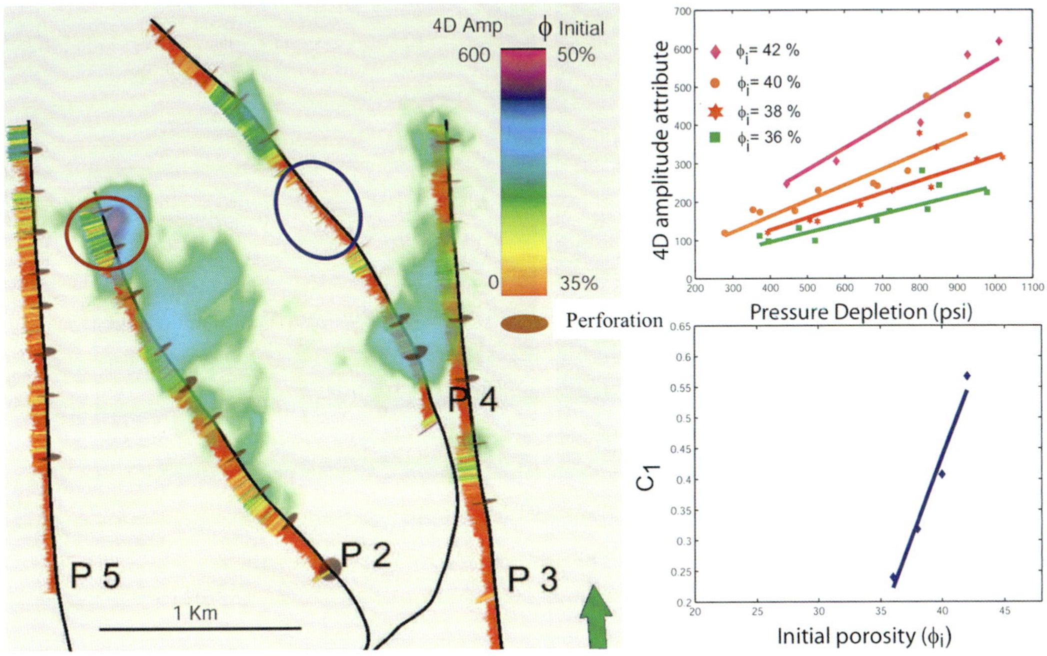

Figure 5.11: a) The porosity log is displayed along the borehole. The background map is the 4D amplitude map for LoFS2minLoFS1. The perforations are the brown disk shapes b) amplitude changes versus pressure depletion, initial porosities are posted in the diagram c) the gradient C1 versus initial porosity, from Corzo et al. (in Press).

The two constants are found by correlating the known pressure distribution as measured in wells to seismic time-lapse attributes from six surveys. Figure 5.11 describes the method. Inside a local area around each perforation in the wells, the mean seismic attribute, the pressure changes and the initial porosity is calculated. Bottom hole sensors at the heel of the well provide the pressure data and the thickness is estimated by combining well observation and seismic impedance data. The four porosity classes are validated using the time-shift across the reservoir By using two seismic time-lapse attribute sets, one based on acoustic impedance changes in the reservoir the other based on time-shifts due to stress redistribution outside the reservoir, C1 and C2 can be determined and it is now possible to predict the pressure change away from the wells using the maps of the seismic attribute changes (given, of course, an initial porosity map), figure 5.12. The number of surveys and the number of independent attribute sets allows us to calculate the confidence of the correlations. The workflow is outlined in figure 5.13.

5.3 TDRM

Kjelstadli et al. (2005) describes how changes in seismic attributes were modeled using the output from the reservoir simulation model at various time steps and compared against observed 4D seismic data in one area of the field. Objective functions describing the misfit between the simulated and measured data were then defined. This was done both for the seismic data and for the production data. Using BP's Top-Down Reservoir Modeling (TDRM) technology, multiple reservoir descriptions were generated that would minimize the objective function. This involved computer-assisted history matching with a genetic algorithm that varied more than 60 parameters in the reservoir simulator to improve the overall history match.

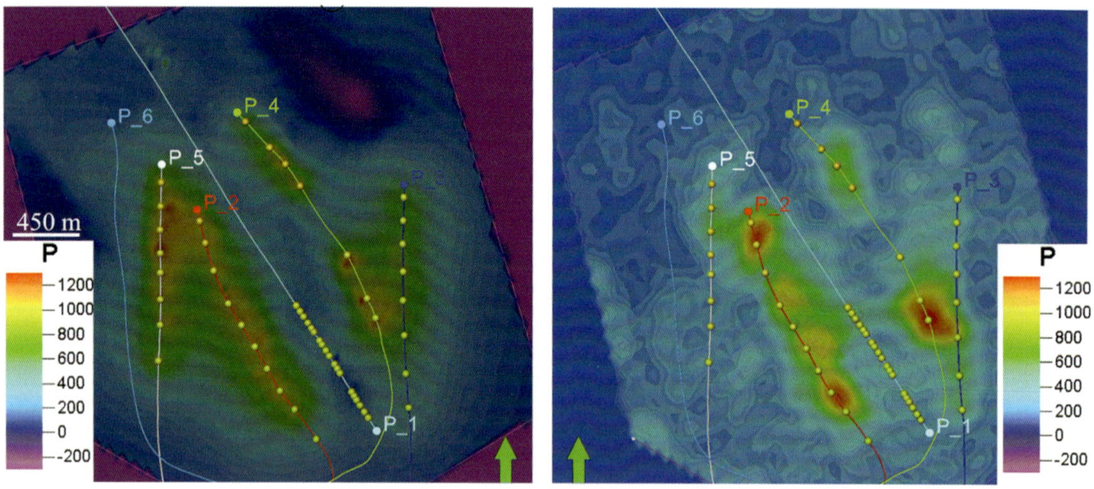

Figure 5.12: Left pressure depletion (psi) from reservoir simulator @ LoFS 4. Right pressure depletion inverted (psi) from seismic @ LoFS 4, from Corzo et al. (in Press).

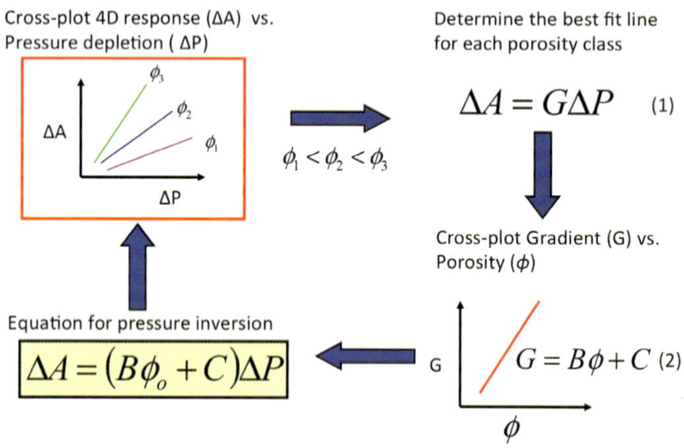

Figure 5.13: Workflow for inversion of pressure from time-lapse attributes, from Corzo et al. (in Press).

In this process several seismic time-lapse attributes were tested, but for this region the attribute set we decided to test were traveltime changes and acoustic impedance changes in the reservoir. Due to the simple geometry and the expected "smoothness" of the depletion in the small region this approach was to be tested on, we ignored the stress arching effect in this test. To represent the compaction energy correctly in the reservoir, it is necessary to incorporate the effects of the stress state redistribution in the overburden. Full field 3D Finite Element Model (FEM) has been used for subsidence predictions (at the platform location) since 1989 and we are well aware of some of the effects through testing out fully coupled models. However, the CPU requirements even for a simplified model are overwhelming. Today, a more rigorous limited coupling approach, **mech2sim**, is available, Pettersen and Kristiansen (2009). We expect that this approach would be practical to use for testing massive numbers of alternative models.

The actual pressure, saturation and porosity at each defined time-step were transformed into elastic properties; bulk density, compressional velocity and shear velocity. These were merged with a background model and the seismic response was calculated by convolving with a representative seismic wavelet. The modeled (synthetic) seismic time-lapse could then be calculated in exactly the same way as we do for the actual seismic observations. The acoustic impedance changes were estimated after projecting the seismic amplitude into relative acoustic impedances using colored inversion. In this projection the reservoir in this region appeared to be defined by local negative values, the reservoir was initially less dense than the surrounding rock. During depletion the impedances increased. The zero crossing above and below corresponded approximately with the top and base reservoir. The traveltime differences, TTD, are calculated directly using the actual velocities V and thicknesses ΔZ.

$$TTD = 2\sum_{i=1}^{n}(\Delta Z_{i2} / V_{ip2} - \Delta Z_{i1} / V_{ip1}) \tag{5.3}$$

Where ΔZ is the total thickness and the summation is performed over the stack of reservoir layers. The time interval corresponded to two of the available towed streamer surveys, 1992 −2002. In this case, these two attributes are perceived to be strongly correlated, so the interest in using two sets is more related to the robustness of calculating the time-lapse response. Geomechanical effects in the overburden were not included but the thickness was corrected for the appropriate reduction in porosity.

The seismic match quality, R_{seis}, was calculated using a map-based correlation technique. Direct point to point comparison is not appropriate as a basis for a match quality indicator, as the results would be too heavily distorted by minor spatial inconsistencies and scaling effects associated with the modeling. The seismic match quality was defined as

$$MQ_{seis} = 1 - R_{seis} \tag{5.4}$$

A low value then represents a good match, while completely anti-correlated results will be given a value of 2. A seismic quality map may be incorporated in the process. We used coherency maps to weight down time-lapse effects close to a fault, as these tended to be anomalous. In total, 2400 seismic attribute maps were generated.

For the production data, a conventional RMS measure between simulated and measured quantities was used for match quality. These quantities included monthly-cumulated water production, gas-oil ratios measured pressure data and monthly oil production rates.

The intention was to run a massive number of realizations through the process and we elected therefore to focus on one of the regions of the field, see figure. 5.14. The preparation for the rigorous testing of the alternative properties of this region included running the full field model with all wells to establish appropriate boundary conditions. The history matching process was done using BP's TDRM technology, Williams et al. (2004), which makes use of computer assisted history matching techniques to find multiple alternative solutions based on match quality. A set of parameters and ranges were tested, a genetic algorithm optimizer was used to modify the parameters and to find a set of reservoir models that minimized the objective function. Some of the parameters that were tested were porosity multipliers (compaction), permeability multipliers, fault transmissibility, vertical transmissibility, aquifer strength, compaction trends and skin along the (horizontal) wells. Further details on the parameters tested and how the objective functions were combined can be found in Kjelstadli et al. (2005).

To better understand how the optimizer would respond to a conventional history matching process, a separate run was done with well data only, figure 5.16. Quite a few "widely" different reservoir models appear to provide a reasonable match to well data, however only a subset of these would also honor the seismic time-lapse data, figure 5.17 and figure 5.18.

A main advantage of having multiple alternative reservoir models with similar match quality is that they can provide a realistic range of future predictions. Future incremental oil production, from a suitable set of realistic models significantly different to each other, were picked from the well, only matched sets, and compared with a set from the well and a seismic matched set. When used in prediction mode, the

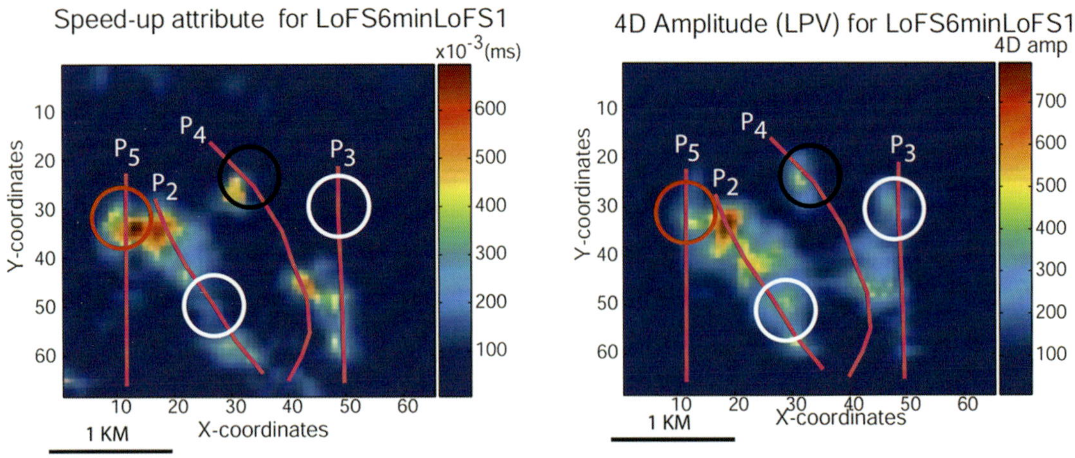

Figure 5.14: Speed-up (left side) and amplitude (right side) attributes for the time-lapse volumes: LoFS6minLoFS1. The circles point out where the two attributes are different. From Corzo et al. (in Presss)

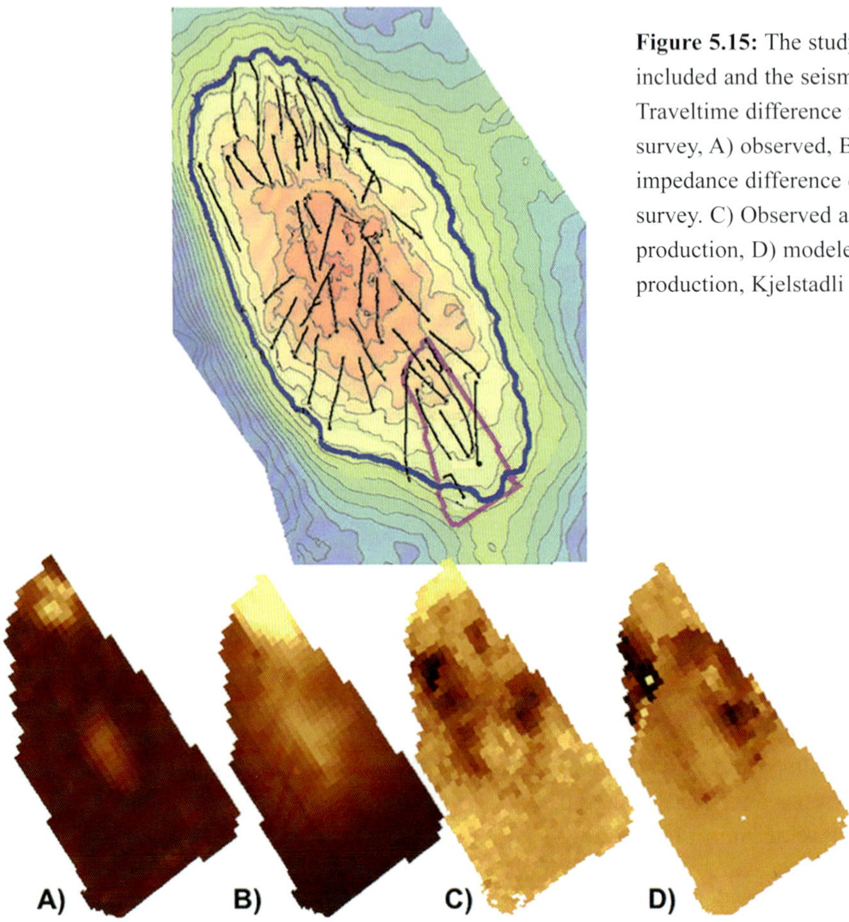

Figure 5.15: The study area, the wells included and the seismic attributes used. Traveltime difference maps from the streamer survey, A) observed, B) modeled and acoustic impedance difference data from the LoFS survey. C) Observed after 9 months of production, D) modeled after 9 months of production, Kjelstadli et al. (2005).

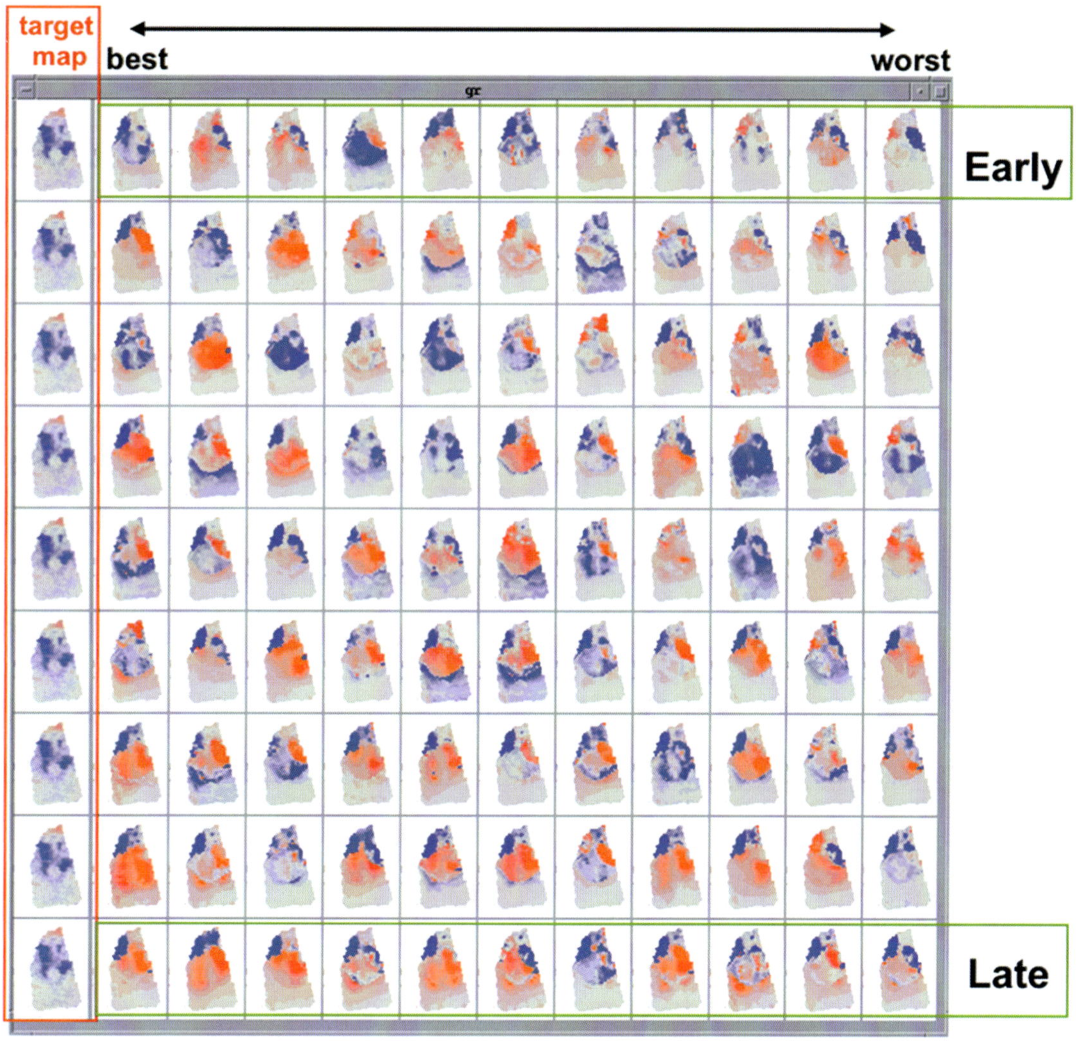

Figure 5.16: Match to well data only, sorted by objective function. Left column is observed seismic response. On the right, 99 maps are simulated responses. Although the well data match is good. None of the maps in the lower row have good semblance with the seismic data. Seismic match quality is not included in this run, Kjelstadli et al. (2005).

history matches that included the additional seismic constraints provided a narrower uncertainty range on the resulting forecasts than did the models that only honored the well data, figure 5.19.

In a subsequent study to the one discussed, we looked into considerations around the selection of appropriate attribute sets when we have frequent surveys available, Lane et al. (2006). A target attribute map refers to time-lapse seismic observations. We are looking for a relevant set of data that corresponds best with the production induced effects, but that also complement each other by bringing new, additional information, which might help in constraining the matching process. Our experience indicates that a history match that honors multiple independent 4D target maps helps in reducing the uncertainty ranges based on history and will result in a narrower range of predicted outcomes. One approach for producing independent target maps is to use maps where the monitor survey is different to the preceding

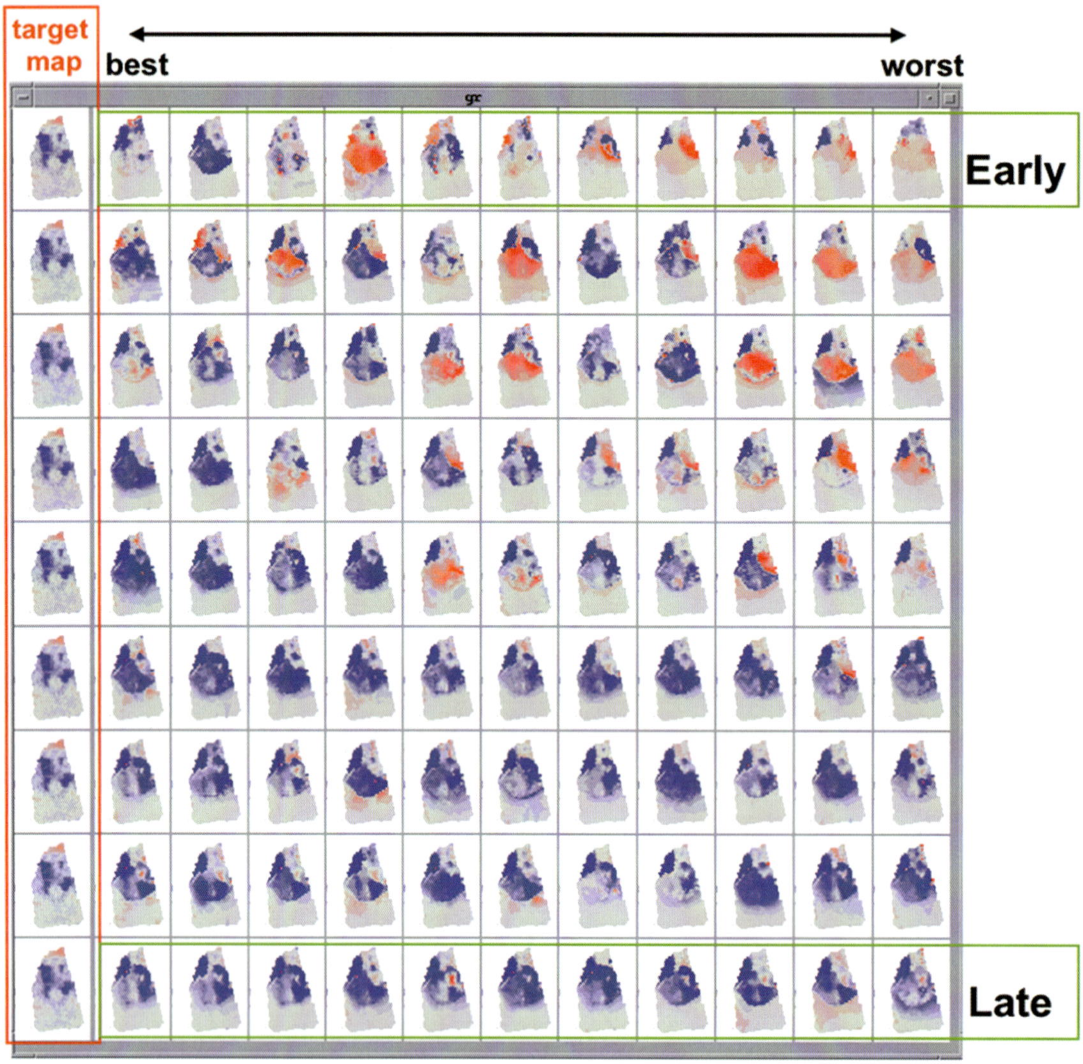

Figure 5.17: Match to well and seismic data. Sorted by objective function. Left column is observed seismic response. On the right, 99 maps are simulated responses. Now the best well matched cases also show reasonable correlations with the modeled and actual seismic response, Kjelstadli et al. (2005).

survey, rather than to a common baseline. These maps will not overlap in time and the production effects appear to better dissolve spatially. To test out whether this has merit, let us take a closer look at the two alternative target attribute sets.

The target attribute sets in figure 5.20, which are all related back to the baseline survey, appear to be highly correlated, in fact the correlation values range from 0.87 to 0.95. On the positive side, this confirms that observed changes are real and not just noise, on the other hand, we probably only have one independent piece of spatial information for constraining the reservoir model during a history match. The attribute shown is "the largest positive value" in a short depth interval across the reservoir, estimated from the difference volume between relative acoustic impedance volumes. The positive value will reflect hardening due to production.

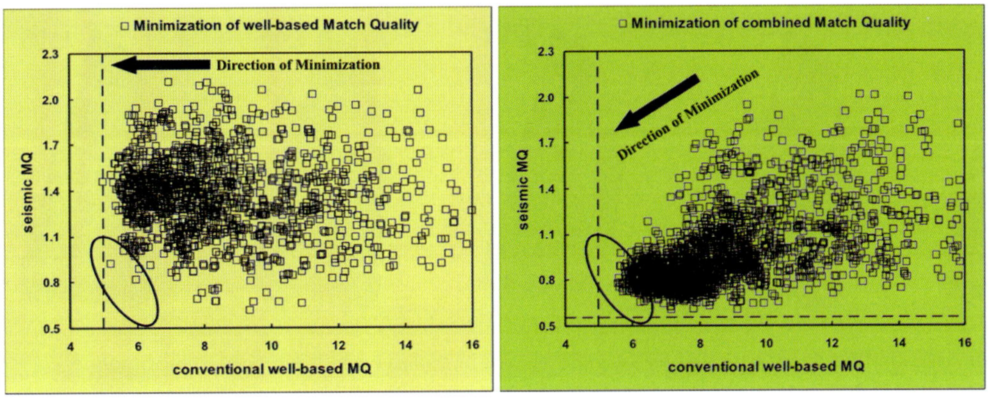

Figure 5.18: Left: well based match quality. Right: combined (well and seismic), from Walker et al. (2006).

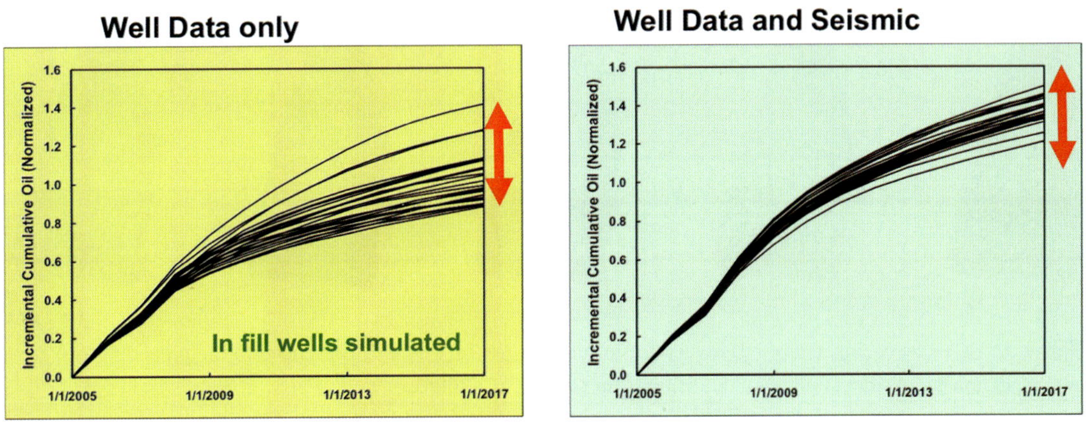

Figure 5.19: Forward model prediction. Ran depletion and one-well infill scenarios in prediction mode. Twenty-four "best" models for well-data-only and combined well+seismic matches. Infill well on-line 1/07, predictions run under THP control to 1/17. Narrower prediction range for both depletion and infill scenarios when seismic data is included. Reduced uncertainty in predicted impact of infill well, from Walker et al. (2006).

The smaller time increments target attribute sets reflecting the incremental changes between the surveys, figure 5.21, provide a weaker reservoir signal and lower S/N. We might have potentially more "independent" maps to choose from. These maps appear to be less correlated, but there are some concerns about how much "independent" information is contained in each map. L12-L23 and L23-L34 map correlation values are ~0.65; the magnitude of change differs, but the spatial pattern is similar. Tables 1 an 2 below summarize the various correlation factors of the target attribute maps in the two cases. An alternative approach might be to remove one or more surveys for example L2 and use L13 instead, and as Lane et al. (2005) show in their paper, the most "complementary" target attribute map set from this area seems to be the combination L14, L45 and L56, figure 5.21 and Table 3.

It is obvious that other types of target attribute set could have been more dissimilar and as such have discrimination "power". If we look at the production history of this region of the field, figure 5.22, production has been very stable up to monitor survey 4. With closer investigation it turns out that between monitor surveys 4 and 5, one of the wells has been shut in. This was expected to result in re-pressurization close

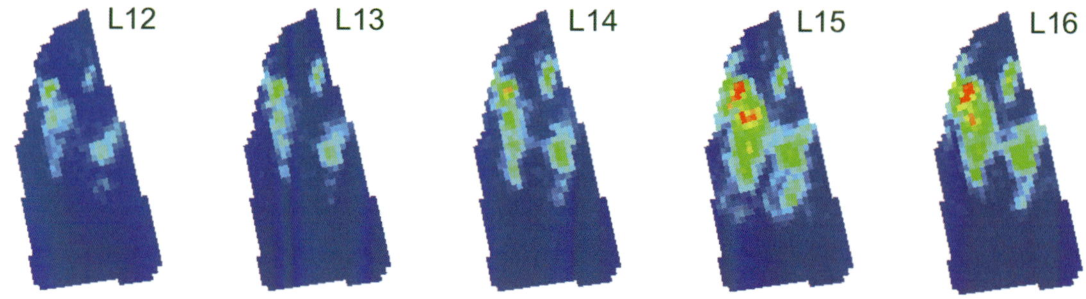

Figure 5.20: Acoustic impedance based target attribute set. All maps has been differenced to the baseline LoFS1 survey (overlapping time windows), modified from Lane et al. (2006).

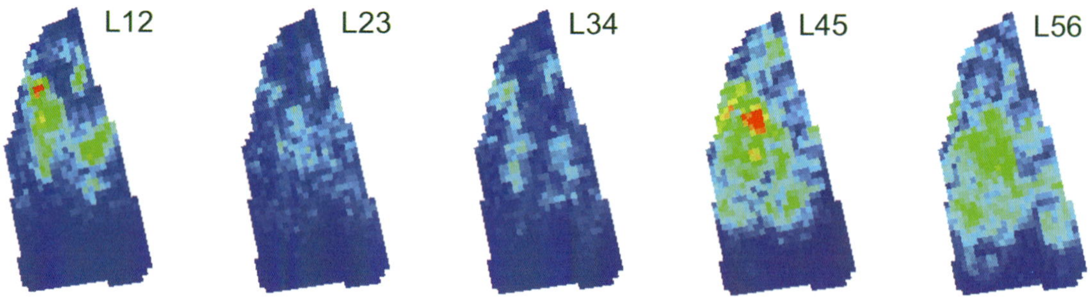

Figure 5.21: Acoustic impedance based target attribute set. All LPV maps differenced to the preceding LoFS survey (independent time windows), modified from Lane et al. (2006).

to the well. The relative acoustic impedance change at this pressure regime (above bubble-point) that is insensitive to pressure builds up but other attributes, i.e., related to overburden stress redistribution, could have picked up this effect.

5.4 Practical integration

In the course of day-to-day integration, the Valhall reservoir management team has implemented a practically full field history matching workflow, building on the findings of several dedicated studies, like the ones mentioned above, van Gestel et al. (2011). The workflow is outlined in figure 5.23.

The field is split up into several areas containing one or more sectors, and each area is worked in some detail as part of a dedicated focus to verify, refine and possibly update the model to support ongoing well planning or well intervention work in the actual area. The findings from this work are later implemented in the full field model.

The input data to this flow is the seismic time-lapse data and the output from the reservoir model for the same time-steps. Using the rock physics model and the process described in Askim, (2003) and Kjelstadli et al. (2005), the reservoir pressure, saturation, porosity and thickness changes is converted into a synthetic seismic trace, using the same wavelet represented in the observed data. In doing do so we may use exactly the same software to generate the attribute volumes or maps. A separate flow is

Correlation	L12	L13	L14	L15	L16
L12	1.000				
L13	0.930	1.000			
L14	0.905	0.967	1.000		
L15	0.915	0.870	0.890	1.000	
L16	0.914	0.935	0.945	0.949	1.000

Table 1: Correlation of target attribute maps in figure 5.20, modified from Lane et al. (2006)

Correlation	L12	L23	L34	L45	L56
L12	1.000				
L23	0.663	1.000			
L34	0.737	0.646	1.000		
L45	0.694	0.592	0.552	1.000	
L56	0.404	0.643	0.486	0.509	1.000

Table 2: Correlation of target attribute maps in figure 5.21, modified from Lane et al. (2006)

Correlation	L14	L45	L56
L14	1.000		
L45	0.540	1.000	
L56	0.309	0.509	1.000

Table 3: Correlation of the 3 most dissimilar target attribute maps, modified from Lane et al. (2006)

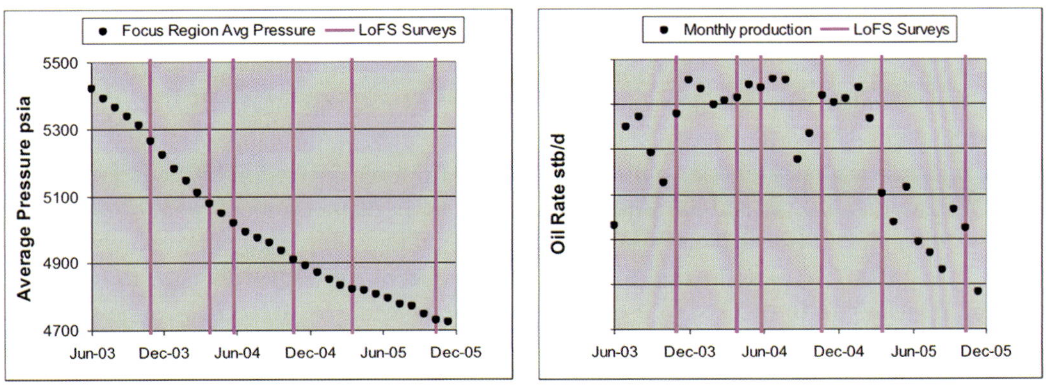

Figure 5.22: The depletion history in the wells during the acquisition of the first 6 surveys, ref Lane et all (2006).

used to analyze and match the compacting induced changes observed above the reservoir to the modeled compaction. The primary time-lapse effect during depletion in the reservoir is an increase in acoustic impedance, or reduction in traveltime, due to reservoir compaction. The secondary effects are due to saturation changes. During production below bubble point, an increase in the gas-oil ratio will result in a "softer" reservoir, which is reversed when the local depletion stops or we inject water into the area. Around the water injectors we also observed an amplitude increase effect due to increased water saturation (van Gestel et al. 2008a). This could be either related to saturation changes or to water weakening, which is additional compaction in the chalk due to increased water saturation, Tjetland et al. (2007).

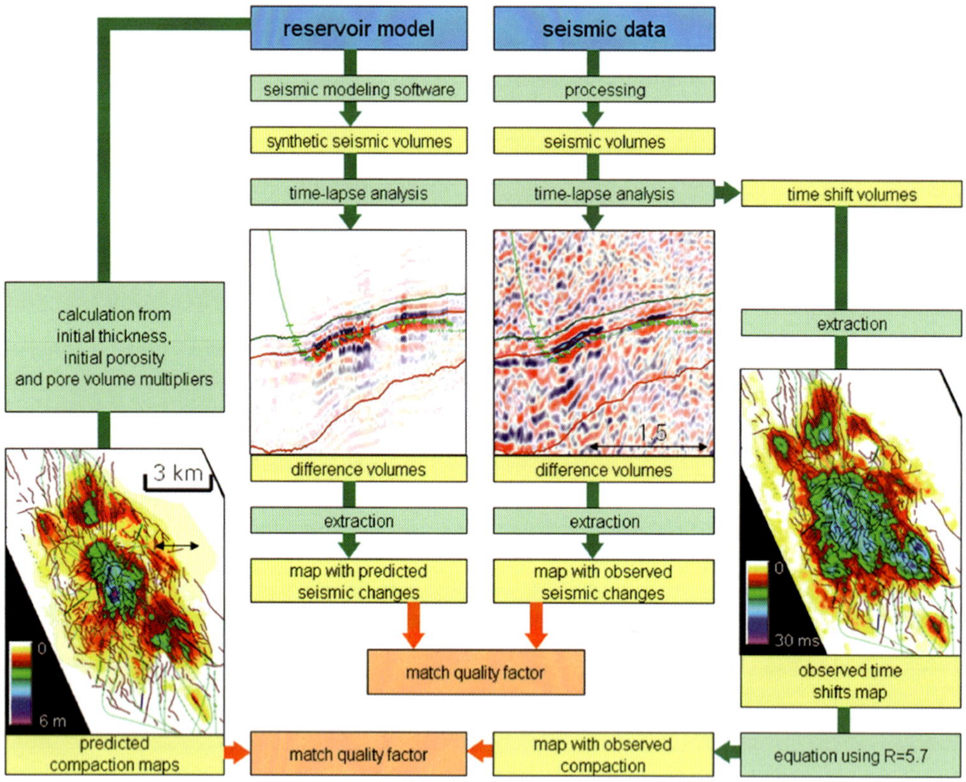

Figure 5.23: The workflow for comparison of life of field seismic data to the reservoir model. The input is in blue, modules are in green, products are in yellow and the final output is in orange, from van Gestel et al. (2011).

The traveltime changes between subsequent time-lapse surveys are calculated and we extract and make a map of the time-shift difference by, extracting the maximum or normalized average time-shift in a short window just above the reservoir. In the standard workflow, we have adopted the R-factor concept as described by Hatchell, Kawar and Saviski (2005). A full field geomechanical model combined with dedicated seismic modeling (**mech2seis**) were used to verify that a simplified approach could be used in this area of the field. The same model, Kristiansen and Plischke (2010), was used to calculate the compaction in the reservoir to be used to estimate the R-factor value. A reasonable value seems to be R = 5.7. Special consideration must be taken in areas where we have abrupt changes in reservoir thickness. But comparison with full field geo-mechanical based results shows that these simplistic analyses are a good first-order approximation, due to the frequency of the time-lapse data. Some further details of the workflow may be found in van Gestel et al. (2011).

The observed and modeled data are loaded into the seismic interpretation system for more detailed analysis, or into the reservoir engineers' preferred visualization engine and as documents in a local web-based documentation system, figure 5.24. Given the outcome of these analyses, a set of relevant parameter ranges is agreed to be tested, these could be thickness, porosity and transmissibility across faults and even relative permeability. Subsequently, a series of simulator runs are initiated, the results are loaded in via the modeling flow and matched to observed data. A similar map-based correlation approach to the one used in the TDRM workflow is used to calculate match quality to assist the history matching process.

In the following we will have a look at a few simple examples to illustrate how the process works.

5.4.1 Well connectivity

Figure 5.25 summarizes the findings from a 3×2 km wide Sub-basin in the northwestern part of the field. Well 2 in this area was converted into an injector, intended to support Well 1. Shortly after injection started, and a few hundred bbls of water had been injected, water breakthrough between Wells 1 and 2 was observed. Even thought the injected volume was quite small, we observed strong amplitude changes. Based on the pressure and the gas-oil ratios observed, we interpreted this effect to be due to gas going back into solution, driven by the increased pressure. Based on the time-lapse response, rightmost caption in figure 5.25, there is no or limited communication between wells 2 and 3.

Forty alternative scenarios, which matched the history of the well data, were tested *versus* the seismic time-lapse data, which included streamer data difference and the first 6 LoFS surveys. The best and worst matched cases are shown in figure 5.25, the primary target attribute used an amplitude extraction at the reservoir level.

5.4.2 Testing relative permeability curves

The gas/oil relative permeability curve in a low permeability field like Valhall controls how good a well is draining areas away from the well. Depletion away from a well may be deduced from observations when drilling a new well in the same area, however, this is often not very conclusive because localized effects, i.e., associated with fractures, may complicate the situation. Time-lapse data may provide enough detail to assess the situation properly. The gas-oil relative permeability has significant impact on depletion planning and ultimately, the total recovery of a field.

To study the impact of various gas-oil relative permeability (rel-perm) scenarios, a range of model with varying rel-perm relations was tested. In this case we elected to use the compaction-induced chang-

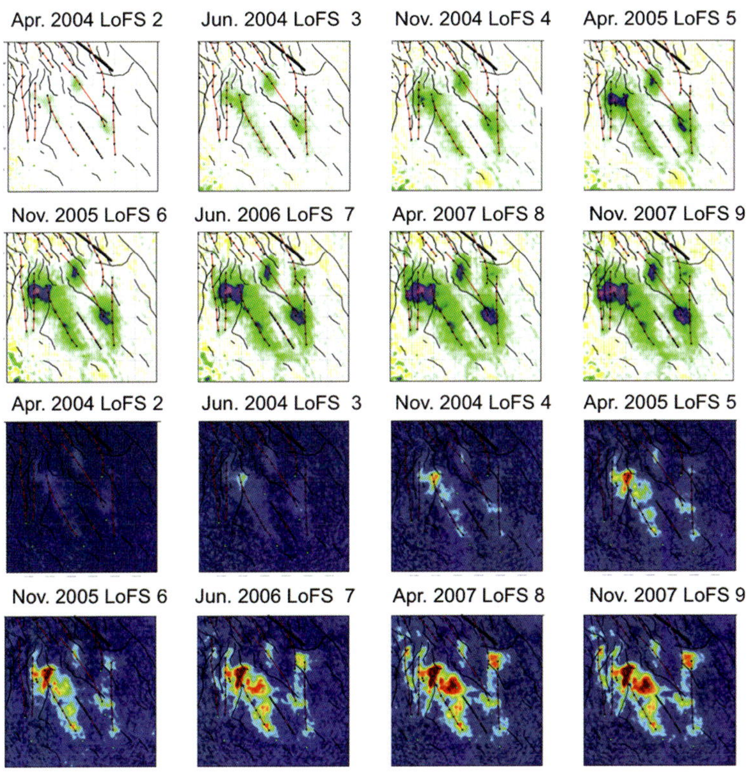

Figure 5.24: Example of maps loaded into the web-based documentation browser for sharing across disciplines, from van Gestel et al. (2011).

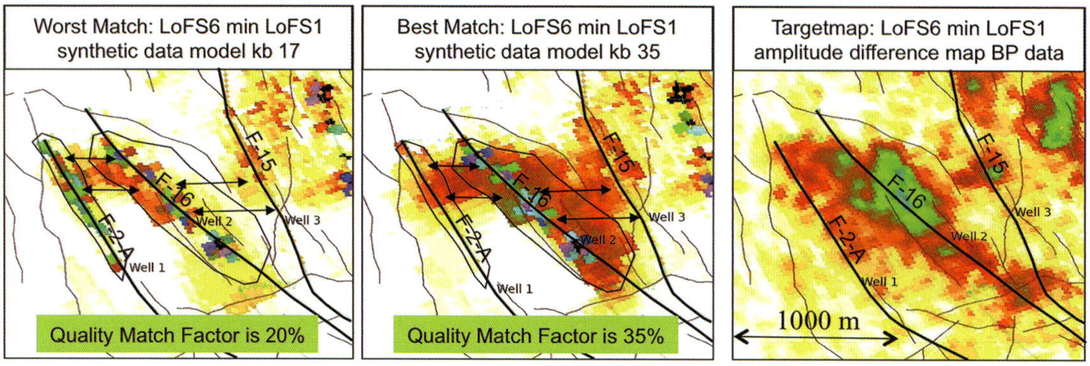

Figure 5.25: The findings from a 3×2 km wide sub-basin in the northwestern part of the field. Well F-16 was converted into an injector, supposed to support Well F-2. On the left, the model with the lowest match, in the center the model with the best match and on the right, the observed seismic time-lapse attribute, from van Gestel et al. (2011).

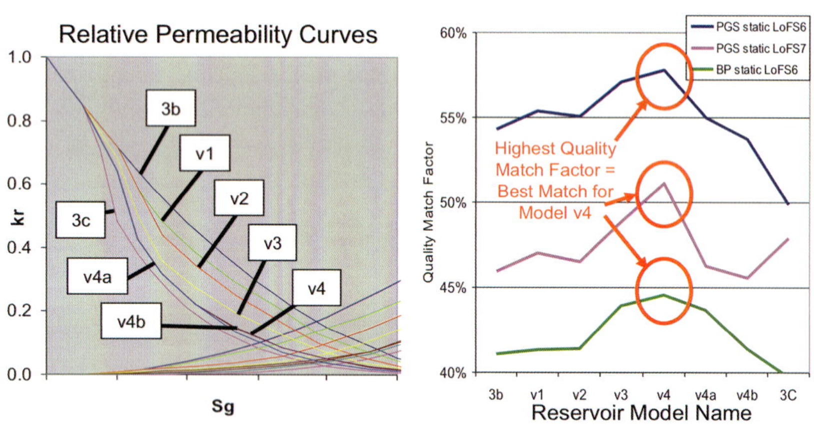

Figure 5.26: Gas-oil relative permeability curves for the 8 different reservoir models (left) and the match quality factors for the compaction response for the 8 different reservoir models (right). We show the results for two different time ranges. In both cases reservoir Model 4 has the best match, from van Gestel et al. (2011).

es in the overburden. This tends to be a very robust target attribute, even in the presence of some gas in the overburden. A quality match factor was calculated, which gave model 5 the highest score, figure 5.26. In figure 5.27 we see a comparison between observed and calculated compaction response for two different models. The results of this study were used in combination with other data sources to select the new G/O rel-perm curves for the full field reservoir model for the Valhall Field.

The third example of the use of his workflow is linked to several wells on the flank of the field. The data quality is generally very good in this area; despite this there were significant local mismatches, which could clearly influence the drilling of infill wells. The continuous depletion and the high-quality time-lapse data gave some leverage for alternative parameters and ranges to be tested. These included variations in skin factor along the wells, adjustment and subsequent testing of the transmissibility of faults, changes and lateral distribution of thickness and porosity of the reservoir. The inversion techniques, similar to the ones outlined by Corsa, figure 5. 14, were very useful complementary elements of

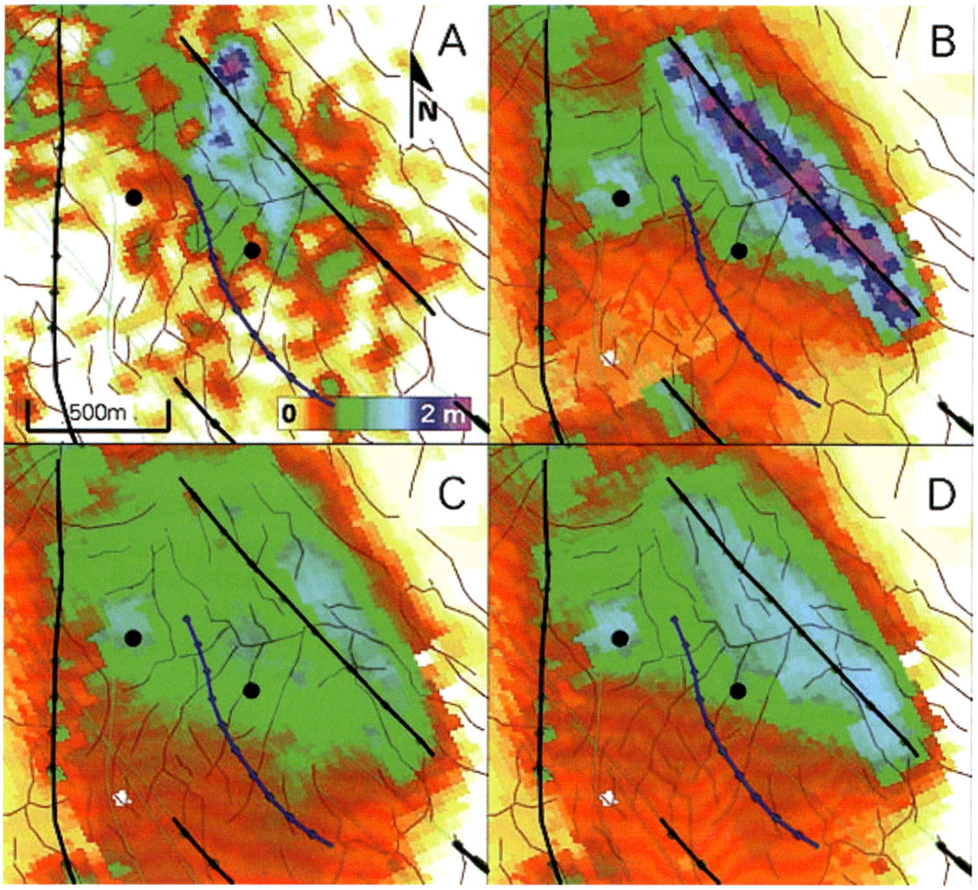

Figure 5.27: Comparison between: A) the observed compaction response from L1-6, (covering October 2003 to April 2006), B) the response of reservoir model 8, C) reservoir model 1 and D) reservoir Model 5 (best). Black lines are horizontal production wells with green diamonds at the location of the perforations. The blue line is the injection well, with blue diamonds at the location of the injection perforations, brown lines are faults, from van Gestel et al. (2011).

this update. Finally, the sensitivity of "improved" models was tested using the reservoir / attribute target map matching and figure 5.28 shows one example of the improvement achieved.

5.4.3 The seismic time-lapse responses of an injection test

Frequent seismic surveying, which is an inherent part of the LoFS concept, allows us to capture the effects at a pace where the various effects can more easily be identified. Continuous sampling of the production induced 4D seismic response increases confidence in the observations and interpretations.

In the early test stages of implementing the water-flood, a sub-optimal producer, figure 5.29 was converted into a water injection well. The well is localized in part of the field characterized by local compartmentalization. The injection started after the base line survey was acquired. The three first monitor surveys show a robust and gradually growing 4D response, although only a small volume of water (2–400.000 bbls) was injected. In the 4th monitor survey, the seismic response due to injection has apparently disappeared around the injecting wells. The seismic response re-appears in the subsequent surveys. During the acquisition of the 4th monitor, the rates and pressure were increased as part

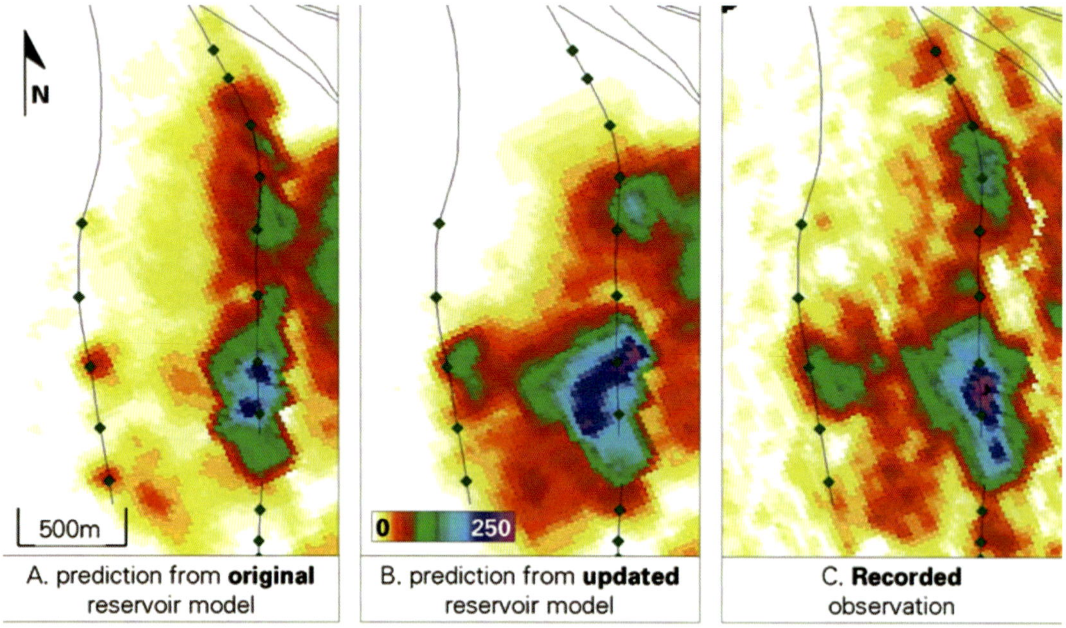

Figure 5.28: Comparison between the original (a) and the updated (b) predicted reservoir model responses that can be compared to the observed response (c). In all figures, the LoFS (survey) 7 minus LoFS 1 amplitude difference response is shown. Black lines are horizontal wells and green diamonds are locations of perforations, from van Gestel et al. (2011).

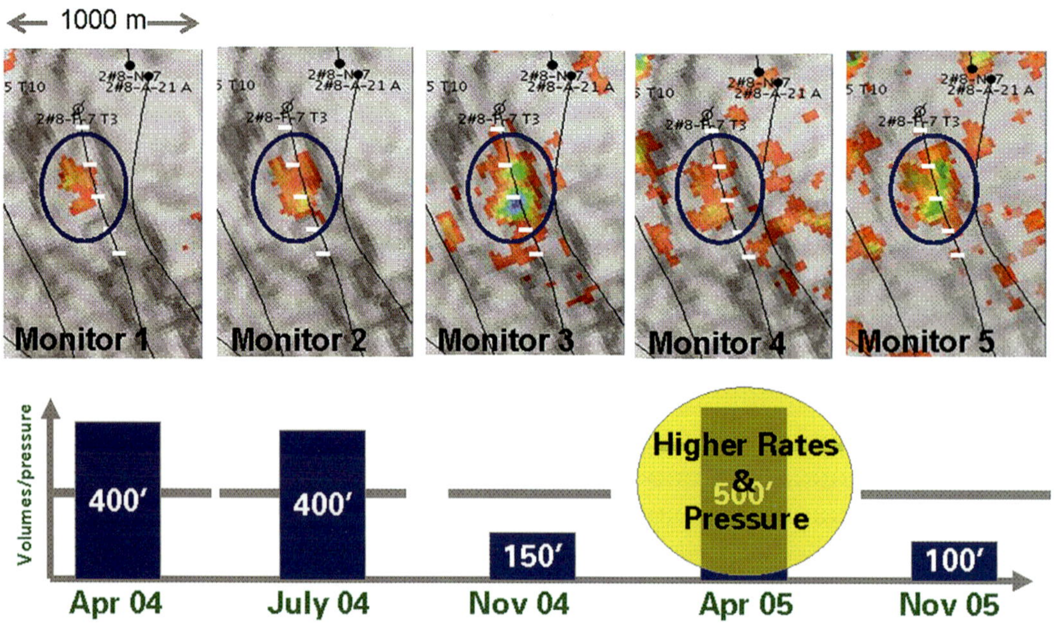

Figure 5.29: The sensitivity of the Valhall LoFS system demonstrated by significant 4D seismic responses due to a small volume of injected water and injection pressure, Barkved et al. (2005).

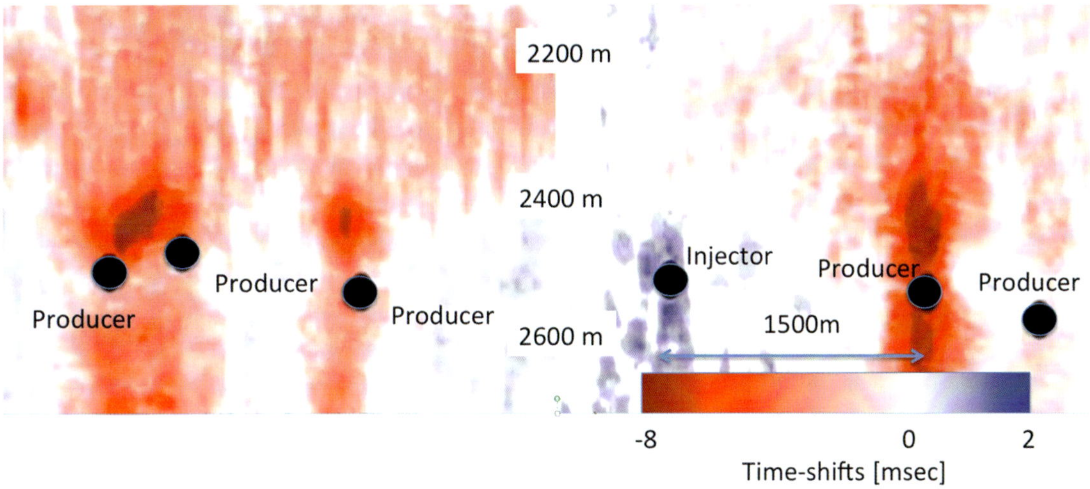

Figure 5.30: Pressure from the injector appears to reverse the time-shift signal in the overburden..

of a 2-week injectivity test. The original response re-appears in subsequent surveys when the pressure is reduced again, Barkved et al. (2008). There is another intriguing time-lapse observation around this well. We expect that higher pressure due to injection of water will arrest the compaction. Around this well we apparently had some pressure-up effects and a reversal of the time-shift effects was also observed, figure 5.30.

5.5 Conclusion

In this chapter we looked at examples of using "dynamic" 4D imaging for improving the static model. The 4D seismic response can help in improving the reservoir model by highlighting the faults that need to be included. Often, the reservoir layers "stand out" due to changes in impedance and could be used as an additional confirmation of the reservoir thickness away from well control. We have revisited generic workflows that have been modified to address reservoir specific issues, i.e., inverting for reservoir pressure. The TDRM workflow allows for the objective analysis of a wide range of reservoir models, which increases the predictive quality of the models. We have also shared a few examples of how the frequent time-lapse data from the Valhall LoFS array has helped in reducing non-uniqueness in the model. Seismic surveillance has become essential for the day-to-day reservoir management at Valhall.

6 Case Study: Seismic Surveillance for New Wells Delivery

"A tree's a tree. How many more do you need to look at?" - *Ronald Reagan*

The highest value of 4D seismic is often linked to the identification of a partly undrained compartment or reducing uncertainties that make a well economic to drill. In practice, the seismic data are often worked in much more detailed for supporting the delivery of a new well than in supporting the reservoir model. In a mature field reducing the risk of drilling problems and building the most robust trajectory for reducing risk could be a labor-intensive task. As we will see in the examples in this chapter, using seismic surveillance for well delivery support is often helped in that the "questions" are well defined. Do we have pay thickness as expected? Any risk of faulting? Risk for pressure variation along the well? Is the reservoir entry robust?

6.1 Introduction

Infill drilling is by far the most important activity for capturing the 4D seismic value. There are several well-documented cases where the seismic time-lapse signal has helped to define new, valuable infill targets. This is also the reason why we have recovery factors far beyond 40% in some of the North Sea fields. Time-lapse seismic helps in high grading new targets through better definition of the container, reduces the risk in delivering the agreed reserves and allows safer and more accurate delivery of the new wells. If the pressure uncertainty is too large for a planned horizontal well, we could risk making a target uneconomic already in the planning, as we will have less chance of delivering an acceptable length. Pressure readings taken while drilling are regularly used to avoid to high-pressure differentials. The 4D seismic images are important to advise where to take the readings and provide context for expected trends. This guidance is important to save time and to avoid exposing the bit to undesired pressure regimes.

In the following we will have a look at what the issues are and how they could be addressed, by looking more closely at two different situations at Valhall, infill drilling in the partly depleted down flank area and in the depleted crestal area of the field, figure 6.1.

Kristiansen (2004), Kristiansen (2007), Strachan et al. (2005) provide further details on drilling history, challenges and performance at Valhall.

6.1.1 Chasing "thin" pay in the flank

Drilling depleted reservoirs is challenging, especially when the formation is weak. Drilling such a reservoir normally requires a long horizontal section to make an economic well; this may imply high torque and drag. Reservoir heterogeneities like small-scale faulting may increase or reduce communication across "compartments" and make the pressure differential along the well too high for drilling. Drilling close to the oil-water contact may also increase the risk of needing to sidetrack or early TD of the well to avoid high water production.

The Valhall Flank area covers a large area, where the reservoir is thin, 5–15 m. During 2002 and 2003, two new wellhead platforms were installed in the field, one 5000 m to the south of the field center and the other about 5000 m to the north. The very high seismic quality in the flank area provided a reasonable basis for defining the geometry of the new long horizontal wells. In general, the wells

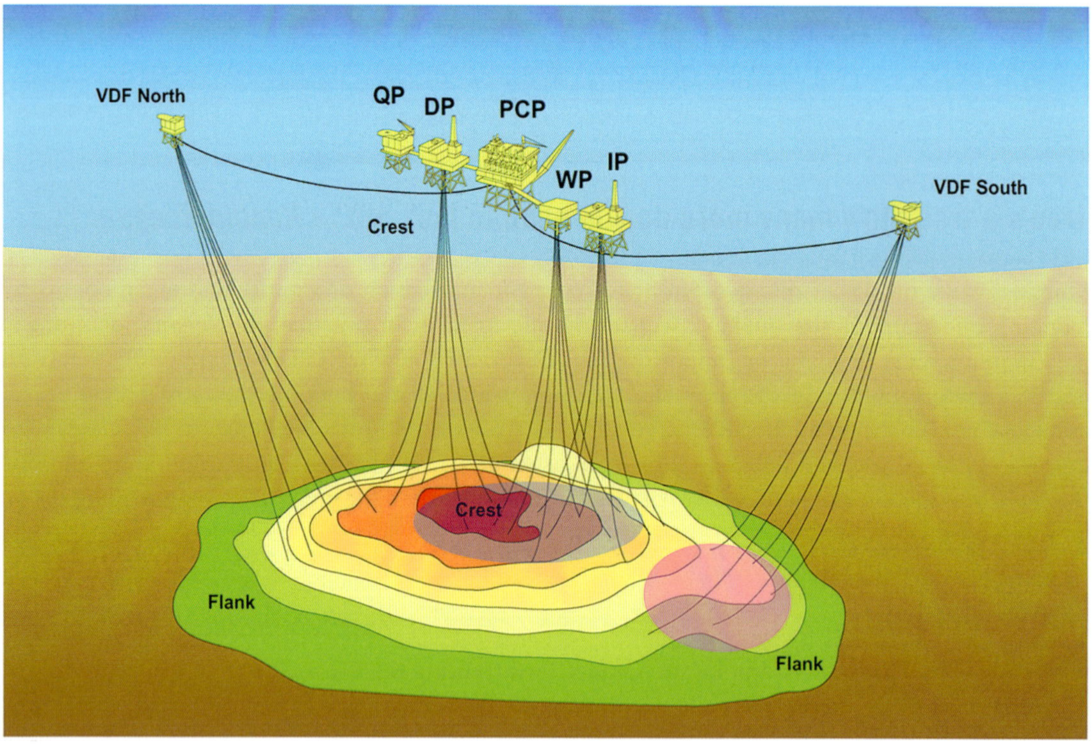

Figure 6.1: The drilling of new wells started up from the drilling platform (DP) in 1982. Since 1996, the wells have been drilled from the wellhead platform (WP), the Injection Platform (IP) or from the North Flank (VDF-N) and the South Flank (VDF-S), Barkved et al. (2005).

would enter the reservoir in areas with modest depletion, but then most wells would be drilled towards the depleted area. Thickness varies rapidly, primarily controlled by syn-depositional faulting. Based on experience from producing from thin pay, there is a significant difference in performance if we can place the wells in areas thicker than ten meters. At the same time, the pressure differential had to be within acceptable bounds along the trajectory.

Reservoir pressure profiles for the drilling program are provided by the reservoir simulator and obviously, using 4D seismic time-lapse for history matching is an important element for obtaining the best possible pressures, figure 6.2. However, details not represented in the simulator or alternative interpretations are important to consider. The flank-drilling program had an impressive performance, Strachan et al. (2005). To ensure maximum impact from drilling the first well on the next well in the same area, we split the drilling campaign into phases. The well was delivered from a jack-up platform that first drilled four wells from the southern platform and then moved to the north to drill four wells before returning to the south.

The first 4D seismic data became available just as the drilling was about to start, and a reasonably good match between the 4D response and the seismic simulator was established. Only minor adjustments to the well trajectories defined for the first phase of drilling were necessary. Figure 6.3 shows a comparison between reservoir pressure and the changes in acoustic impedance across the reservoir. The small seismic changes toward the northern end of well 3 were interpreted to be real. The plan was changed, and the trajectory for well 3 was extended into the area, this resulted in a few hundred meters of extra well length and added value to the well.

Pressure readings taken while drilling added new points to the database. In addition, we obtained a pressure point when the new wells were stimulated. The high-quality pressure points was then cross-plotted with seismic time-lapse attributes, as shown if figure 6.4. In this case we have used the changes in acoustic impedance, which show very good correlation to pressure as long as we are above the bub-

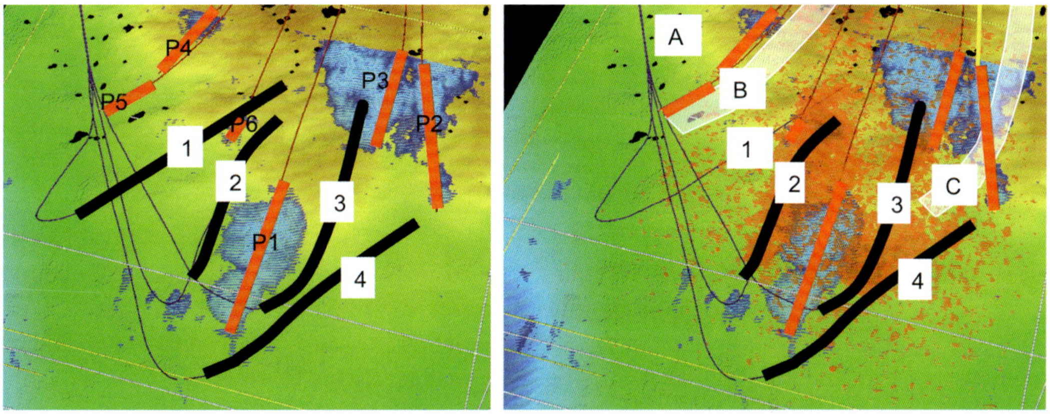

Figure 6.2 Left: 4D attribute from the 92-02 streamer data (blue), draped across the structure. Right: 4D attribute map from the LoFS array, reflecting 9 months of production (orange, posted on top of the map shown to the left. Red wells are "old" producers drilled from the field center. Wells 1, 2, 3 and 4 were drilled in the first phase. A, B and C were drilled during the second phase, Barkved et al. (2005).

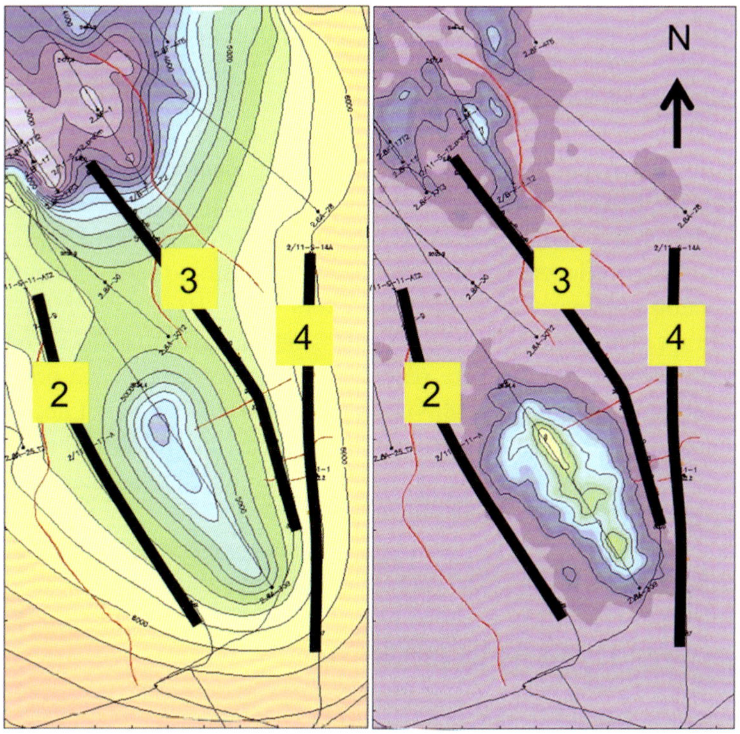

Figure 6.3: Comparison between simulated reservoir pressure, left, and changes in acoustic impedance, right, across the reservoir between 1992 and 2002, Barkved et al. (2005).

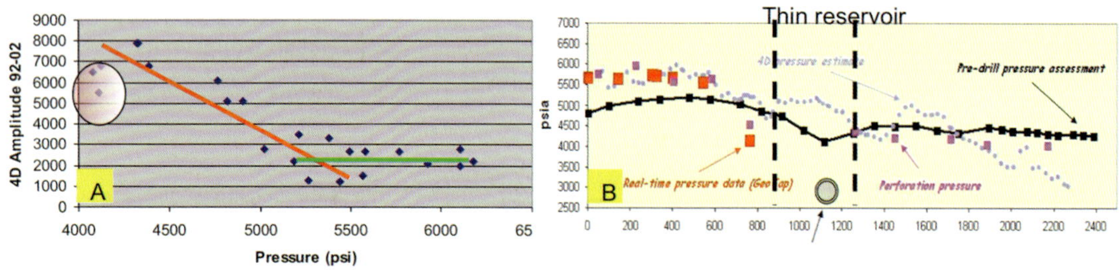

Figure 6.4: A) cross-plot between pressure readings and changes in acoustic impedance (92–02). B) Prediction from simulator (P50) shown as a black line, predictions from seismic time-lapse data, blue dotted line and recordings in the well, red points, Barkved et al. (2005).

ble point, ~4000 psi. As a result of this observation, the pressure estimates from the seismic time-lapse data were included in the pressure predictions from the reservoir simulator and used in the assurance process pre-drilling. Figure 6.4 shows that in this case the actual pressure points (red) closely followed the seismic derived pressure point, apart from where the new well was to cross an older well, at this point the reservoir was known to be thin. The exact minimum thickness has not been established, but it appears that when the reservoir is thicker than 10 meters we have a good correlation in a high-quality area. The explanation for the technique working at this thickness range is that the acoustic impedance of the reservoir and the cap rock are quite similar, while we have a significant increase in acoustic impedance below the reservoir.

6.1.2 Moving into the difficult crestal area

The crestal area of the Valhall field used to be defined as a seismically challenged area, of poor seismic quality. This situation greatly improved when we got the permanent seismic array in place. Furthermore, it turned out that the dynamic images were much more conclusive than static images and especially the traveltime changes appeared to be a robust parameter. This is because we often reduce the impact of systematic noise, like certain (intrabed) multiples, during the subtraction process.

The crestal area has been subject to significantly more depletion than the flank area. A new water injector was planned to help re-pressurize one of the depleted segments, figure 6.5. To the east of the planned well we had one of the field's best horizontal producers C, and to the NW, there was a mediocre old vertical producer A. Well B used to be a producer. The depletion signal appears to reflect hardening of the reservoir, implying that compaction dominated over the increase in GOR. By closely following the time-lapse signal we noticed an abrupt change in the depletion rate. This appeared to reflect a baffle zone between well A and the planned new injector. Pressure readings in A indicated that the pressure differential could be an issue when drilling the full length. Moving the well eastward was not necessarily a good idea, as we were rapidly depleting this area. The best solution appeared to be to drill the well just to the east of the baffle zone, while keeping to the west of defined N-S trending faults. The well was successfully drilled and has since been one of the field's best injectors.

Flatebø et al. (2010) share in some details of how the well delivery team makes use of advanced technology in the planning and delivery of new wells. The team attributed the ability to deliver their well A, figure 6.6, 32.5 days ahead of plan, to the integrated planning process and drilling performance. Seismic surveillance data provided an essential basis for their achievement.

The first challenge they faced was the presence of a shallow pressure anomaly linked to the presence of shallow gas. This shallow anomaly is reflected in the shallow data by an abnormally low seismic velocity, which appears in short wavelength velocity models based on using Full Waveform Inversion

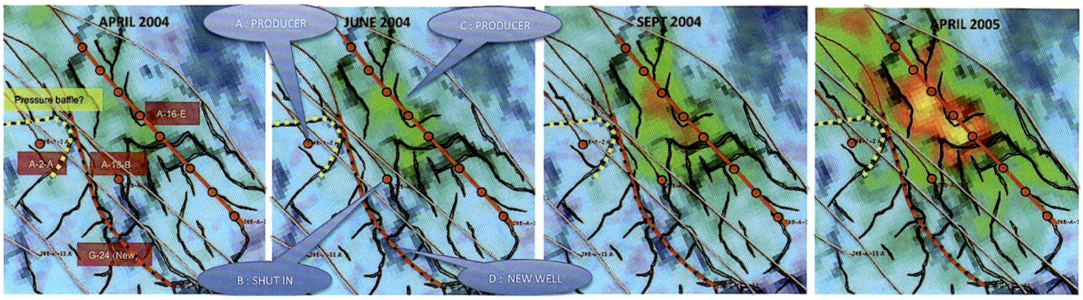

Figure 6.5: Accumulated changes in traveltime across the reservoir (RTS) from subsequent LoFS surveys. A, B and C are existing wells, D is a planned water injector. Red dots show the position of existing perforations. The black/yellow line shows a possible pressure baffle. The area of the map is 3×3 km, yellow colors indicate 6 msec. north is up, Barkved et al. (2005).

techniques (FWI), Barkved (2010). Using these high-resolution FWI seismic velocity volumes from recent LoFS surveys, the team designed a well trajectory that allowed them to avoid the anomaly until a depth where the "worst" case could be easily managed with the weight of the mud.

The transport section in the overburden had to cross an area of significant compaction induced stress state changes. Stress changes in the shale overburden above the differentially compacting chalk reservoir are most pronounced close to the reservoir. A history-matched full field finite element based model, Kristiansen and Plischke (2010), was used to calculate stress, strains and displacements. The outputs are used to test the well bore stability of alternative trajectories. The model is typically validated against 4D seismic, i.e., by verifying the localization of stress arching and stress pillars and the possible impact of large-scale faulting.

In the case of planning well A, the team noticed that the preferred trajectory passed an area of increased vertical stress (a stress pillar), which resulted in a very limited mud weight window. The subsidence induced stress variation is visualized over the field by calculating the deviatoric stress invariant, Figure 6.6.

To minimize the risk of overburden well bore stability issues during drilling and early well life collapse, the team selected to move the reservoir entry point about 500 m closer to the platform. This additional 500 m transport section in the reservoir was designed such that it could later be completed to support the water-flood program. Not only did they avoid drilling into a zone with potential troublesome stress regimes, they also significantly reduced the inclination of the well in the area close to the presence of strong stress, and secured an upside water injection opportunity. The well was drilled down to top reservoir in 30 days, 22 days ahead of schedule.

To reduce the uncertainties in the reservoir thickness the team used the 4D seismic signal, still the actual thickness at the reservoir entry was 15 m less than that of the offset wells, which quite consistently proved to be a 40 m thick Tor formation. Seismic data quality in this area was considered mediocre, being influenced by gas in the overburden, which also reduced the seismic resolution. The 4D seismic response combined with the production history to fill the gaps prior to the first seismic survey resulted in an estimated compaction of 5 m, the remaining less than 5-10 m of apparent thickness was attributed to the presence of a known fault close to entry. The team made a very important observation while drilling. The average porosity in the well was about 8 porosity units less that in the previous well B drilled into the area, figure 6.7, which again was more than five porosity units less than the average porosity seen prior to any drainage in the area. A reduction from 45% to 32% in a 40 m thick reservoir works out to be 5.2 m, which is on the high side of pre-drilling prediction. Later investigations confirmed that there were a depositional bias to the lowest porosity observed.

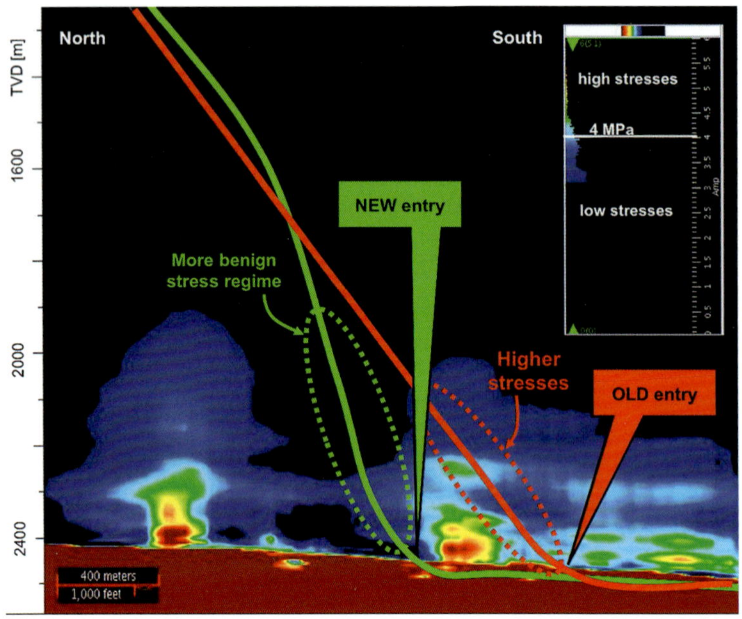

Figure 6.6: Initial revision of the well trajectory goes through an area of increased overburden stresses with a high wellbore deviation whereas the final revision is moved about 500m north, into a more benign stress regime, and with a reduced deviation. From Flatebø et al. (2010).

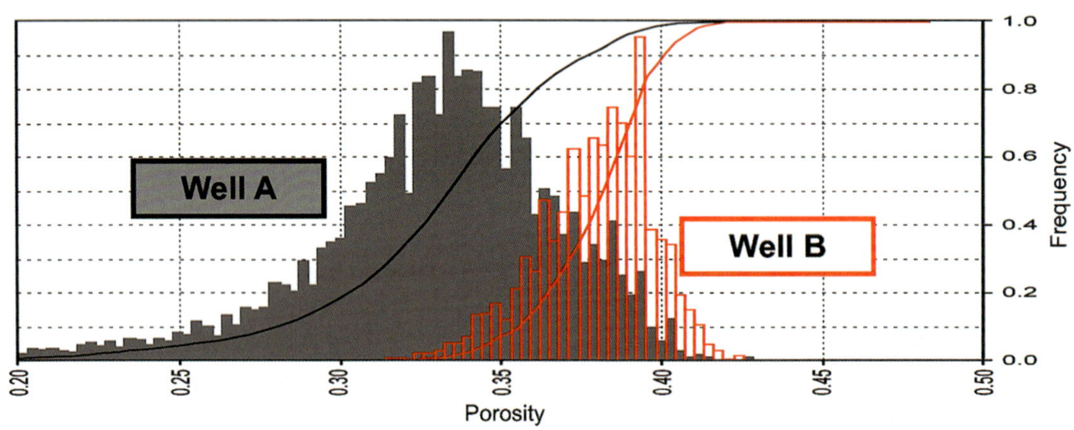

Figure 6.7: Histogram showing the frequency distribution of calculated porosity for Well A and Well B, from Flatebø et al. (2010).

The planned well was to be drilled between two horizontal wells, Well B (injector) and Well C (producer), as a partial extension to the injector line drive strategy, Figure 6.8. The LoFS surveillance data is a response to a complex depletion and re-pressurization process. Well B to the east is an injector. The seismic time-lapse data reflects a clear hardening response. The shape and areal extent of the anomaly correlates well with the history matched modeled water saturation and pressure change, van Gestel et al. (2008). An anomalous western extension two-thirds of the way along the well was flagged as possible higher-pressure compartments, due to the conformance with structure/fault. A possible risk that this could be associated with injected water was also flagged. The increased pressures were confirmed by pressure readings while drilling data and interpreted to be associated with a possible fracture system, resulting in a localized increase in permeability. No sign of injected water was observed.

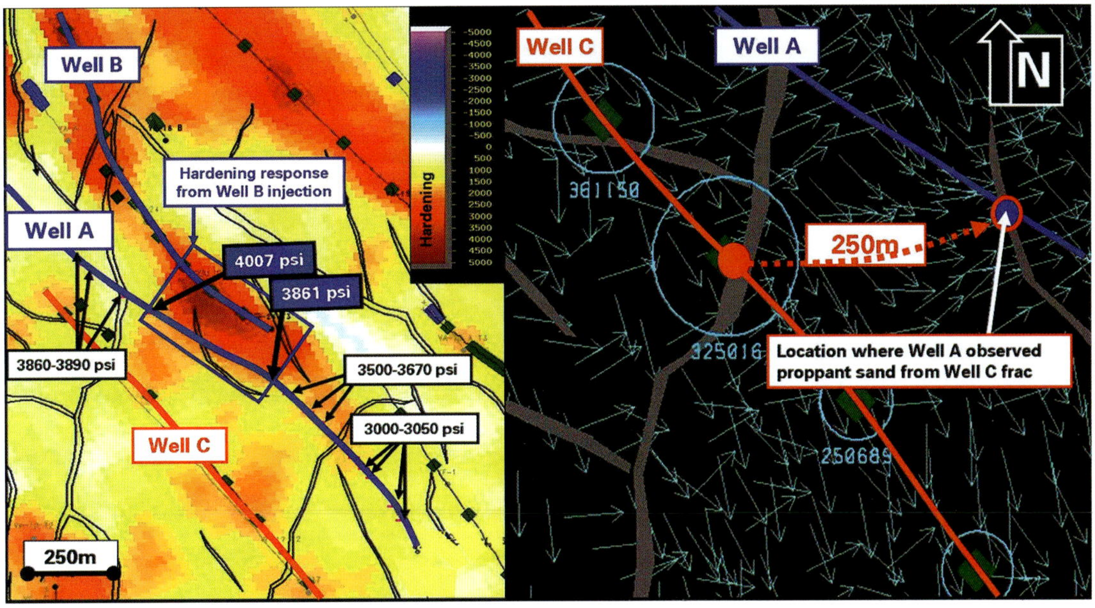

Figure 6.8 A: LoFS1-11 impedance map, Testrak™ pressure points for Well A are annotated. B: illustrated are the principal horizontal stress directions in the reservoir at the time of drilling Well C, the estimated fracking radius from Well B stimulation and the location where Well A produced proppant, from Flatebø et al. (2010).

Setting realistic expectations is one of the main reasons for the well delivery team's success in this case. Advanced modeling and analysis was translated into a common picture for the well that all the involved stakeholders bought into. Seismic surveillance has an important role in providing context, but also helps greatly in constraining the pre-drilling prediction.

Planning and drilling new wells probably provides some of the best learning opportunities in the industry. During the drilling of well A, a few grains of fracture proppant appeared with the drilling cuttings. They were identified to originate from well C, 250 m to the west of well A. A pressure sink was also associated with the location of the proppant. The teams grasped this great opportunity for testing out the capabilities of the geo-mechanical modeling. Figure 6.8 B illustrates the principal horizontal stress directions in the reservoir at the time of stimulating Well C. The second northernmost fracture stimulation was considered poor, believed to create a long and narrow proppant fracture. The notable 'vector highway', figure 6.8 B towards the location in Well A where fracture proppant was observed, supports this.

6.2 Conclusion

Seismic surveillance data from towed marine streamer surveys and from the LoFS array continue to provide key data for the planning of new wells. Time-shifts in the overburden are used to support geomechanical modeling, which provides essential guidance in avoiding unwanted effects due to stress redistribution. Improved structural information and detailed insight into the presence of the reservoir and drainage has been important in planning the well in the reservoir. The value of seismic time-lapse information will increase as more water is injected into the reservoir and the risk of drilling into a swept reservoir increases.

7 Case Study: Seismic Surveillance for Base Management

"There's a fine line between fishing and just standing on the shore like an idiot."
- *Steven Wright*

Combining seismic time-lapse response, production data and forward modeling may provide useful insights about what is going on in the wells and the reservoir. The permanent installation at Valhall has allowed for frequent surveying, and has led to closer links between the geophysicists and the team responsible for optimizing production in the short term. In-well observations and 4D seismic are an ideal combination for understanding the detailed hydraulics of the wells and the reservoir. If it is impossible to enter a well we might be able to find out which part of the well is contributing to injection or production. In this chapter we start with an introduction and then discuss two approaches to linking production and injection in the well to seismic time-lapse observations. We will then discuss the use of seismic surveillance for well optimization and well intervention work, before we conclude with the realization from seismic surveillance data that sometimes we are better off doing nothing, contrary to what the initial observations on the well indicate.

7.1 Introduction

Managing the base production includes optimizing the reservoir performance through controlling the individual well's production & injection performance. Prediction, well surveillance and interventions, and system optimizations are all part of this. The key contribution from seismic surveillance is to provide the spatial context for well observations and predictions and to assist in the planning of intervention programs. The seismic surveillance data are used to assist in making operational decisions, by simply providing insight into how the water and hydrocarbons are moving in the reservoir.

7.1.1 Well surveillance

In defining the field specific surveillance plan we need to have a good understanding of the risks, uncertainties and the opportunities of the field and the individual wells, Grose (SPE 2007). A good starting point is to articulate these issues into tangible elements like sweep efficiency, production optimization, integrity of the wells and the system, compartmentalization, etc. This helps greatly in defining what information we need to capture, how often we need the data and how different surveillance schemes could be combined to deliver the best possible unique interpretations. The key outcome is to support safe operation, improve production and assist in recovering as much as possible of the hydrocarbons in place. It is as important to understand the trends and spatial context of the data, as it is to interpret the individual snapshots through time

The essential data in any well surveillance scheme are the basic production or injection volumes. Other fundamental data to gather in the well include, pressure, fluid composition, water cut, rates and pressure, water chemistry, tracers, gravity, and zonal contribution along the wells. Time-lapse petrophysical logs and production logs are great tools for understanding the reservoir performance. More details on how these tools can contribute and the importance of integration and planning the surveillance program may be found in Grose (2007).

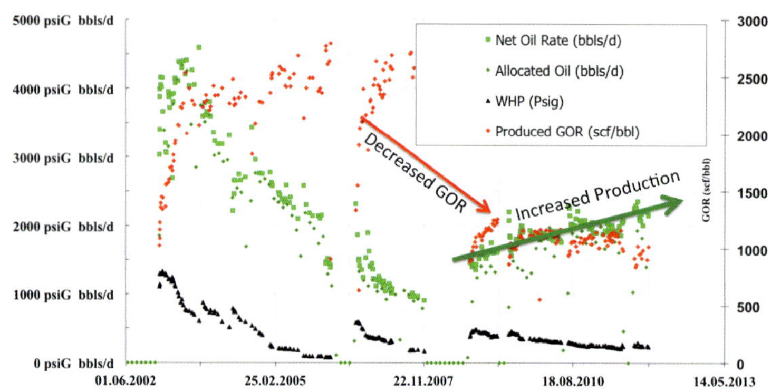

Figure 7.1: Producing well PB 2's response to waterflood.

Griffin et al. (2007) discuss the importance of surveillance for optimizing the water-flood performance for the Valhall field. They identified the main surveillance technologies to be:

- Continuous data gathering at the wellhead, production volumes, samples, well allocation and integrity data, i.e., using ion chromatograph at the offshore lab
- Time-lapse data gathering includes seismic surveys, cased hole logging, production logs (PLTs), saturations logs and compaction logging) in a monitor well, injection of tracers in injectors and pressure surveys.
- Completion based data gathering, using downhole gauges for temperature, in well geophones installed in injectors above the reservoir.
- Data gathering from drilling has been optimized by adopting a key well strategy to gather more data in wells.

A key element is not only to gather the data, but also to rigorously and punctually enter it in a production database to enable easy access to the well operators and to decision makers. A dedicated workflow is defined to help to extract changes in trends which, i.e., reveal the impact of an injector on a producer, figure 7.1.

According to Schlumberger Oilfield Glossary, a production log is "a record of one or more *in-situ* measurements that describe the nature and behavior of fluids in or around the borehole during production or injection. Production logs are run for the purpose of analyzing dynamic well performance and the productivity or injectivity of different zones, diagnosing problem wells or monitoring the results of a stimulation or completion." "The earliest production logs consisted in temperature logs (1930s) and flowmeters (1940s), to which were soon added fluid-density and capacitance logs (1950s). Flow-rate measurements were gradually improved by the development of tracer logs and improvement to the basic spinner flowmeter."

"This technology developed for near-vertical wells and single or biphasic flow and could be misleading in highly deviated and especially horizontal wells. New techniques were developed starting in the 1980s."

At Valhall, acquiring conventional production logs is a significant challenge. This is due to limitations in bed space, risk to the well, limited access due to restrictions (deformations or scale) combined with the probability of being able to record conclusive information due to low flow rates or potential chalk influx. The conventional production logs that have been acquired lately show results that are close to what can be deduced from analysis of the depletion around the wells from the time-lapse seismic images generated by the LoFS system. These observations resulted in the development of a new way of presenting the 4D seismic response, which will be discussed in the next section. Presenting and linking the seismic time-lapse observation in this manner has greatly helped in facilitating discussions linked to well performance across the disciplines involved. The displays we use as part of this a have a strong similarity to the bar diagrams sometimes used in production logging and are sometimes referred to informally as "seismic PLT display". The seismic time-lapse observations are normally made at the surface and are interpreted based on the response due to injection or depletion around or between wells. **That is, the methods should not be viewed as a replacement for a production log but as complementary information.**

7.2 Linking production and log data to 4D seismic response

Most of the recent Valhall wells are drilled horizontally and are completed using propped fractures at a spacing of about 150–200 m. This distance is larger than the spatial resolution of the time-lapse images that can therefore be used to resolve the individual contributions of the different perforations. A significant part of the Valhall field has only one main reservoir layer with a very strong time-lapse response. These unique parameters combined with the presence of the LoFS system allow us to study the detailed contributions from each propped fracture along the well bore and led us to the implementation of a dedicated workflow for this.

We realized that production induced 4D seismic changes around individual perforations in a horizontal well can be transformed into the fractional contributions and presented in a similar way as the inflow measured with Production Logging Tools (PLT), figure 7.3. We refer to this way of presenting seismic time-lapse data as Seismic Production Tracker Plot (Seismic PTP). Linking the 4D seismic response directly to the production greatly adds to the integration of the seismic 4D signal across the disciplines. Using a relative rather than absolute measure reduces the requirement for absolute measurements initially and is better aligned with the strength of the seismic measurements. Presenting seismic 4D changes in terms of the contributions of the individual perforations was initially developed to support well interventions and activities aiming at improving productivity. However, it has turned out to be a very useful attribute for co-visualization of dynamic information; for providing context for planning new wells and for reservoir management decisions, in addition to the routine use for identifying and implementing well interventions options.

The seismic PTPs cannot replace conventional PLTs, as the methods are not mutually exclusive. Conventional PLTs measure the flow in or out of the well; while the seismic PTPs are based on measuring changes around the well bore. Seismic PTPs are independent of access to the platform and the wells and allow for much more frequent estimates, which then allows for analysis of trends and changes over time. Combining these trends with the actual production data may be used for additional confidence in the observations and gives an improved basis for understanding the well's performance. In the past the term Seismic PLT has been used for what we now refer to as seismic Production Tracking Plot, seismic PTP. This has led to a number of confusions, and we suggest dropping this term.

7.2.1 Workflow

We started by calibrating the collected PLTs to the time-lapse responses. We compared various attributes, areas of influence and time ranges to the data of the conventional PLTs. Figure 7.2 shows a standard presentation of the data from a conventional PLT and Figure 7.3 shows the comparison between conventional PLT results and estimated response for three different wells in the field, two of them are producers, one of them is an injector. Later, two additional wells were logged, one showing a similar response. In the second, the PLT data were unfortunately not conclusive.

The methodology applied up to now relies on a linear relationship between the seismic response and the produced volumes. The seismic 4D response that had the best match to the conventional PLTs was the Sum of Signed Amplitudes (SSA), calculated in a window across the reservoir from the Acoustic Impedance (AI) volume. Figure 7.4 shows a seismic PTP calculated on the basis of difference between the baseline and subsequent monitoring surveys. We will be looking for trends that can reveal problems with one or a group of perforations and will further investigate in the well, i.e., using a coil tubing tool.

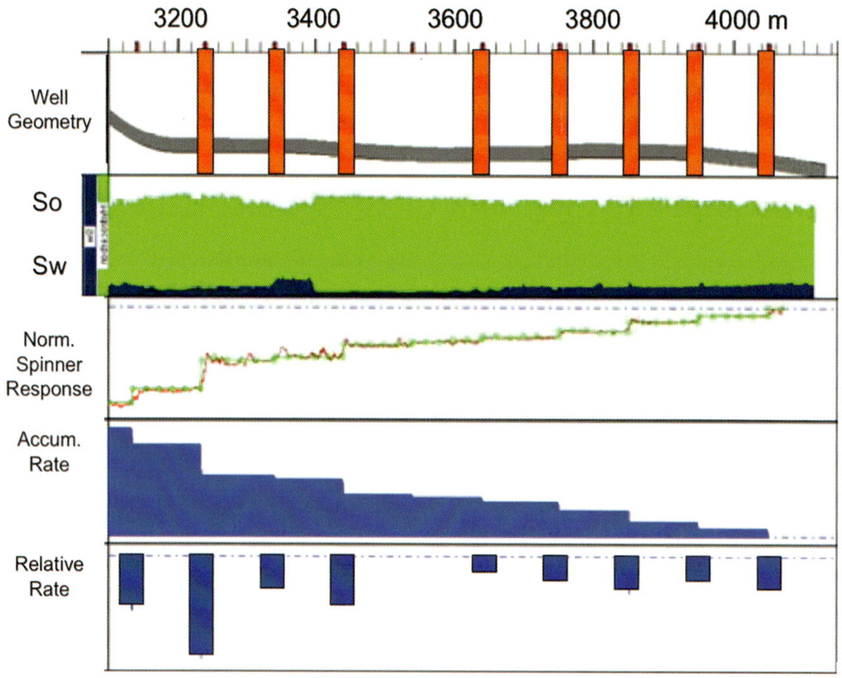

Figure 7.2: Standard presentation of a conventional PLT. The red bar in the upper window indicates perforations. The blue bars show the flow profile along the well, and the relative rate for each perforation below. The example is from a water injector. From Barkved et al. (2009).

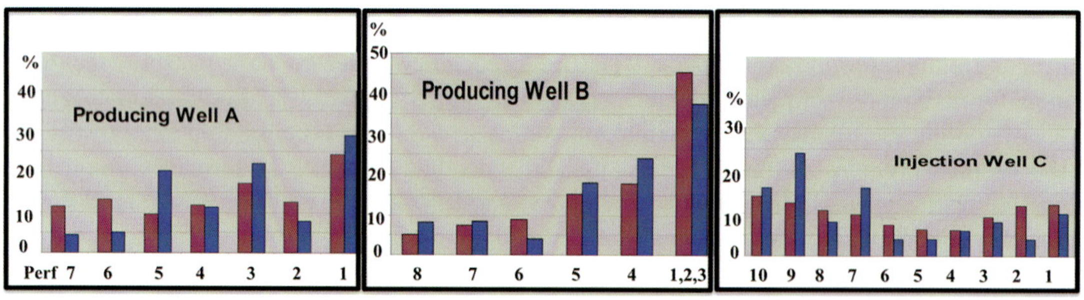

Figure 7.3: The comparison between conventional PLT and calculated PTPs is shown for three different wells in the field. Blue bars show conventional PLTs; brown bars show seismic PTPs. The horizontal axis is the measured depth; a bar is given for each perforation location. From Barkved et al. (2009).

7.2.2 Integration & validation

Seismic PTPs have been important in the integration of the LoFS data with the team responsible for optimizing production and injection levels from the existing well stock. It is important to understand what part of the well is contributing and whether any regions remain unswept. This understanding allows the team to target well interventions to improve production. Since casing deformation or chalk influx is a common phenomenon at Valhall, Seismic PTPs also provide additional insights to reduce the risk of the intervention work.

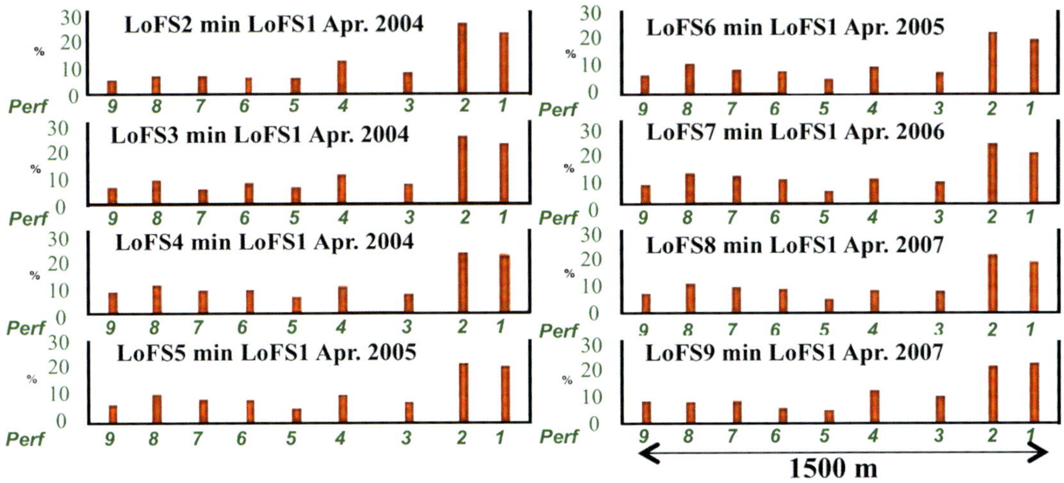

Figure 7.4: The seismic PTPs generated for a well in the South Flank of the field. The different seismic PTPs show a consistent response over time, which increases confidence in the signal. From Barkved et al. (2009).

Seismic PTPs provide an excellent tool to simplify the conversations between the geophysicists who are familiar with the various maps and extractions of the time-lapse data and the engineers who are more familiar with the well performance. Figure 7.5 shows an example from a recent integration session. Seismic PTPs provide a quick estimation of the contribution of the various perforations. The geophysicist can provide additional information about confidence and uncertainty of the measurement by further analysis.

It has turned out to be useful to also present the output from the reservoir simulator in a similar manner. The modeled production log are extracted directly from the flow simulator, while the synthetic PTPs are derived by applying the exactly same routine as for the seismic PTPs computations to the synthetic seismic data generated from the flow simulator parameters at the actual time-step. This alternative way of presenting the simulator output allows for effective 3D display scenes used by the multi-disciplinary model to validate the relevance of the static and dynamic reservoir model. Mismatches between model and observations are easily pointed out and help to facilitate the constructive discussions aiming at understanding the reservoir and optimizing the performance of existing and new wells.

Seismic PTPs are now routinely constructed for all relevant wells and from all monitoring surveys acquired at Valhall. The results are used to condition the flow simulator and to visualize the production and injection inflow for each perforation along a horizontal well as observed by the 4D seismic signal. While the concept is based on linking production observation directly to seismic 4D changes, it has also been useful in validating model output. This has turned out to be a useful way of quantitative matching of the flow model and the 4D seismic; however the greatest benefit is often in using the results as a basis for multi-disciplinary discussions and analysis.

7.3 Where does the production come from

Production induced 4D seismic responses may be linked to the activities in the wells, as we discussed in the previous section. In deriving the seismic PTPs, we assumed that the individual perforation was responsible for draining a symmetric area around the well. This might not always be the case when we have multiple wells in a complex field and we may have to verify which wells drain or inject into what

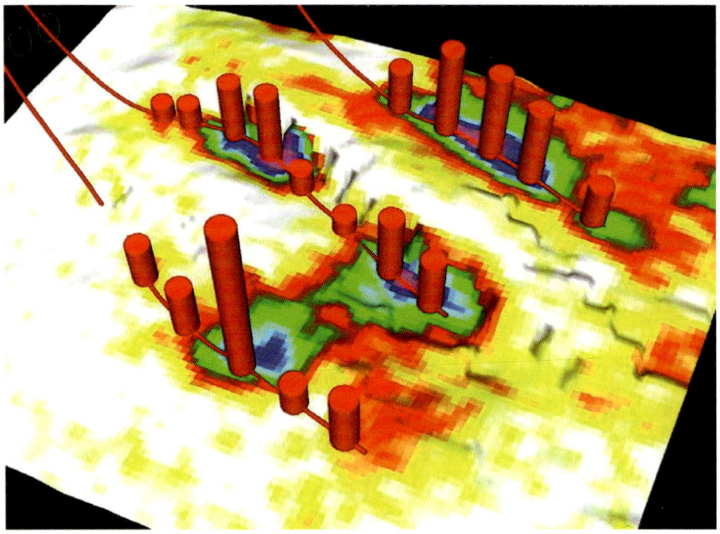

Figure 7.5: Coviz display showing the part of the North Flank of the Valhall field with three horizontal wells and their associated "seismic PTPs" above the LoFS time-lapse amplitude difference indicated the fractional inflow for the individual perforations. From Barkved et al. (2009).

area. If we have several seismic time-lapse signals, we would expect a correlation between changes in the 4D response and, for example, the accumulated production or injection over same time period in a well. Very often the wells are seeing a similar production scheme but over time there might be variations in rates, a well could be shut in for a period of time or start production later than the others, etc. Issues relating to maintenance may result in that the injector wells and the producers may be active or shut in for different periods, if we then have frequent seismic time-lapse survey, we might see a different patterns emerging over time from tabulating the total production/injection for the same period, Huang and MacBeth (2009), Huang et al. (2010).

Assuming that at any specific location in the reservoir the 4D response is a direct function of the production and injection history from wells connected to that location. Modifying the proposed pressure-saturation equation of MacBeth et al. (2006) to only well volumes, the pressure-dominated 4D signature evaluated over a fixed time interval can be related to the cumulative sum of the produced or injected volumes from these wells (Huang and MacBeth 2009; Huang et al. 2010). If we do have several time-lapse surveys, we might form a series of the changes for a specific point, comprising the changes in seismic attributes for all possible combinations of time-lapse signals, ΔA. Similarly, we might form times series of all combination of production/injection volumes ΔVj for the wells from the same time interval, ΔT_k. Huang and MacBeth (2009) formulate this as:

$$\Delta A_k(x,y) = G\Delta V_j(\Delta T_k) \tag{7.1}$$

or in the more general form as:

$$\Delta A_k(x,y) = G\left[\sum_{i=1}^{M} B_0 \Delta V_{ik}^{0} + \sum_{i=1}^{M} B_w \Delta V_{ik}^{w} + \sum_{i=1}^{M} B_w \Delta V_{ik}^{w}\right]$$

here Bo and Bw are the formation volume factors for oil and water, respectively, ΔVo the oil volume difference and ΔVw the produced or injected water volume difference. For n surveys there are n(n-1)/2 possible combinations of differences. The methodology was tested on a sequence of 10 repeated surveys from the Valhall LoFS array. This leads to a sequence of 45 different measures that can be made from

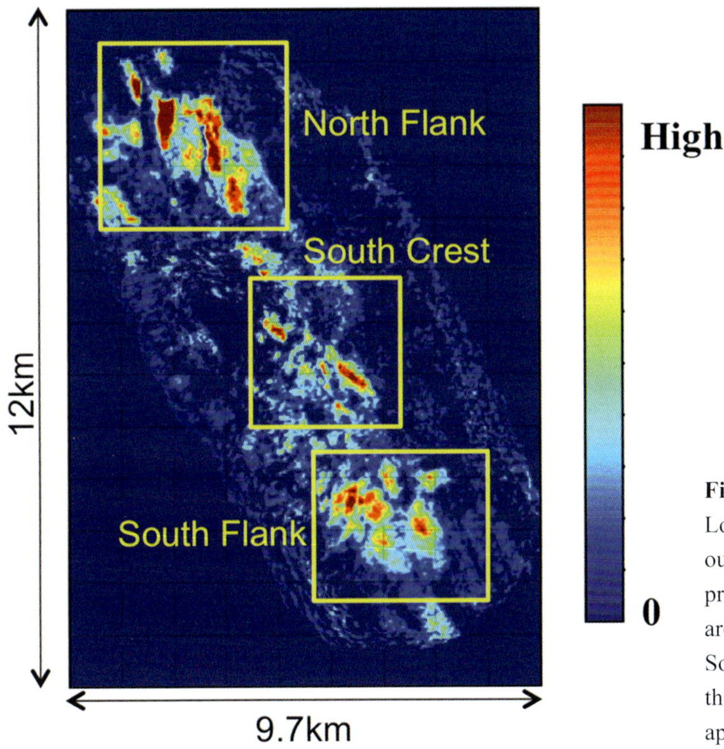

Figure 7.6: Map of AI change between LoFS10 and the baseline, showing the outline of the Valhall Field. Three major production regions in the Valhall Field are identified: the North Flank, the South Flank and the South Crest, where the technique proposed in this paper is applied. From Huang et al. (2011).

one set of seismic time-lapse attributes or from one set of measures in the well and we can analyze how these measures from different sets correlate.

The approach was tested in three different areas of the field and two different sets of seismic time-lapse data were used. These were the changes in acoustic impedance across (AI_Diff) the reservoir and the changes in two-way traveltimes (TT_Diff) at top reservoir level. Here we will only discuss the north flank case, see Huang et al. (2011) for more details.

		L1	L2	L3	L4	L5	L6	L7	L8	L9	L10
Nov 2009	L1										
Apr 2004	L2	L2L1									
Jun 2004	L3	L3L1	L3L2								
Nov 2004	L4	L4L1	L4L2	L4L3							
Apr 2005	L5	L5L1	L5L2	L5L3	L5L4						
Nov 2005	L6	L6L1	L6L2	L6L3	L6L4	L6L5					
Jun 2006	L7	L&L1	L7L2	L7L3	L7L4	L7L5	L7L6				
Apr 2007	L8	L8L1	L8L3	L8L3	L8L4	L8L5	L8L6	L8L7			
Dec 2007	L9	L9L1	L9L3	L9L3	L9L4	L9L6	L9L6	L9L7	L9L8		
Apr 2008	L10	L10L1	L10L3	L10L3	L10L4	L10L6	L10L6	L10L7	L10L8	L10L9	

Table 1: The order in which time lapse survey attribute sets difference pairs are created in order to generate the sequence defined in the main text. Each sequence is correlated with a similarly defined sequence for the well activity.

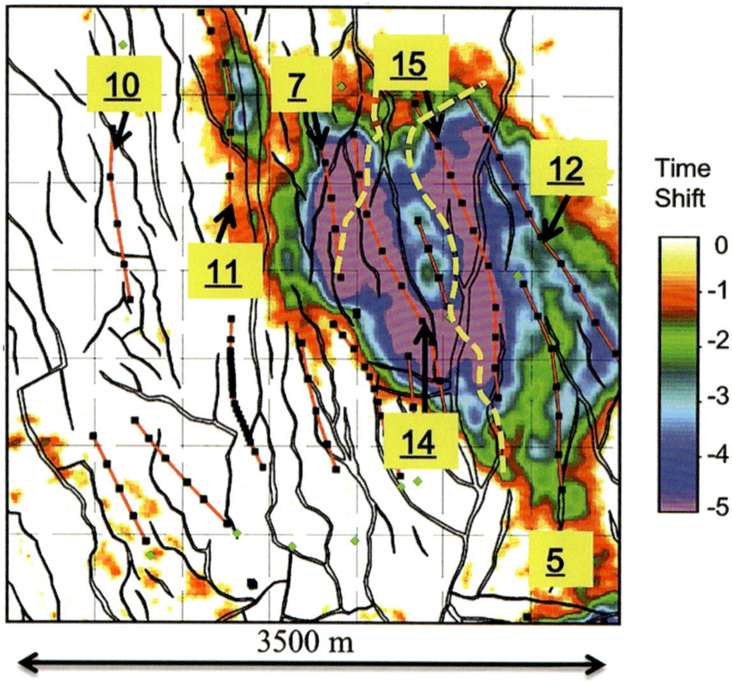

Figure 7.7: Major production wells (in red) on the North Flank superimposed on mapped time shifts between LoFS10 and the baseline survey. Black squares indicate perforations and black lines are interpreted faults. Dotted yellow lines are the common boundaries of separation between the N-14 and N-15 responses obtained by applying the technique in this paper. From Huang et al. (2011).

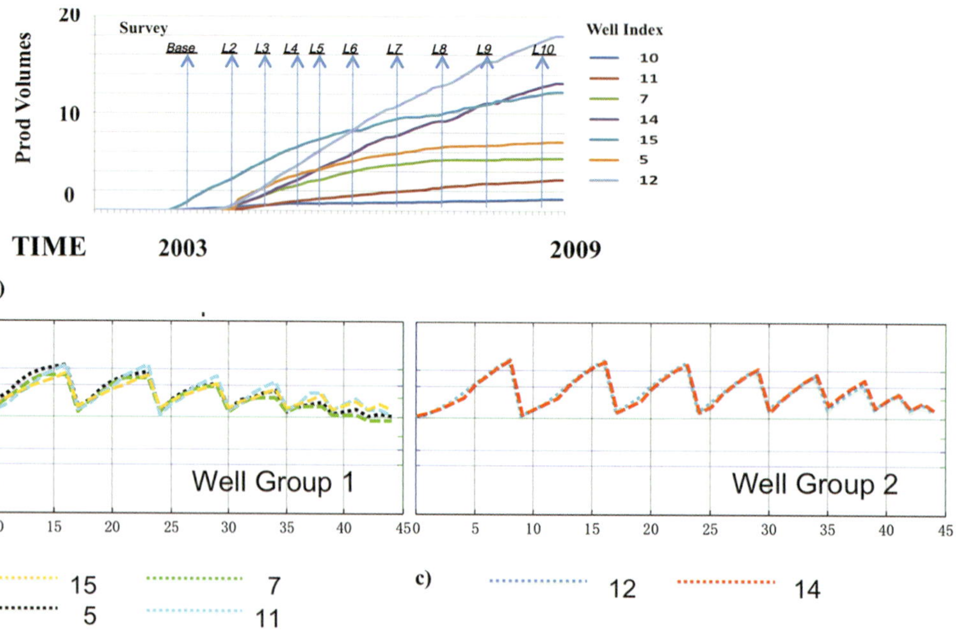

Figure 7.8: (a) Cumulative fluid volumes produced from wells in the North Flank. The plots are converted into the time sequences of cumulative volumes. Wells with similarly sequenced cumulative volumes are shown together: (b) Group 1 – N-7, N-15, and N-5; and (c) Group 2 – N-11, N-12, and N-14. From Huang et al. (2011).

Figure 7.7 shows how the traveltime difference between survey 10 (L10) and survey 1 (L1). The well location is plotted on the top of the attribute map and we can see that the 4D response of the wells overlap and interfere with each other. Plotting the cumulative volumes of each well producing from the area, Figure 7.8, we see that they group into two distinct groups of wells based on their production profile. The wells in group 1 are N-11, N-12 and N-14 and the wells in group 2 are N-5, N-7 and N-15, the characteristics of each of the groups are tabulated in Table 2.

We then calculate a measure (NCC) for how the individual well response and the seismic attribute set correlates, using a different combination of time intervals as shown above. Figure 7.9 a and b shows a map of the NNC statistics relating to the well activities for group 1 and group 2 wells. The threshold values are set such that the maps do not reflect the lower correlation factors across the groups and we clearly see that by using this technique we have managed to separate two sets of characteristic production behaviors and distinguished them laterally.

Two main groups of wells are also found for the southern flank. This is apparently linked to the sequence in which the well was drilled and the fact that there was a drilling break. The study in the south flank is summarized in figure 7.10 and revealed another intriguing insight. There is a noticeable difference if we compare the mapped NCC factors using a time-shift above reservoir map *versus* what we see when using the acoustic impedance difference map, figure 7.10, lower two images. The most likely explanation for this is that the time-shift difference map reflects compaction. This whole area is under normal depletion and we also expect compaction to be a dominant effect in acoustic impedance difference maps. However, if gas comes out of solution or if the production is stopped for a period of time, the acoustic impedance changes might respond to this but it is unlikely that the compaction induced effects in the overburden will be affected.

If we look more closely at the localized anomaly pointed to by the arrow in the lower left-hand map in figure 7.10, we see that while the changes in traveltime above the reservoir appear to continue to changes consistent with the established statistic correlation to the well activity, while the acoustic impedance does not have the expected behavior and gives poor correlation. The small area appears to be at a local structural high and there is also a high-quality hydraulic fracture crossing the well at this position. In figure 7.11 we see from left to right, acoustic impedance traces, from survey 1, 1,2,3,....... 9 plotted on top of each other at this point. To the right we see the log pointing out the reservoir, and the color code for the traces. The trace represents relative acoustic impedance and has been adjusted to the same top reservoir datum. The reservoir is between the two black lines and is about 25 m thick. In the overburden, we see a continuous decrease in acoustic impedance for consecutive traces. In the reservoir we see continuous hardening until survey 7. The situation then appears to change and we see a softening appearing during survey 8, this is confirmed by survey 9 and following surveys. We interpret this to be due to gas accumulated in the most porous reservoir at a localized structural high. The effects were initially only on a few traces but extend laterally in subsequent surveys as expected.

	Group 1			**Group 2**		
	N-5	N-15	N-7	N-11	N12	N14
N-5	1					
N-15	0.958	1				
N-7	0.992	0.945	1			
N-11	0.916	0.926	0.912	1		
N-12	0.875	0.905	0.870	0.988	1	
N-14	0.835	0.878	0.830	0.972	0.993	1

Table 2: The matrix of NCCs calculated between time sequences of cumulative produced volumes at the wells from the North Flank. The wells are divided loosely into two groups based on the magnitude of their cross-correlations. The correlations between wells in groups 1 and 2 are lower than any of the intra-group correlations. From Huang et al. (2011).

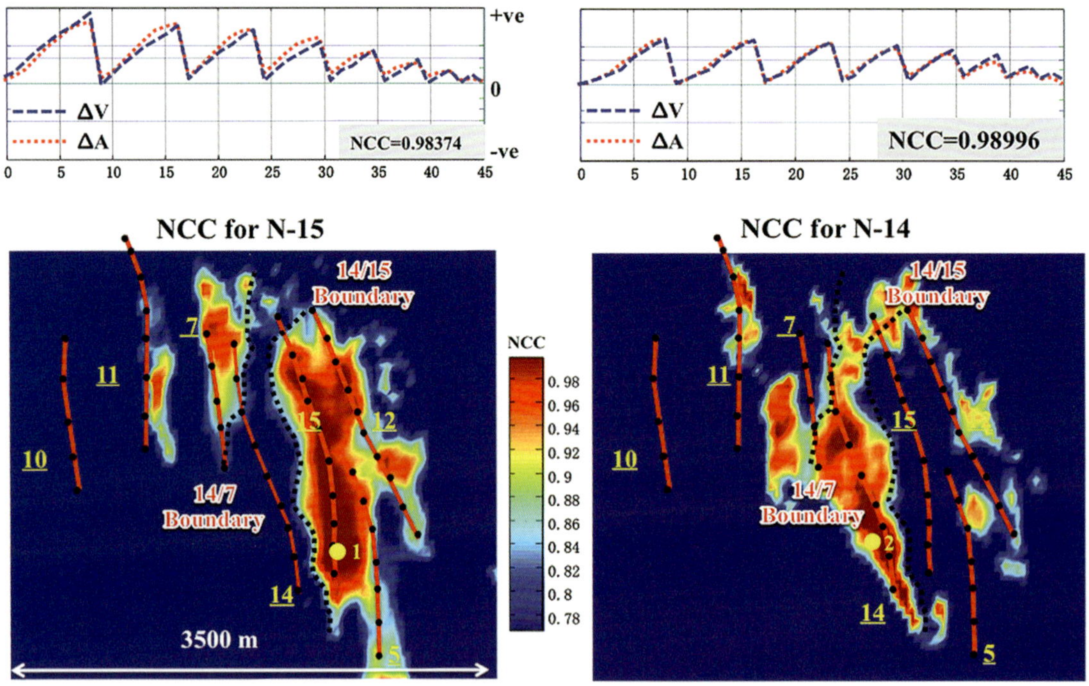

Figure 7.9: (a) The time sequence of seismic attribute changes ΔA at point 1 (left figure) and 2 (right) compared to the well groups incremental cumulative volumes Δ V for the same time sequences. (b) (left) Mapped NCC factor generated using the well activity of N-15. Dotted lines are interpreted boundaries between the areas of influence from N-15 and N-14 and between N-14 and N-7. (b) (right) Mapped NCC factor generated using the well activity of N-14. From Huang et al. (2011).

7.4 Business impact

The first indication of possible performance issues in producing wells is normally reflected in production rate, rate changes or pressure behavior. In case these are indications of sub-optimal behavior, the next step will be to find the cause. This may not be straightforward. In some cases we might not even have measurements of the individual wells and if the well is horizontal, we might have to run a production log to assess the situation. Especially in offshore wells this may be costly and is not without risk. In the case of a weak formation, we might have got sand into the well bore or some of the perforations might have been clogged up. Measuring the thickness by drilling horizontal wells is also challenging. This implies that even if we make a sophisticated single well flow model, the interpretation of the flow regimes in the well might be non-unique.

Seismic time-lapse data can be very useful in providing the context for where the well is producing from and also to identify the risk in doing interventions in the well. In a weak formation we might even have deformation of the liner. These are likely to occur where there are faults or rapid thickness changes.

Figure 7.12 shows cross-sections through three wells with different production behaviors. Well A reflects a homogenous reservoir section being optimally drained. Well B did not produce anything for the toe section, the last quarter of the horizontal well. All other data indicate that we have the presence of a reservoir in an up-thrown block, but there is apparently no communication to it. The well in this section is below the possible reservoir and we might try to frack into it from below or we may drill an infill well to drain this section.

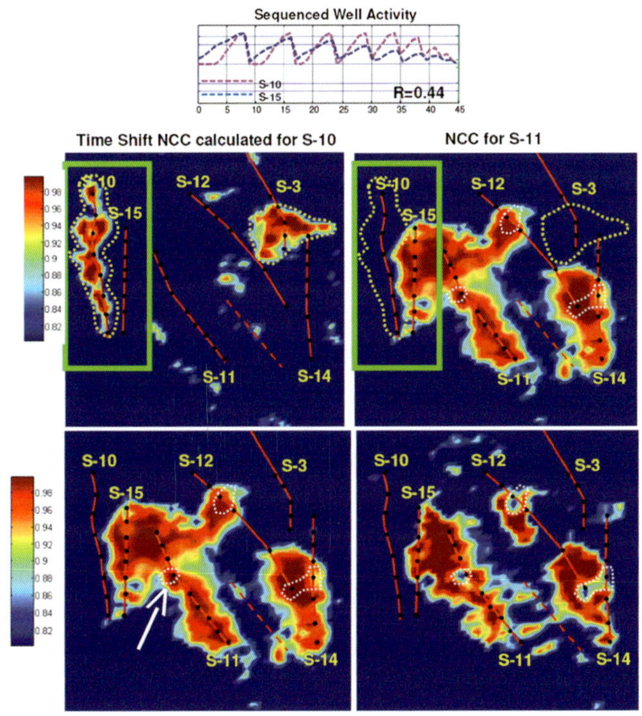

Figure 7.10: Example for the South Flank area. Top sequenced well activity for two groups of wells. Middle: maps showing correlations to Group 1 wells (left) and Group 2 wells (right). Bottom: NCC factor map based on correlating well activity set for group 2 wells to traveltime difference attribute sets (left) and to the acoustic impedance difference sets (right). The white arrow in the lower left-hand map points to a possible localized gas accumulation. Black dots denote perforations in the wells, shown as red for the horizontal section of the reservoir. From Huang et al. (2011).

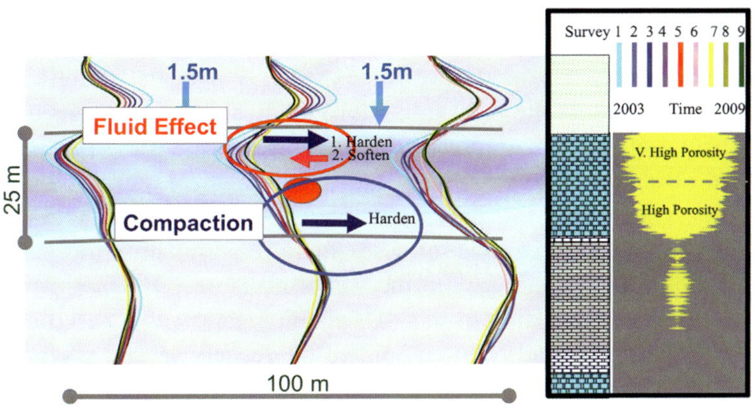

Figure 7.11: Acoustic impedance from color inversion from subsequent seismic surveys, see the text for explanation. The traces are selected from the location shown at the end of the arrow in figure 7.10.

Well C is more complicated. The first 400 m of this well were drilled in non or thin reservoir rock and it was decided not to complete it, the next 800 m are in good pay, before we apparently drilled through a fault and continue for 600 m in what we considered to be a good but possibly thin reservoir. Other data indicated that we could have a restriction in the well and it was decided to do a well intervention to clean it out. Based on the seismic response, we flagged a potential risk for restriction in the well after passing the fault at 400 m and also after about 1200 m. At these two points the operator will be very careful about applying too much force on the tool. A second risk was defined to be the apparent lack of drainage of the last 600 m. Given the potential pressure differential measured by the permanent sensor positioned where we enter the reservoir section and what the model predicted by closing off the section, we needed to have the right procedures in place to manage this while cleaning out the well bore. Although this might be a

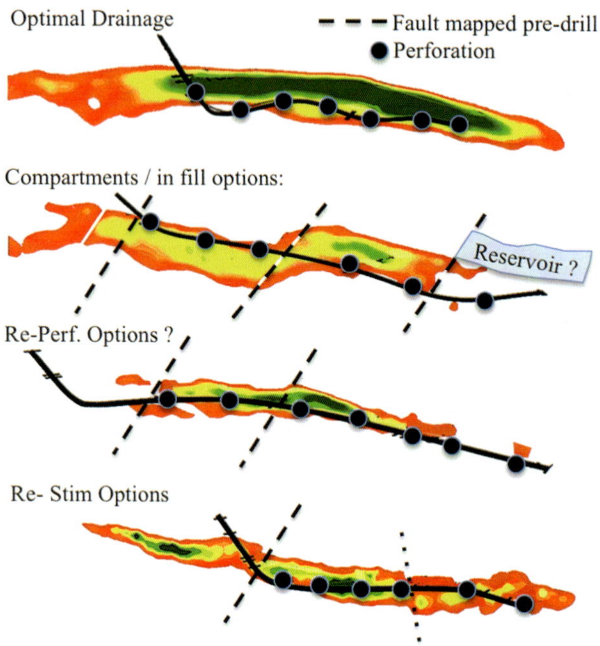

Figure 7.12: cross-sections through three wells with different production behaviors, the color reflects changes in acoustic impedance, red indicates small changes and while green is large changes. From Barkved et al. (2005).

generic risk for this type of operation, being able to locate where it is most likely to occur makes a difference to the focus of the operation. The conclusion of the well intervention was that the well had been clogged up 600 m before TD and there was a higher pressure toward TD, which was mitigated through the proper preparation.

The seismic from Well D were a bit surprising, as the well unexpectedly appears to drain quite well an area to the north of where the well enters the reservoir. Close study showed that part of this drainage is actually linked to an offset well. Our main concern for this well was in the last part of the horizontal section, where we had observed quite significant faulting and where parts of the well penetrated quite poor reservoir quality. We see that this area is depleted and the proposed action was to run a production log to acquire more information to use for assessing a re-stimulation job.

To understand the contribution to the production between the wells, maps or 3D visualization is needed. One of the first well intervention jobs benefiting from a 4D seismic image was a well draining an area down flank of the field center. This well had been completed by densely spaced perforation and acid stimulation. The well had been producing well but a re-stimulation job had already been done. Now the production engineer proposed a second job as their models indicated that the well was underperforming, mostly due to declining pressures and this was interpreted as due to fracture closure. A meeting brought together the geologist, the geophysicist, the reservoir engineer as well as the production and intervention engineers, all with a different experiences and images of the subsurface settings in mind of how the subsurface condition is, the wells hydraulics and how this could impact the planned work.

The availability of a 4D seismic map, figure 7.13, completely changed the process. The confidence in the 4D image was established by showing the consistency with the 19 perforation clusters and the strength of the time-lapse attribute. The 4D images provided a common reference to the reservoirs conditions and "everyone" could see that the area was unevenly drained; furthermore it appeared that the toe of the well was less well stimulated. This could be linked to the way this acid stimulation was run and a new scheme included use wax beads to divert the acid. The work resulted in a rate improve-

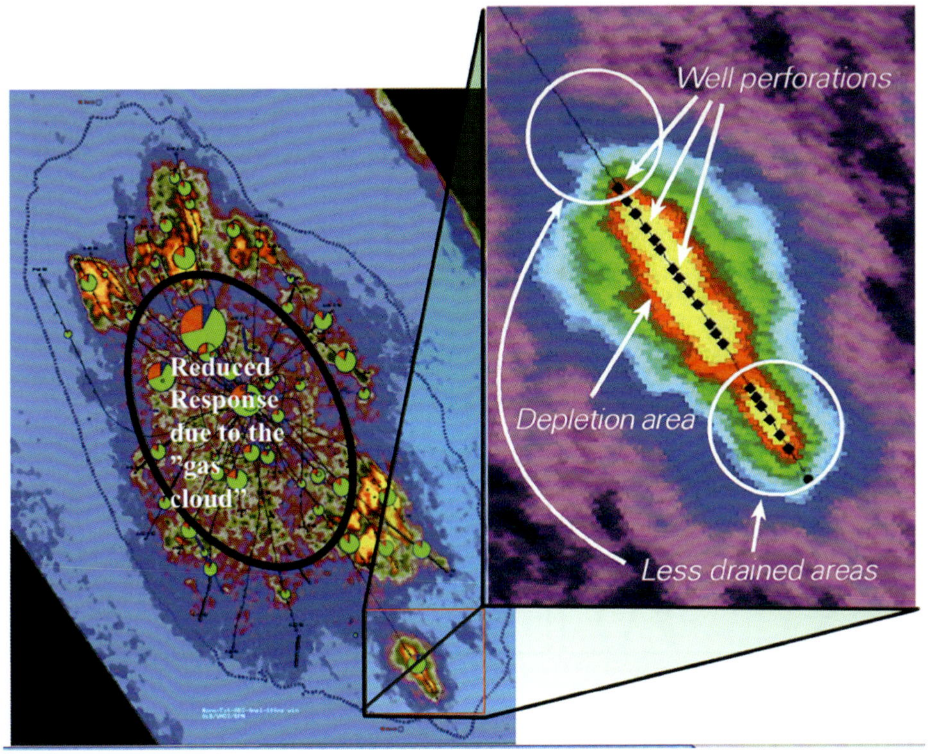

Figure 7.13: 4D seismic images used to guide re-stimulation of horizontal well, From Barkved et al. (2005).

ment of 800 bopd. The well intervention cost 470,000 USD, which at the actual oil price meant 12–15 days of production to pay back, an additional 100 days would pay for the 4D seismic as well.

7.5 The right action may be doing nothing

One of the unexpected outcomes from consulting with the 4D seismic on production issues has been the conclusion to do "nothing". In a few cases there have been indications that the loss in production was most likely due to a permanent failure. If suddenly the production in a 1000 b/d well disappears, it would not go unnoticed. If there were a chance to restore it with a cleanup operation, this would normally be a very good business case. However, this may need re-prioritization of well intervention with longer-term well surveillance operations and could also require expensive last-minute mobilization of equipment. The seismic observation would not be conclusive but it will inform us about the chances of a successful operation and help in making a sound business decision.

One of the main drivers of the LoFS installation at Valhall was to support the water flood program. A new well, called WID, was drilled in the North Crest Basin and put on injection in April 2008. The well was on injection of about 10,000 b/d and had injected 3.4 mmbbl by January 2010, Van Gestel (2010). LoFS data were used intensively in the planning and drilling of both the overburden part of the well and when picking the final depth of the well, Kristiansen and Flatebø **(2009).**

Soon after the start of injection of water into the new well, figure 7.14, we recorded higher water cut in one of the producers, PD1. Based on the ion analysis this appeared to be seawater. This created serious concerns for early water breakthrough from WID and plans were made to investigate the well bore and potentially shut in the toe perforations of the injector. LoFS 11 was the first survey acquired

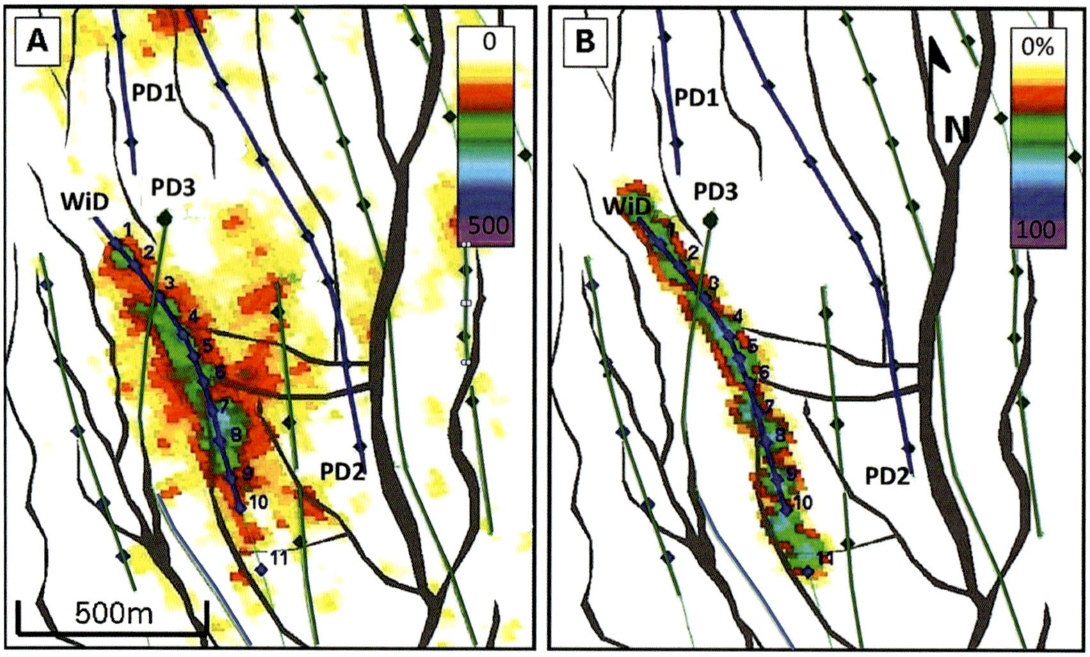

Figure 7.14 A: Observed amplitude difference map between LoFS 11 and LoFS 10. B: Modeled changes in water saturation between LoFS 11 and LoFS 10. The blue lines are the wells and the green diamonds are the locations of the perforations, grey lines are the mapped faults. From van Gestel et al. (2010).

after the well came on injection. The data for this survey was processed and analyzed within weeks to allow for interpretation and integration of observations. The LoFS 11 data shows a clear seismic response of the water injection around this well, figure 7.14. We observed an increase in acoustic impedance or a hardening response as predicted and observed in the other water injection wells. The area of the seismic response had a very good match to the area where we expected water saturation increases. This shows even distribution of the water injection along the well bore, with all perforations contributing.

When the LoFS observations were analyzed and matched with the simulator response, it was realized that the observations were more or less as predicted, even in the case when the majority of the water went into the matrix and the observation did not necessarily require direct communication through a fault into the PD1 well. We decided to defer the expensive well operations and thereby save unnecessary work. LoFS 11 observations also confirmed that the packers between perforations 10 and 11 are not leaking, as perforation 11 was closed in at the start of injection and the reservoir model required an update in this area. The LOFS 4D images flagged a potential risk in injecting higher volumes into an area close to an old shut-in producer, which combined with its proppant-filled hydro frocks, could act as a conduit and feed early water to the TD2 well. The contextual information from the LoFS data help us in making a better decision by continuing as planned, saving one million dollars in work that could actually reduce the sweep efficiency of the area.

7.6 Conclusions

We have shown examples of how seismic surveillance provides the spatial context for well observations and predictions and to assist in the planning of intervention programs.

Seismic time-lapse data has been used to assist in making operational decisions by simply providing insight into how the water and hydrocarbons are moving in the reservoir. We have also discussed two new techniques that have been developed to assist in projecting the seismic time-lapse information into parameters more easily adopted by production engineers.

8 Passive and other Emerging Technologies

"Any sufficiently advanced technology is indistinguishable from magic."
- *Arthur C. Clarke*

Noise is there for free. We are interested in two aspects of passive recording; one linked to the failure of the rock in the subsurface, the other linked to using the industrial or natural noise present to deduce subsurface parameters. In this section we will review a few intriguing technologies that have been tested on Valhall. They are listed in chronological order, and include passive monitoring, shear wave splitting, detection of fault re-reactivation and AVOA. We have also included innovative analysis of recordings from the active source that probably will make it into routine usage in the future. Finally, we will close out with some interesting observations of combining passive monitoring and S-wave splitting analysis from an onshore field.

8.1 Introduction

Our interest in passive monitoring is linked to the information this can provide about deformation processes. Information on stress state changes can provide important parameters for reservoir management and drilling, and provide valuable insight into hydraulic fracturing, production and injection mechanisms and into drill cutting re-injection mechanisms.

One of our key sources of information is acoustic emissions, or micro-seismic events. Understanding where these take place help us to identify possible weak zones in the subsurface and to introduce mitigation actions if needed.

Mapping of the subsurface structure and properties is mainly done on data acquired using an active source. However, various types of industrial or natural noise can also inform us about the sub-surface.

The recorded micro-seismicity can be used to localize and characterize induced hydro fracturing. The early applications of passive monitoring using downhole sensors were for onshore gas storage reservoirs. In 1993, Deflandre et al. (1993) reported on a research project that Total, Elf and GdF were involved in, where the objective was to determine whether the withdrawal of gas inside a reservoir created any micro-seismic activity.

Microseismic signals are typically recorded and processed in the same way as earthquake signals. First, continuous signals are analyzed with detection algorithms to determine when an impulsive energy source has occurred, by identifying the event. To locate where the events originate we need three-component (or triaxial) sensors and velocity models, which express the Vp in case we have widely separated instruments. Alternatively we can use the Vp/Vs ratio observed from arrival times of the different phases and ray-path orientation from polarization or moveout. The radiation pattern from an event can also be analyzed to determine the seismic moment tensor related to the stress state changes that resulted in the seismic deformation. The moment tensor can then be interpreted in terms of the shear *versus* dilation or compressional nature of the deformation and we can obtain an indication of whether the fracture is opening or closing. The fault orientation and approximate principal stress directions can also be estimated from these analyses.

The possible applications of passive micro-seismic monitoring, Maxwell (2001):

- Well stimulation or hydraulic fracture mapping and characterization
- Localization of possible weak intervals in the subsurface
- For fault mapping and characterization
- Mapping of fluid movement
- Mapping thermal fronts
- Mapping of drill cuttings re-injection DCRI domains

During the last 15 years a significant microseismic monitoring business has evolved around tight gas/shale gas development onshore in the US. The early testing started in the 1970s, and a few important experiments were funded by the DOE, Sleefe et al. (1993). The Cotton Valley Consortium project in 1997, Ruthledge et al. (2004), triggered the commercial services around microseismic monitoring. Interestingly, the engineers in the companies drove the interest, while the geophysicist generally followed half-heartedly. Geomechanical insight is essential in interpreting the observations during a massive hydraulic fracturing job, as the micro-seismicity we record is associated with the geomechanical response of the rock during the stimulation. Being able to distinguish between shear failure and dilation is directly linked to the quality of the stimulation.

8.2 Passive recordings from Valhall

The Valhall field was produced by pressure depletion until 2006, when a water injection program started to provide sweep and pressure support. The effect of pressure drawdown in the reservoir is compaction in the reservoir (~ 10 m) and stress unloading and subsidence in the overburden. The deformation process associated with the stress unloading effects is expected to be a source of acoustic emissions, figure 8.1.

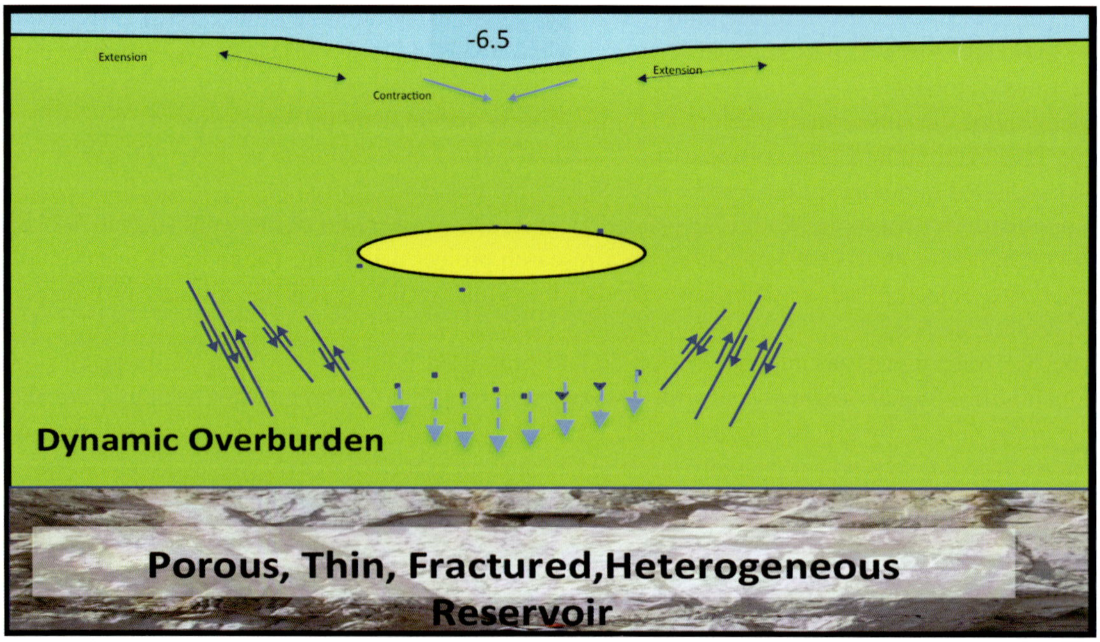

Figure 8.1: Production induced compaction resulting in stress redistributing and subsidence may induce fault reactivation in the overburden. Barkved et al. (2003c).

8.2.1 Early tests

In June 1992 we installed a seismic observation system at the seabed at Valhall. The objectives were: to use this for undershooting the platform (active source) during a 3D acquisition and to test viability for recording acoustic emissions from what we consider a passive source or micro-seismicity. Unfortunately, the system disappeared during a foggy night, it later appeared in Hirtshals – a fishing harbor in Denmark The active source part of the project had to be abandoned after just a few lines had been acquired. However, a significant number of events were recorded and a few of them were interpreted to have originated in the subsurface, possibly linked to the ongoing drilling of a nearby well, figure 8.2.

In 1993, Phillips tested a triaxial downhole geophone in one of their wells for monitoring of possible ambient micro-seismicity associated with the discrete shear failure of the reservoir rock, ref Rutledge et al. (1994). During 104 hours of monitoring they observed 572 discrete events, which were reported as shear-failure micro-earthquakes. Events were reported to originate up to 1000 meters from the borehole.

These encouraging results led to new experiments where a modified six-level 3C tool (CGG SST500) with modified geophones packages was installed for 57 days at 2200 m, 250 m above the reservoir in a well at Valhall, ref. Dyer et al. (1999). Close to 600 events were detected, of which 324 could be reliably located, figure 8.3. The event rate was approximately 10% of what was observed at

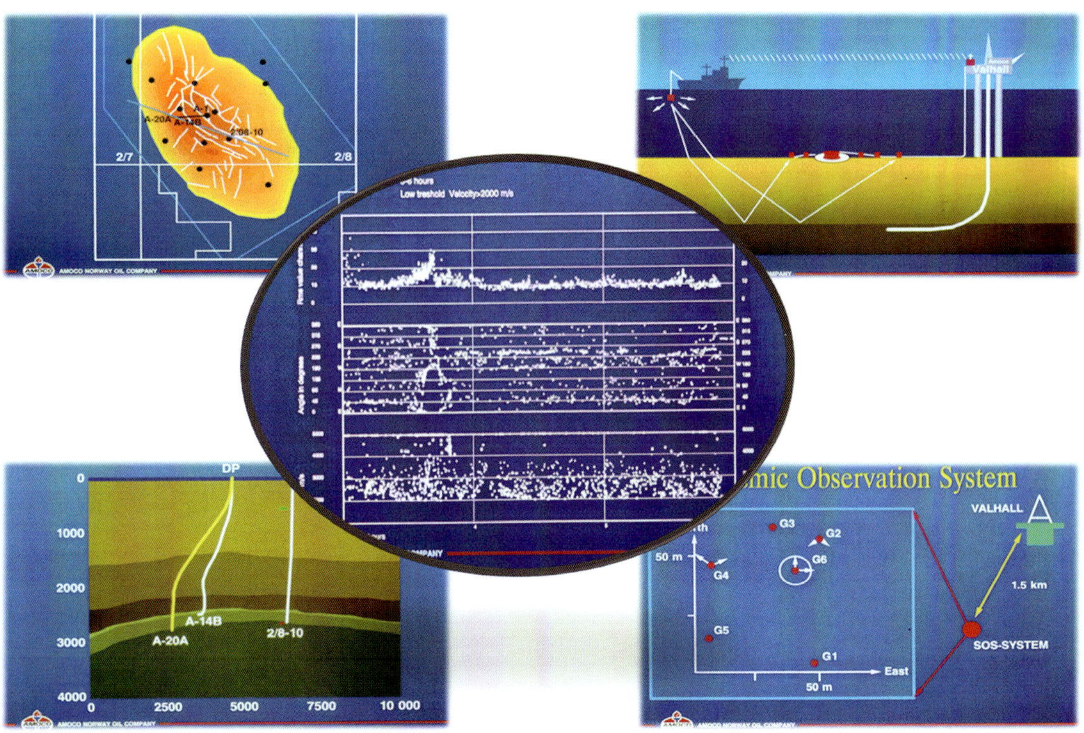

Figure 8.2: The outline of the Valhall Field (top left), and the location of the profile shown. The system contains three 3C and three 1C geophones, which were mounted within a 50×50 m steel frame that was placed on the seabed at 70 m depth. The circular image in the middle shows six hours of data, from the top is the RMS value of one of the geophones, in the middle is the azimuthal polar orientation of the event 0–360 and the lower graph shows the apparent velocity, ranging from 2000 m/s to 5000 m/sec, Berteussen et al. (1993).

Ekofisk, where the same tool was installed across the reservoir, Maxwell et al. (1998). All of the events, except those related to drilling activities, were located to within a 50 m thick zone, about 50–100 m above the reservoir. These observations had a direct impact on how we designed the wells at Valhall, Kristiansen et al. (2000). The presence of the microseismic events was attributed to a weak zone or slip planes at these intervals, which would tend to focus the deformations associated with compacting induced stress unloading effects. Because of this, the casing program was modified to include a heavy wall liner across this interval, which should reduce the impact of the deformation. Since these changes were introduced in 1999, the well failure rate at this level has been reduced significantly.

The dataset was later donated to the University of Leeds and was studied in more detail. They were able to reduce the location uncertainty and to map two faults within the Valhall cap rock and near the crest of the reservoir. The orientations of these faults agree well with those obtained from interpretation of surface 3D seismic provided by BP. De Meersman et al. (2007) conclude: "Stresses in the microseismic source area probably recharge in a cyclical fashion. Stresses build up in response to oil production and are dissipated through microseismic activity. These changes are likely to drive the short-term variations in seismic anisotropy and the reactivation of microseismic source mechanisms." The anisotropy has been deduced from shear-wave splitting analysis. The shear-wave splitting had apparent temporal variations, which are most likely associated with fracturing, and the observations could be explained by two (conjugate) sets of fractures whose relative dominance are linked to the compaction in the reservoir.

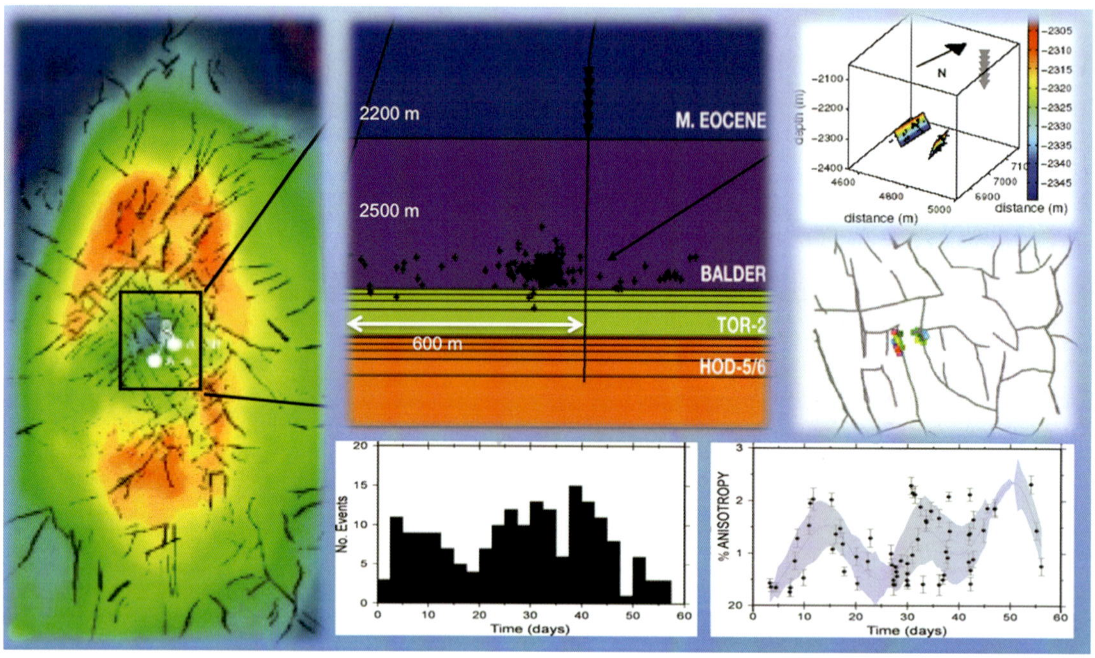

Fig 8.3: From the left, then clockwise: map showing the location of the well where the passive data was recorded. Most of the events were located to between 50–100 m above the reservoir. The precise location showed that a significant number of the events were located very close to two (conjugate fault) plans, which matched well with existing interpretation. The two images at the bottom show the number of events per day and how the anisotropy apparently varied as a function of time; there is an apparent lag of 11 days between the peak number of events and the maximum anisotropy observed. Modified from Teanby et al. (2004) and De Meersman et al. (2009).

Shear-wave splitting is generally the best indicator of seismic anisotropy, as the anisotropy materializes itself in difference in velocity between the fast and slow mode of the S-waves. Shear-wave splitting of the passive recordings showed delays up 30 msec, which corresponds to anisotropy of 3%, given that the anisotropy is uniformly distributed. Closer inspection of the data indicates a time dependency in the properties, Caley (2001). Figure 8.4 shows an example of a fast and slow S-wave event before and after shear-wave splitting corrections.

Zoback and Zinke (2002) also found evidence of a normal faulting stress state at Valhall, using data from the passive seismic monitoring experiment. An inverse/composite focal plane mechanism of these micro earthquakes is consistent with a normal faulting stress regime, figure 8.5.

8.2.2 Permanent installation of electrical systems in drill cutting re-injection wells

Today's completions with down-hole pressure and temperature gauges mounted on tubing may be expanded to include in-well seismic sensors. These sensors can then be used for active VSP recordings with a seismic source at the surface or for continuous recordings of micro-seismic events. The ability to utilize these sensors to record seismic data in the presence of production/injection noise will be critical to most applications.

During re-completion of the Valhall A-25B T3 well at the end of July 2001, a permanent in-well seismic monitoring system was deployed; figure 8.6 Barkved et al. (2002). This well was initially completed as a horizontal deep flank producer but was later converted to a drill cutting re-injector due to poor performance. Drill cutting are slurrified and injected at limited volumes and rates, as needed. During

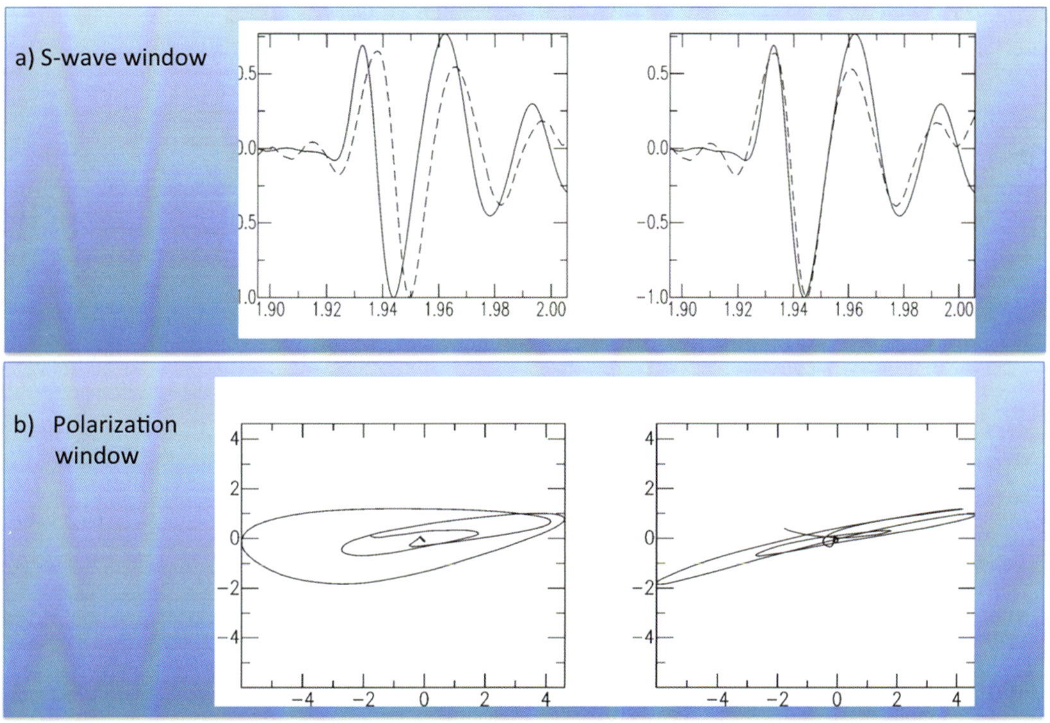

Figure 8.4: Shear-wave splitting analysis a) isolated and slow shear-wave, before and after splitting corrections and b) horizontal particle motion, before and after splitting correction. Notice that the correction linearizes the particle motion. From Caley et al. (2001).

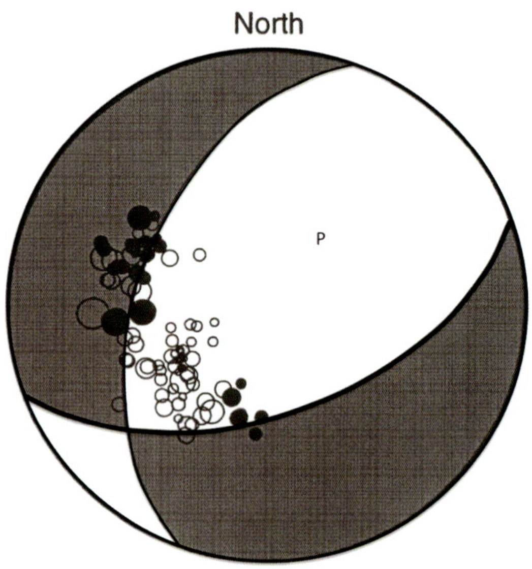

Figure 8.5: Inverse/composite focal mechanism for micro earthquakes recorded at Valhall. The focal mechanism is an attempt to best fit both the polarity data (where open circles denote dilatational and filled circles denote compressional P-wave first motions) and S-to-P amplitude ratio (indicated by the size of the symbols). While there is considerable uncertainty in the focal plane mechanism, it is certainly suggestive of normal faulting. Modified from Zoback and Zinke (2002).

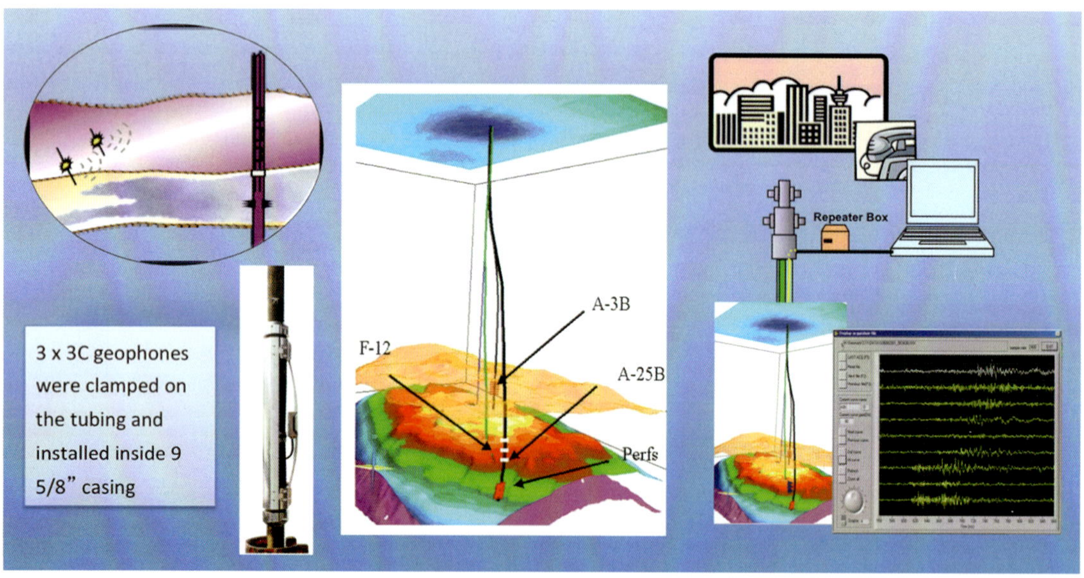

Figure 8.6: Installation of electrical geophones in a DRCI well. From Barkved et al. (2002).

drilling, the injection will normally last for 2–3 hours twice a day. The main objectives for installing the seismic array in this well were:

- To validate the feasibility of recording seismicity in an active well
- To monitor the waste injection domain
- To monitor the micro-seismic activity related to compaction of the reservoir.

Three tools, each with 3C geophones, were clamped to the tubing and successfully installed at approx. 50 meters separation as deep as possible into the cap-rock. The selection of final geophone depths was driven by the objectives but guided by well-bore design. A dial-up connection was installed to supervise the acquisition system from the shore.

Due to a failure in an electrical component of the geophone string, the system was only active for two weeks. During this period the system was on automated permanent monitoring for five days and was operated manually during three injection tests.

A total of 3260 reliable files were recorded during the whole monitoring period. Four main types of record have been identified and categorized according to their principal characteristics; see Barkved et al. (2002). Figure 8.7 shows an example of an event occurring "near" to the geophone, <750 m.

Twelve "near" seismic events and ten "far" seismic events were identified and analyzed. Preliminary analysis showed reliable seismic arrival times and polarization of four "near" events and six "far" events. On average, the detection rate is only two seismic events per day, reflecting that the well is outside the production area.

Due to the early failure of the system, surface calibration shots were not recorded. Therefore, the geographical orientation of the geophones is partially unknown. A simplistic velocity model using arrival times of the S and P-waves and measured polarization of the events was used to define locations relative to the well bore deviation. The local magnitude of the seismic events recorded ranges between –3.2 and –2, which is consistent with earlier observations, Dyer et al (1999).

The seismic attributes show that the events were not induced by the three injection tests. No seismicity induced by the injection was observed. This may be due to the unfavorable acoustic setting resulting from the actual velocity contrasts between injections at 5200 m MD and the deepest geophone at 4850 m MD. In addition, the maximum injection rate was around 9 bpm and the total volume injected was only around 1900 bbl.

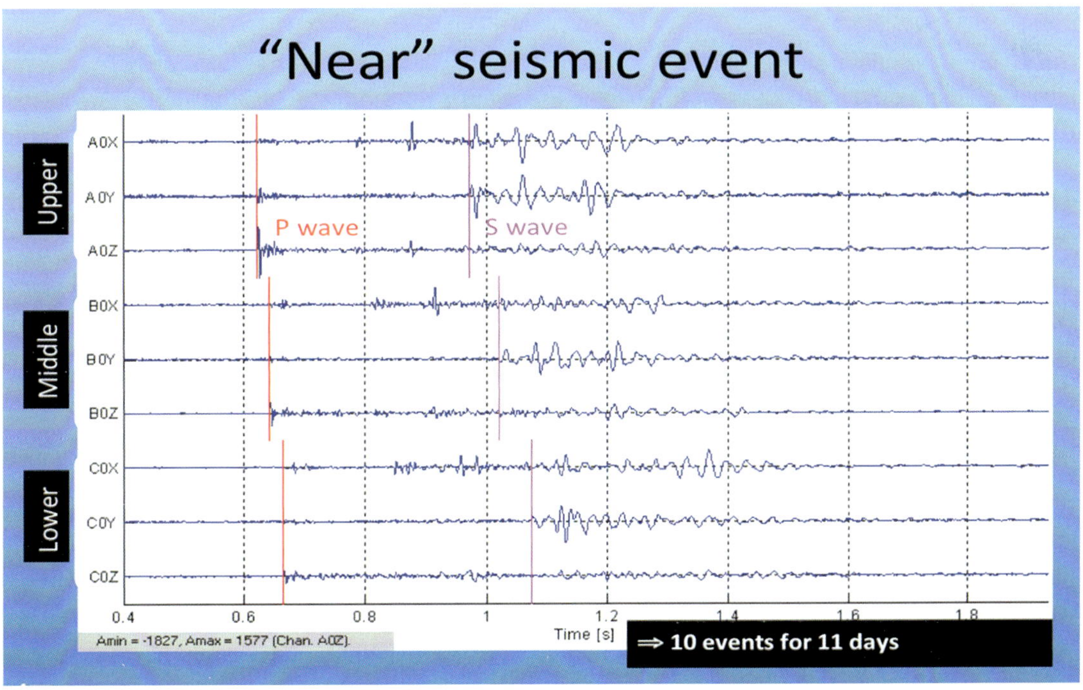

Figure 8.7: A typical near event. Note that the data is raw and unrotated. From Barkved et al. (2002).

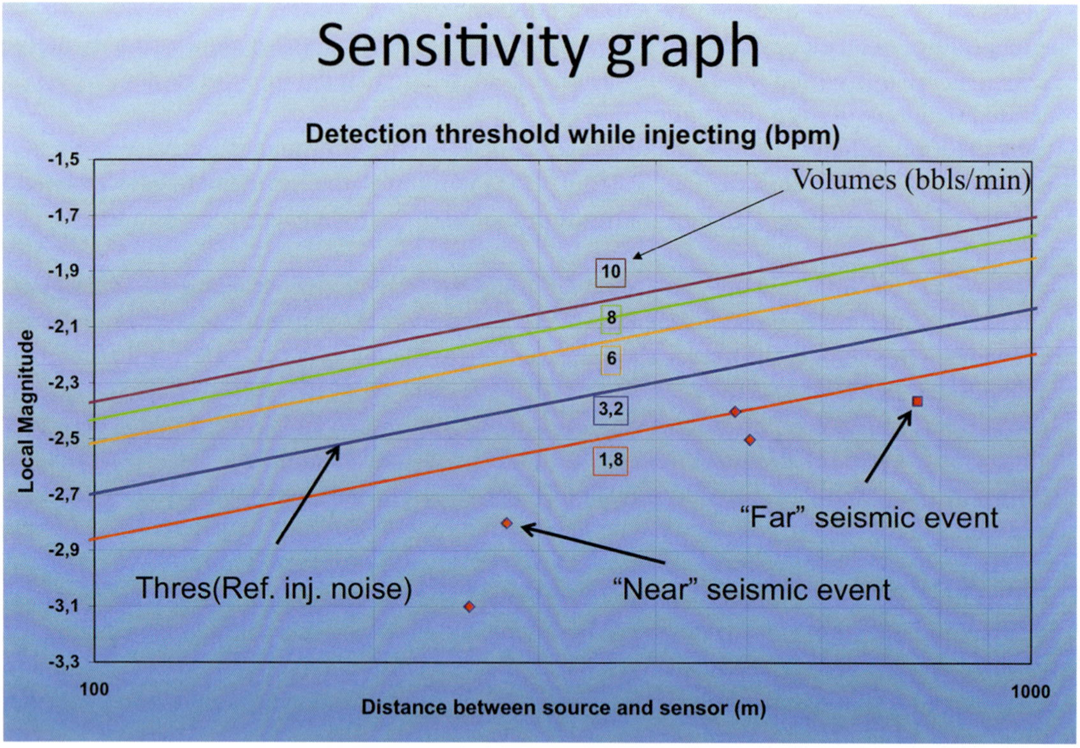

Figure 8.8: During injection it was impossible to detect any of the type of events observed between the injections. From Barkved et al. (2002).

The system was under manual supervision during the injection tests. Seventy records were captured during various stages. The records were analyzed to qualify feasibility for acquiring passive seismicity in an active production/injection well, figure 8.8. This analysis concluded that a linear relationship between the injection rate and associated seismic amplitude recorded on the geophone channels exists. Figure 8.8 illustrates this relation and also emphasizes the differences between the channel responses. The geophone tangential to the well bore is generally less noisy than the radial one and the deepest geophones are always the noisiest.

The results have also been expressed in terms of sensitivity of the network during various injection rates. The linear relation proposed between the injection rate and noise level allows us to express the network sensitivity as the range of magnitude and distance seismic events that have to be detected while injecting at a given flow rate. According to this, the seismic events, which were recorded during non-injection periods, would not have been detected for injection rates above 2 bbls/min.

8.2.3 Permanent installation of optical system in producing well

Permanent borehole seismic is seen as complementary to the permanent seabed seismic array with the following goals:

- 4D seismic calibration. Use the directly recorded downhole signal to calibrate the surface seismic signal. This might best be done in conjunction with instrumented seabed seismic receivers for full-field monitoring, Barkved (2004).
- 4D VSP. With several wells instrumented, use surface sources and downhole sensors for reservoir monitoring.

- Micro-seismic monitoring. Passive monitoring of faulting caused by reservoir compaction and fluid fronts and active monitoring of cuttings injection and well stimulation.

Figure 8.9 shows the details of a permanent fiber-optic in-well seismic system installed at Valhall. The 3-C sensor package is installed in a seismic carrier. The carrier comes pre-loaded into a seismic array spooling unit, which has the fiber-optic cable pre-attached to the seismic carrier. At the well site the seismic carrier is installed on the mandrels at pre-configured distances determined by the fiber cable and tubing spacing. The installation of the world's first successful optical offshore production well system at Valhall involved a significant effort in system design and qualification. This was the result of an extended development effort with Weatherford, in which BP provided significant guidance with respect to in-well deployment risks and requirements. The overall system was designed in part with support from Norwegian Demo 2000 and industry partners; in particular Statoil and Total. Several field trials have been conducted to date, Bostick III (2000); Hornby et al. (2005); Keul et al. (2005); Knudsen et al. (2006).

The objective of this installation was to qualify the type of system and test out the performance and utilization in a flowing injection well, but also to monitor the implementation of the water-flooding scheme started in 2006, by imaging closer to the reservoir. In addition, during the initial stages of the water injection, reservoir compaction was expected to increase due to water-weakening effects. Monitoring of micro-seismicity related to this compaction and associated subsidence may provide vital information about the sweep-efficiency of the water-injection scheme. In addition, it is desired to use micro-seismic monitoring as a means to monitor waste injection of cuttings and associated fluids. Figure

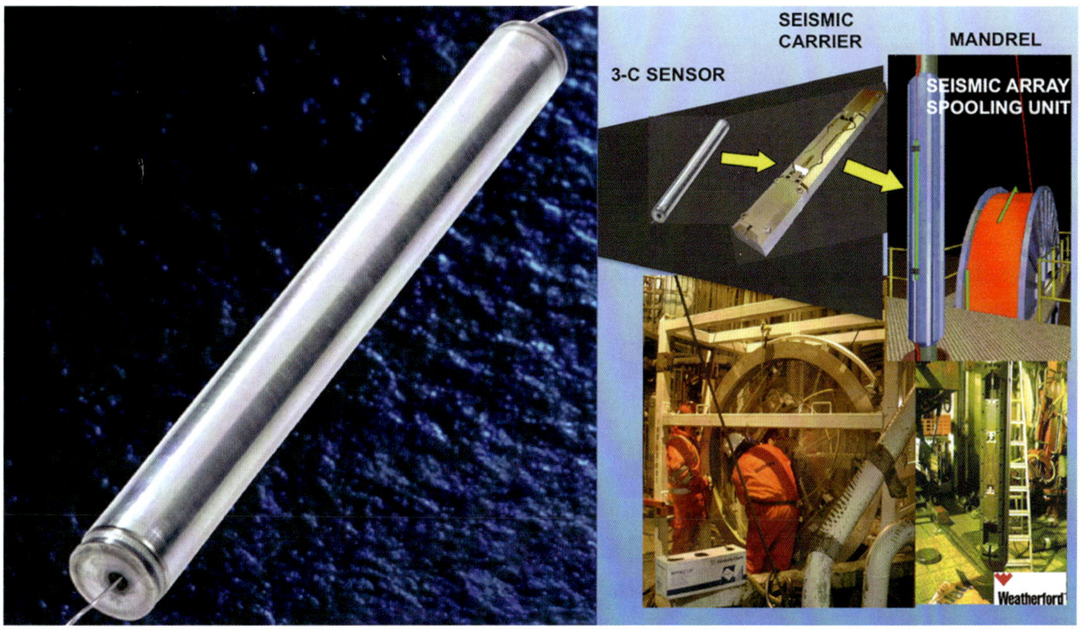

Figure 8.9: The 3C sensors are mounted in a small cylinder, about 1 inch in diameter, this is installed in the seismic carrier, which is mounted into the mandrel, which has already been mounted on the 6-inch line being run into the well. The carrier will eventually "land" up in the 9-inch casing, before the pressure is increased beyond a preset level and the sensor package is released and pushed against the outer casing. From Hornby et al. (2005).

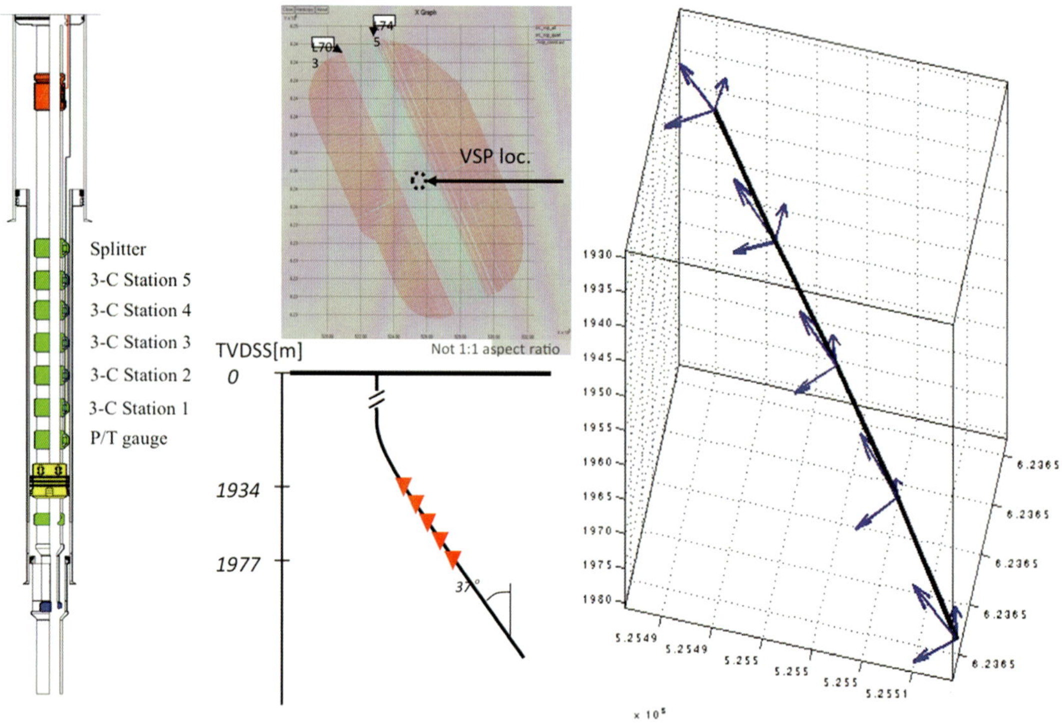

Figure 8.10: The depth and inclination of the tools are known from the well's deviation. One sensor is pointing in the same direction as the liner but the two others will be orthogonal and we do not know what the azimuth is. The sensor therefore needs to be oriented and this is done using seismic signals emitted from a known position at the surface. We use the source geometry for the 3D seismic survey with simultaneous seabed and borehole acquisition. Red lines represent when the well was injecting and green lines are when injection was shut in. From Askim et al. (2008).

8.10 shows the details of the tool and the geometry of the installation. The tool was oriented using data from the one of the Valhall LoFS surveys.

The system has a total of five 3-C fiber-optic sensors spaced 13 m apart along the borehole, using up two of the three available optical fibers, figure 8.10. In addition, the installation includes a pressure/temperature gauge in the reservoir section, which takes up the remaining optical fiber. The results were successful and the 15 channels of seismic sensors and pressure/temperature gauges were fully operational. The system is completely integrated with the Valhall permanent seismic monitoring system and is connected through the Oyo-Geospace GeoRes system. The result is simultaneous acquisition of the entire seabed permanent seismic array along with the downhole sensors. Figure 8.11 shows a comparison between data from a line in the 3D seismic volume, compared with a line from the VSP volume of data, Askim et al. (2008).

Several months of recordings from the well were run through an automated event detection algorithm by ESG. The flow noise makes it difficult to isolate the events. The noise is still present for an intermediate period of 5–10 days after the injection has been stopped and with current parameterization only about 17 high-quality two-phase microseismic events were detected, figure 8.12. In addition, 2792 single-phase events were detected, the majority of which were recorded during non-injection.

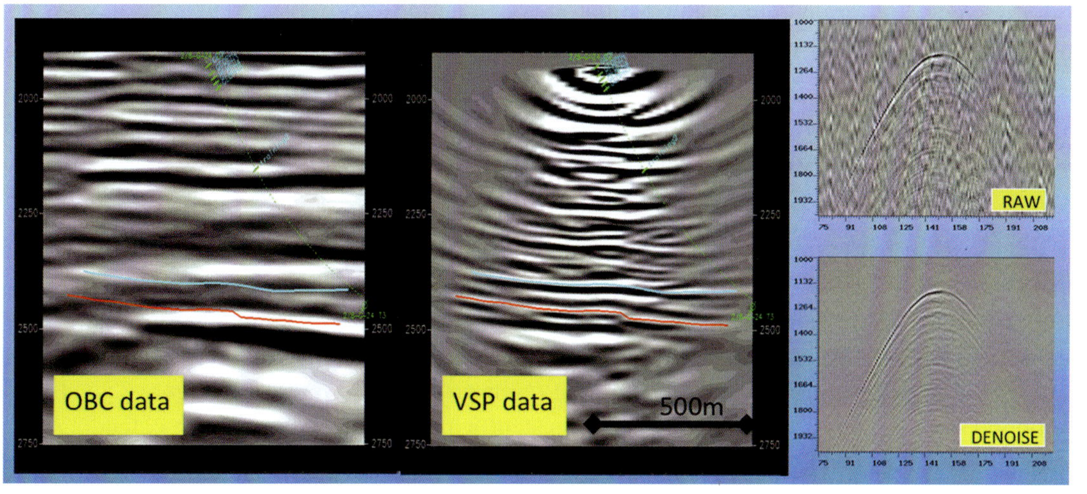

Figure 8.11: A comparison of pre-stack depth migration of the VSP data and seabed seismic (left). The blue line is the reservoir top and the red line is the base. Clearly, the VSP has higher vertical resolution than the surface seismic, yielding good image quality to a radial distance up to 400 m from the borehole. Monitor surveys will focus on time-lapse effects in the reservoir section below the sensors.

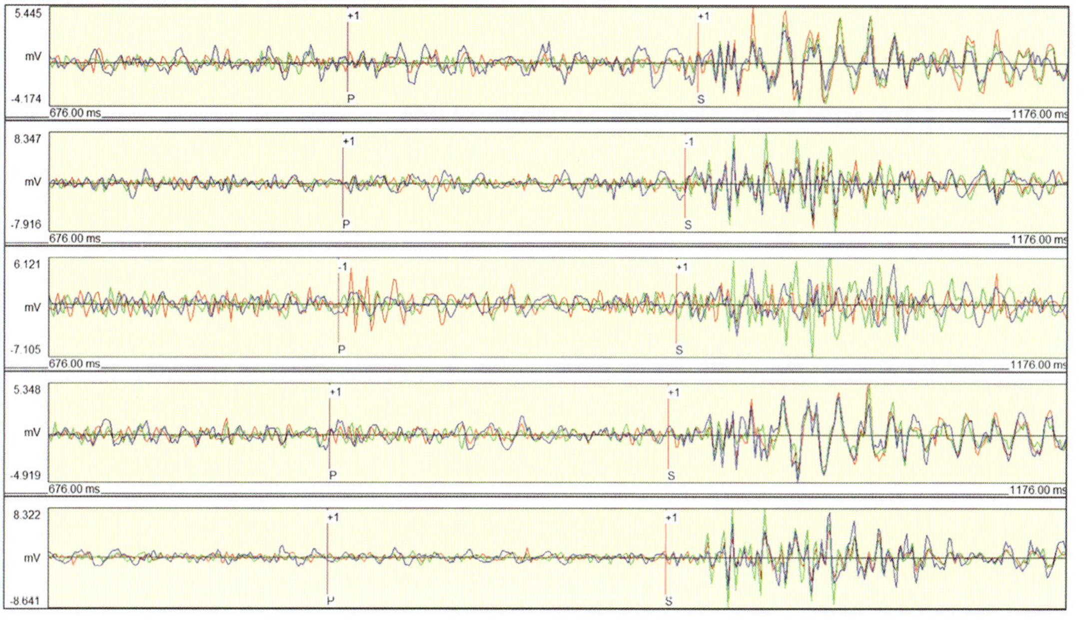

Figure 8.12: This microseismic event from the optical array was recorded on May 12, 2009 at 14:31:36.857. The trace shown above is 500 ms long. The moveout of the P-wave and S-wave is characteristic of a microseismic event. The S-wave moveout (red line) is more pronounced than the P-wave moveout (green line), due to the differences in traveltime and attenuation. The bottom-up moveout indicates that the event is below the array.

The use of a band-pass filter helped significantly in the detection of micro-seismicity, by filtering out low-frequency background noise during non-injection periods. It is possible that these filters can be optimized and additional events might be detected. The background noise levels were quite constant, indicating that there is no significant drift in the system.

8.3 Passive recordings using the LoFS (surface) array

8.3.1 Monitoring hydraulic fracturing in horizontal wells

We have used data from the Life of Field Seismic (LoFS) surface array in conjunction with a migration-style approach to locate microseismic events in Valhall Field during 6.5 hours of hydraulic fracturing. We identified an event distribution that extends roughly 300 m vertically and 200 m horizontally, with sources both within the chalk layer containing the reservoir and the overburden, figure 8.13. However, the temporal distribution of the events is not consistent with the induced fracturing extending into the

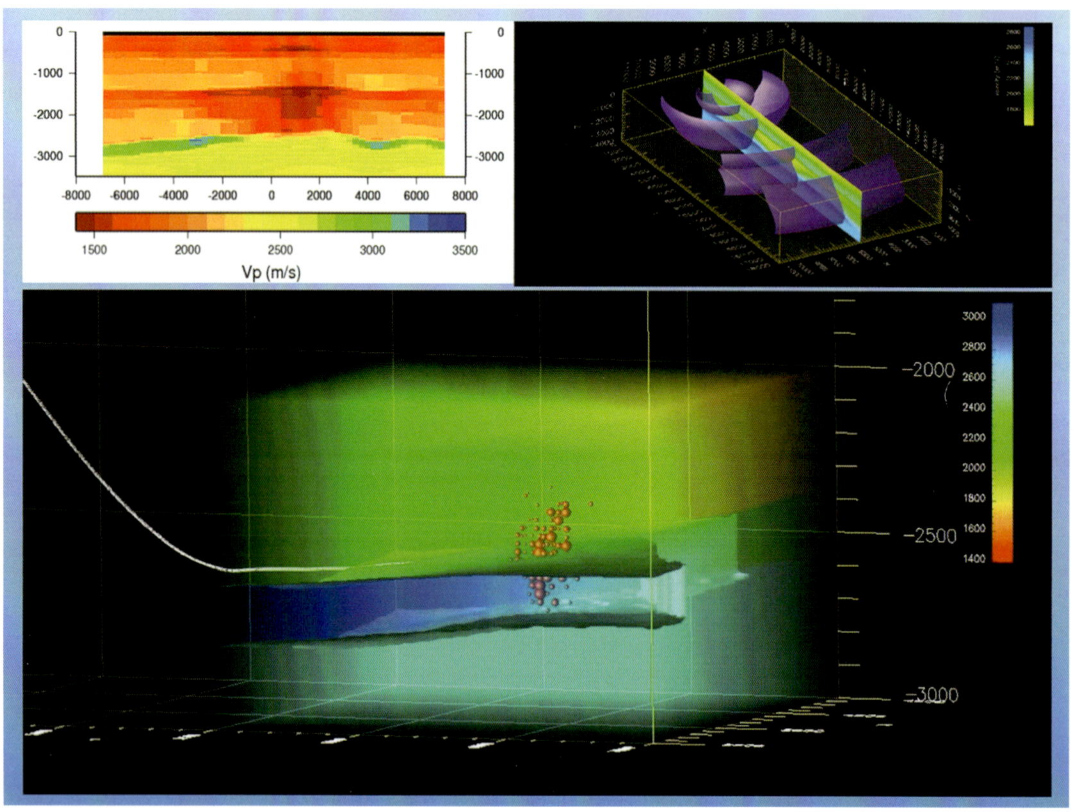

Fig 8.13: Top left shows a cross-section though the P-wave velocity model for the area covered by the Valhall LoFS array. Top right is the corresponding traveltime iso-surfaces (wave fronts) for four time steps for a wave-field propagating through the velocity field. Travel timetables are constructed by initiating wavefronts from each sensor position and interpolating travel times throughout the sub surface. The travel timetables are used to align and stack passive recordings resulting in images that identify the position and time of sources in the sub surface. The bottom figure shows the position of events identified using the imaging procedure. Chambers et al. (2010).

overburden, figure 8.14. Instead, the synchronous increase in event activity with downhole pressure could be interpreted as evidence that elastic wavefronts originating from well activity trigger movement on faults where the *in-situ* stress is very close to that required for failure, Chambers et al. (2010).

8.3.2 Shallow shear waves and shear-wave splitting

We have seen that microseismic events can provide useful insights into the medium they travel through. Before we look at how "noise" can be used to image or characterize the subsurface, it might be useful to look into some intriguing observations from using a controlled source.

The opportunities for passive monitoring are most likely best in the shallow overburden or near the surface. So let us start looking at this level. At Valhall we have an opportunity in that the seabed is subsiding. This will lead to a slight extension at the edge of the subsidence bowl, while the center of the bowl will be under contraction.

Seismic anisotropy is a term used to describe the fact that seismic velocities at a point vary as a function of the orientation the seismic wave's travel. Anisotropy originates from aligned heterogeneities at a scale less than the scale of measurement.

Azimuthal anisotropy can be caused by a combination of several factors. These include variations in horizontal stress, vertical cracks or fractures and lithology. Anisotropy has effects on seismic waves

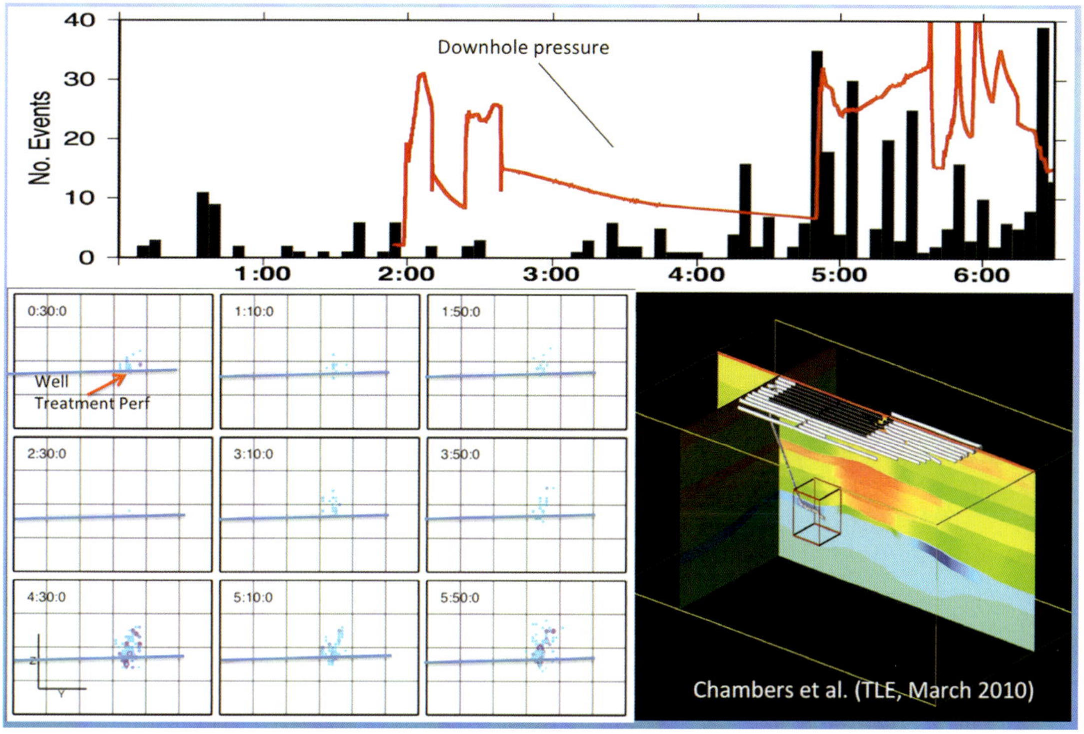

Figure 8.14 Top: downhole pressure reading plotted against number of microseismic events identified in 5 minutes time windows during a hydraulic fracture treatment. Bottom left shows "snapshots" of the event positions in cross-section during the treatment and on lower right the outline of the study area is shown. Events occurred throughout treatment. The main activity pulse occurs after ~4.75 hours and coincident with an increase in the down-hole pressure.

traveling with different azimuths and can result in variations in the traveltime, phase and amplitude, due to azimuthal variations in the velocity and reflection coefficient.

Shear-wave splitting (birefringence) is the phenomenon where the particle motion becomes polarized along the principal axes of the anisotropic layer. In a subsurface exhibiting azimuthal anisotropy, vertically propagating shear-waves split into two perpendicular polarizations, which travel at different velocities, figure 8.15, often referred to as the S1 (fast) wave and the S2 (slow) wave.

When the first 3D OBC survey was acquired across the field, the subsidence at the seabed was only 3 m and the extent of the subsidence bowl was around 3000 m. The maximum extension, estimated by geomechanical modeling, was in the order of 0.01 mm/m. The regional dominant stress orientation is north-north-west (NNW). The differences in horizontal stress magnitude are expected to be small and will change locally when production induced subsidence effects occur, figure 8.16.

When analyzing the 3D OBC data in detail, we exploited some properties of shear-wave splitting to derive two azimuthal anisotropy parameters, the azimuth/direction of the fast S1 polarization and the time delay between the slow S2 wave and the fast S1 wave. Apart from their dependency on the azimuth, these two parameters also depend on the angle of incidence as well as other properties, such as frequency. In this analysis HTI symmetry was assumed and we ignored any vertical deviation of the ray-path, Olofsson et al. (2001).

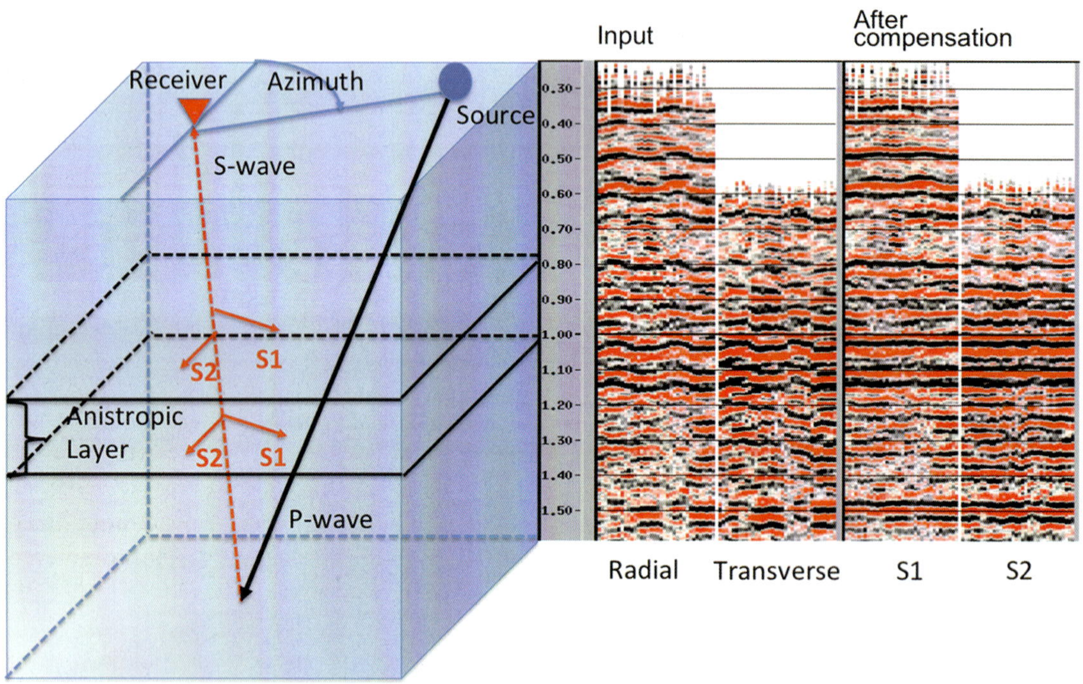

Figure 8.15: Left: shear-wave splitting for a converted wave passing through an azimuthally anisotropic layer. In a subsurface exhibiting azimuthal anisotropy, vertically propagating shear-waves split into two perpendicular polarizations, which travel at different velocities often referred to as the S1 (fast) wave and the S2 (slow) wave. Right: Common Receiver Azimuth gathers, from the radial and transverse horizontal geophones. Each consists of 36 traces, one for every 10 degrees. Before the corrections, we see a wavy appearance reflecting anisotropy. When properly corrected for anisotropy and sorted into the fast and slow modes, the events in the gathers appear flat. From Kommedal et al. (2004).

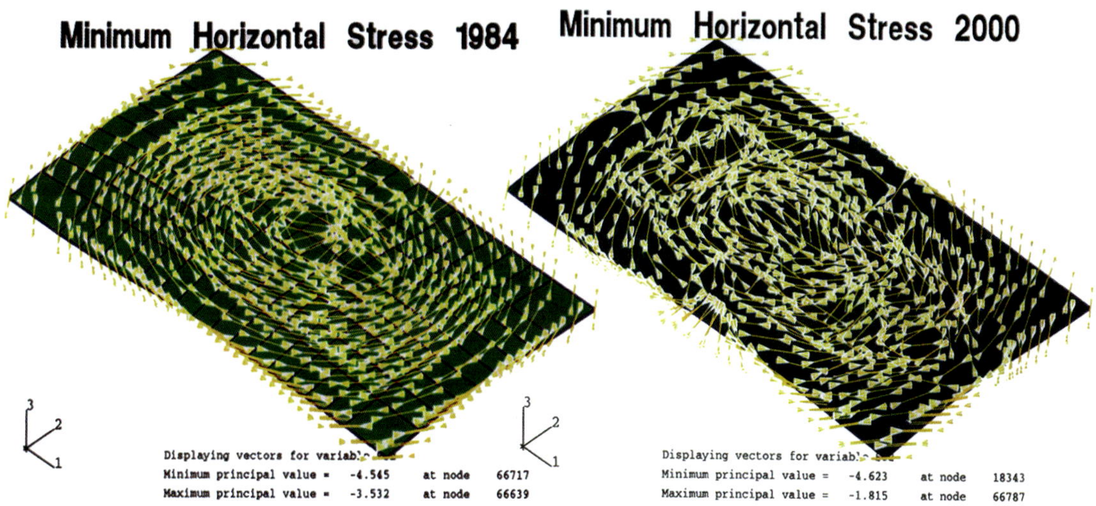

Figure 8.16: The orientation and relative magnitude of the minimum horizontal stress. In 1984, most of the production had come from the most central area (basin) of the field. Modified from Kristiansen (1998).

The first Azimuthal Anisotropy analysis was applied to a sparse OBC survey acquired in 97/98. When the anisotropy result was presented in a map view, figure 8.17, it showed a remarkable similarity to the modeled minimum horizontal stress. In this case, we have mapped the anisotropy parameters into the Valhall LoFS geometry, to allow comparison with a later LoFS survey from 2004. The analysis from the LoFS data set shows the same overall effects. In the most central part, there are less anisotropy effects, consistent with a compressional regime, see figure 8.16, while on the edge of the subsidence bowls we see the effects of the extension. In 2003, we produced more volumes from the flank area of the field, as we also can see in the anisotropy maps. As the thickness is not uniform across the field but has some similarity to basins separated by local thins, we obtain a more complex pattern linked to the development of local stress pillars.

By placing the seismic receivers on the seafloor we are also sensitive to various interface waves. One of these groups is the Scholte waves, a horizontally propagating interface wave that is also seen on the vertical geophones as a low frequency, low velocity event, figure 8.18. These events have properties similar to ground roll as seen on land. The velocity is primarily linked to the S-wave velocity and to a lesser degree to the density. In a typical unconsolidated rock, the depth of penetration is around 100 m but decreases with higher frequency and increases with mode numbers. The observation of these waves was first reported by Kommedal et al. (1997), who used them invert for shallow S-wave properties and S-wave static. In Figure 8.18 we see the results of inverting for shallow velocities.

We can also estimate the azimuthal anisotropy effects from the Scholte waves, Muyzert et al. (2002). The results are comparable to those described above; however, the dominant mode is only capturing the effects down to about 100 m.

Peter Wills investigated Scholte waves further using the Valhall LoFS data. He found that there were considerable time-lapse changes. The link to geomechanics via azimuthal anisotropy was already established but the sensitivity reflected in wave velocity found by this study took us all by surprise. Wills et al. (2008) also showed that there is a strong correlation between velocity changes and stress anisotropy from geomechanical modeling, figure 8.19.

We concluded that the shallow seismic anomalies provide very useful information, which can be used to history match the geomechanical model. The geomechanical model is the basis for a number of

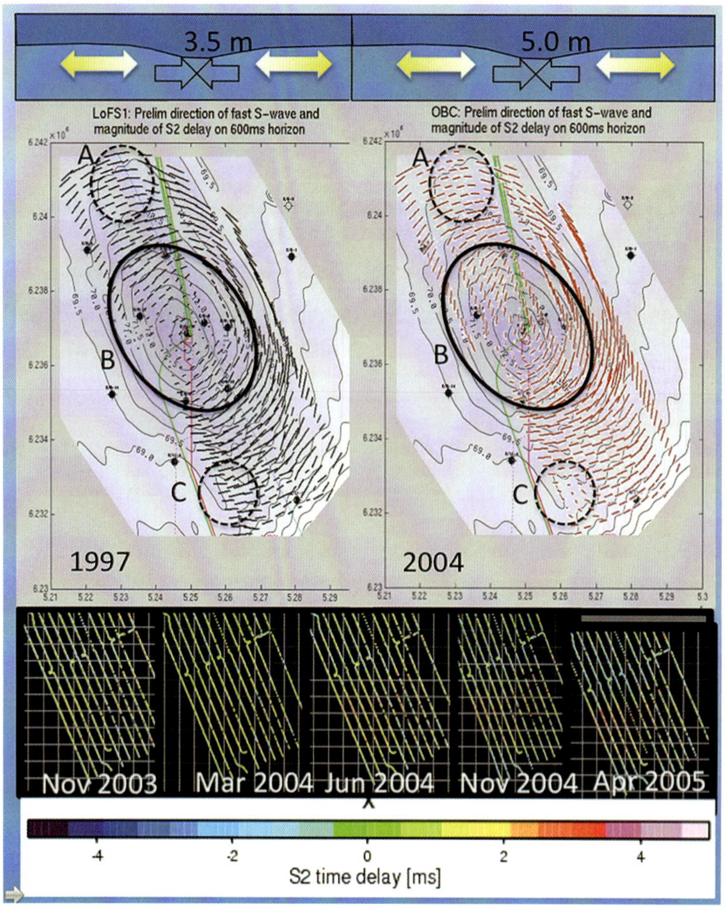

Figure 8.17: Central map shows the azimuthal anisotropy effects in 1997 and 2004. The lines show the S1 axis, while the length indicate the S2 lags. The blue maps and the contours in the back show the depth of the seafloor, notice the circular shape. The caption on the top shows a schematic drawing of the subsidence bowl. Below is series of maps derived from the LoFS surveys from part of the survey. They show how the S2 lags develop over time, Barkved (2005).

important decisions, it is used in well planning, to predict seabed subsidence at the platform location and other infrastructure at the seabed. Apart from also contributing to improved seismic images, the knowledge of the detailed anisotropy has limited implications and is only consulted for possible anomalous behavior. In cases where there could be issues linked to overburden integrity, i.e., shallow re-injection of drill cuttings, these types of surveillance could have merit.

The ultimate value in shear-wave splitting would be if we could isolate the effects deeper in the subsurface, i.e., just above the reservoir or linked to intervals with known drilling issues. We will get back to these challenges but first let us see what passive recording could potentially offer.

8.3.3 Making the most out of noise

When a permanent seismic array is in place, possible passive applications become an intriguing subject. One of the obvious uses is analyzing tele-seismic events; however, these are sporadic and possibly of less value for local monitoring. Local acoustic emission and or micro-seismicity are other applications which are being pursued and have been discussed earlier. However, there are additional sources of seismic energy, not all of them understood and not all very well appreciated

Brad Artman (2005) was one of the first to look into passive recordings from the Valhall LoFS array. He analyzed about 43 hours or 2–3 terabytes of recordings. He focused his study around the hydrophone recordings and found several complete hyperbolas in the data that could correspond to events analogous with active seismic data. However, he also noticed a dominant event arriving from a source about

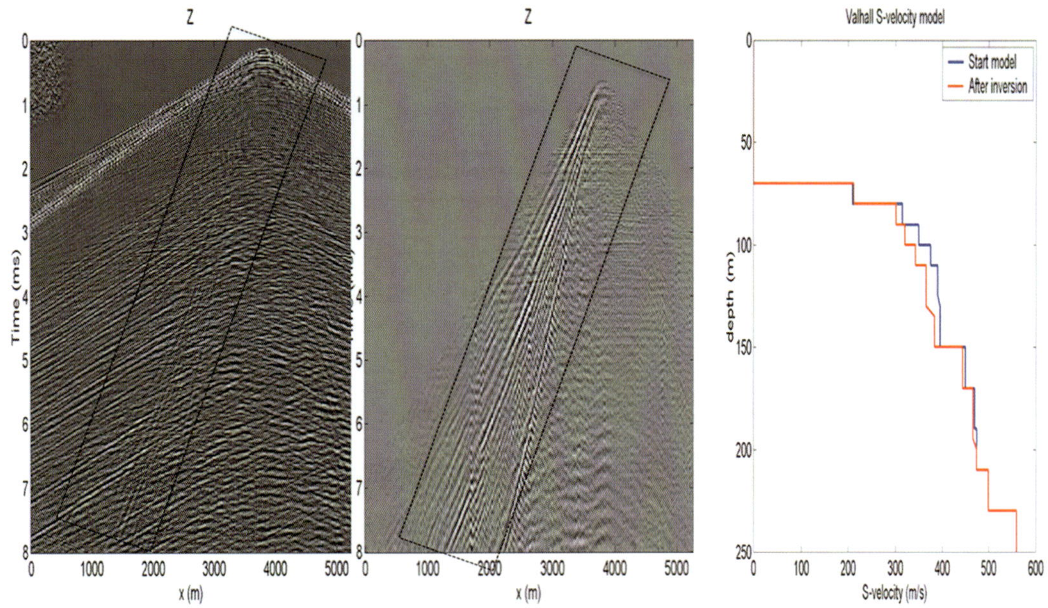

Figure 8.18: Scholte waves as they appear on the vertical geophone sensors to the left, the image in the middle is the same common receiver recording but with a low-frequency pass filter events. The Scholte waves are typically between 3–9 Hz. To the right we see the derived velocity. Note that the Scholte waves are not propagated with the S-wave velocity and a further refinement is needed to link these to the relevant S-wave velocity. From Kommedal et al. (1997),

40 km west of Valhall. This event propagated with a speed of 1450 m/s and had its locus exactly at the Ardmore field in British waters. Brad concluded, based on information from the Department of Trade and Industry, that the possible source of this energy was gas flaring!

Later, other workers had more success with analyzing the Z component. Dellinger and Yu (2007) started to look into the very low frequency band of the passive recordings and were able to find useful energy in the 0–3.5 Hz bandwidth. In that bandwidth he was able to synthesize relatively strong, low frequency Scholte wave virtual sources, figure 8.20.

Landes et al. (2009), extracted Love waves, another type of surface wave that is restricted to horizontal movements orthogonal to the travel direction and inverted for shear-wave velocities and shear-wave statics, figure 8.21. The platforms are apparently the source of the Love waves and they were using a higher bandwidth for this analysis. Their results were comparable to previous observations using active sources.

De Ridder and Dellinger (2011) progressed this work further and have been able to delineate detailed lateral maps that correspond to the morphology revealed by active P-waves sources and the use of full waveform inversion.

The studies to date have been based on relatively short time windows of data. We believe that by expanding these time windows even more robust results may be derived. At this stage, these results might be viewed as academic results, as we are primarily interested in the deeper part of the subsurface. However, the methodology might have merit for monitoring minor changes, i.e., associated with gas seepage or leakage, as suggested by Landroe and Amundsen (2011).

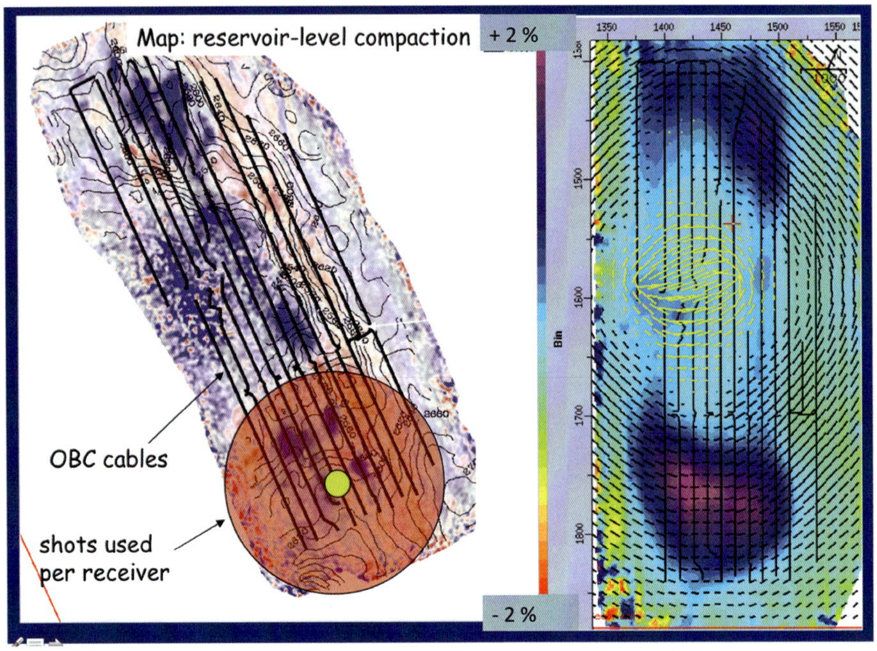

Figure 8.19: Strong correlation between changes in shallow S-wave velocities and geomechanical effects. To the left we see the reservoir compaction taking place between 2003 and 2006. To the right is the change in S-wave velocity posted on sticks representing maximum horizontal strain from the geomechanical simulation. From Wills et al. (2008).

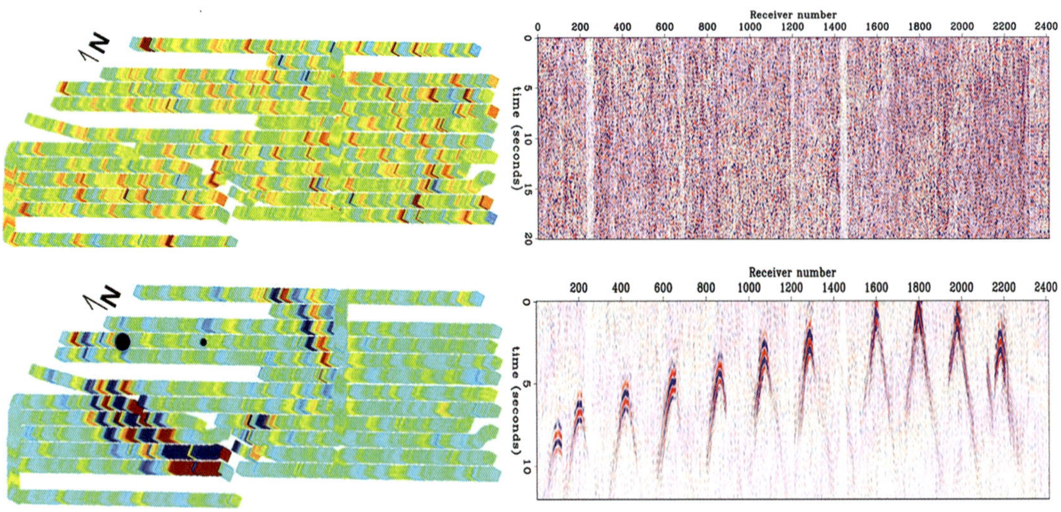

Figure 8.20: "Ocean-wave noise data" (microseism) has been successfully cross-correlated to generate strong virtual point sources. The left panes show a snapshot in time from the Valhall LoFS array, while the right panes show a cross-section from one of the center lines in the array. The moveout velocity is around 400 m/s. This is consistent with what previous analysis of Scholte waves using an active source have generated. From Dellinger and Yu (2007).

8.4 Delineating fault re-activation

Røste et al. (2005) developed a method based on pre-stack data that discriminated between changes in layer thickness associated with production induced reservoir compaction. When applying this technology to a 2D line, he noticed a localized time-lapse anomaly with characteristic time-shift *versus* offset pattern, which crossed the line of common depth points, figure 8.22. Using the angle of the anomaly in the CMP-offset domain, he could back project the anomaly to originate from the tip of a fault at a certain depth. This indicates that it was movement or stress build-up associated with this fault that caused the anomaly. This could be a very useful application for use in planning new wells, as localized stress anomalies tend to impact the applicable mud window.

The angle of this line shown in figure 8.22, can be used to estimate the depth of a buried anomaly creating the anomalous behavior at the reservoir level. The bottom figure shows the relationship between a buried anomaly (marked as a star) and the distortion of straight ray-paths at position x0 for zero offset and position xh for offset 2h. This relationship correlates the distortion zones shown in the top figure, with buried anomalies at depths around 1.3 km and 1.9 km.

8.5 AVOA

We have previously discussed amplitude *versus* offset, and seen how this can be used for separation between fluid and pressure changes in time-lapse seismic data. We used a simplified equation, referred to as the Shuey's equation, Shuey (1985), which suggests that the amplitude *versus* angle of incidence may be expressed by a constant and gradient, see chapter 4.6.2. To calculate the corresponding offset, distance between source and receiver position, at the surface, we need to know the velocity distribution between the interface and the surface. In some cases, this might be a challenging exercise, i.e., below salt, but usually it is straightforward. Given that we have adequate description of the velocity distribution, we may fit the seismic amplitude as a function of incidence angles in a least–squares sense, to find the gradient and the constant needed to describe the elastic properties across an interface. This type of analysis is based on assuming a small contrast in elastic property and isotropy.

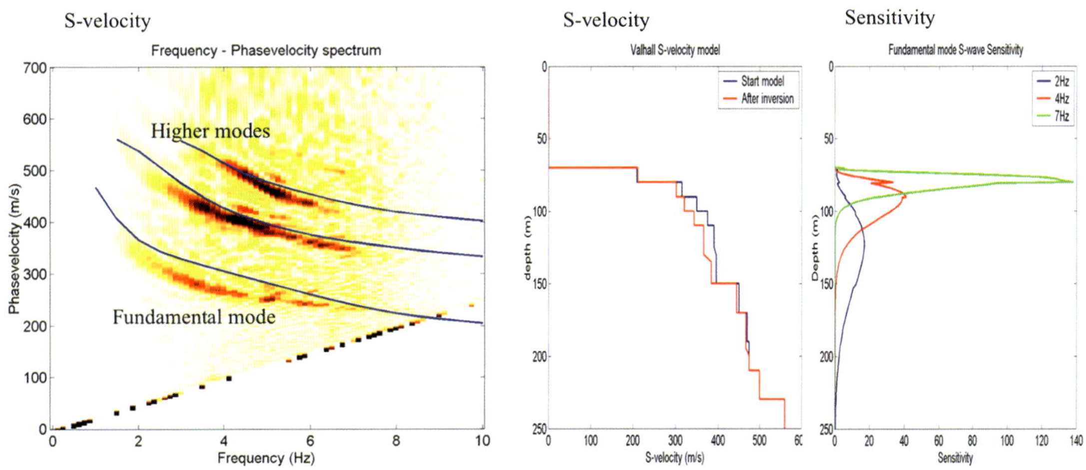

Figure 8.21: a) Fundamental mode and higher mode from active data, Kommedal et al. (2001), b) S-velocity from inversion of passive data (noise) and sensitivity in the inversion, from Landes et al. (2010).

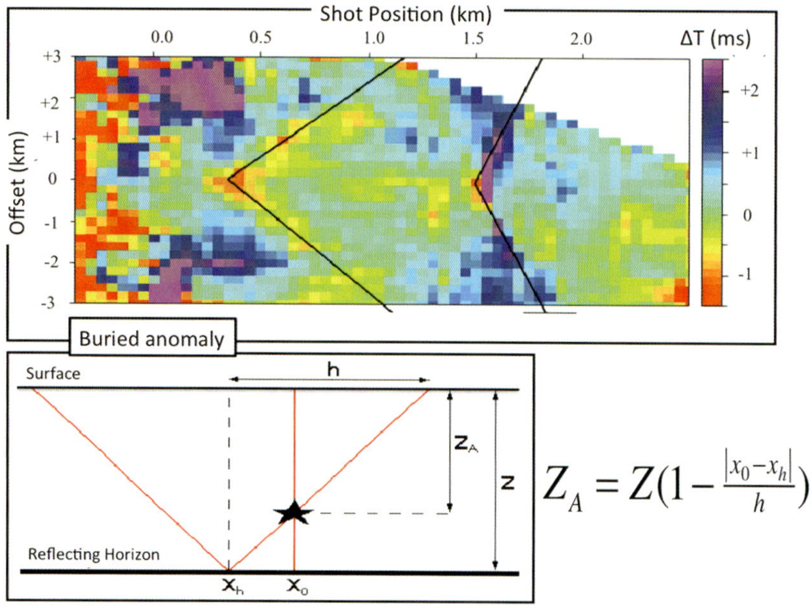

Figure 8.22 : The top figure shows time-shift calculated at top reservoir level displayed as a function of offset and position along the 2D line. An anomaly associated with this level will appear with the same orientation as a CMP, assuming straight ray-paths and horizontal reflectors. From Røste et al. (2005).

In general, rocks are not isotropic. Both the layering effects from deposition and the presence of cracks due to deformations cause apparent anisotropy, Thomsen (1986). We refer to the effects of horizontal layering as polar anisotropy. Thomsen (1988) showed that polar anisotropy has a first-order effect on AVO. This may be an issue we need to keep in mind when applying our standard AVO analysis, especially if we have changes in effective stress. If we have preferred alignment of vertical cracks, we will obtain an apparent azimuthal anisotropy effect.

P-wave amplitude variation as a function of offset and azimuth (AVOA) contains information about the anisotropy effects and we may estimate fracture density and orientations by analyzing this effect. Figure 8.23 shows the geometrical relations for an AVOA data set. Figure 8.24 demonstrates that it is systematic effect that can be observed in real data. A compact form of the P-wave reflectivity for small angles may be written as:

$$R(\theta,\phi) \approx B_0 + (G1 + G2\cos^2\phi)\sin^2\theta \qquad (8.1)$$

The AVOA dependencies can be found via fitting a surface to the amplitude and evaluating this in terms of angle of incidence θ, and azimuth ϕ. B0, G1 and G2 are all constants linked to the actual elastic properties. See the classical Valhall AVOA work by Hall et al (2002) or Xia et al. (2006) for a detailed description of these and for practical advice on implementation.

The technique was applied to Valhall LOFS data sets 1 and 2, Xia et al.'s interpretation of the implied fracture density and orientation at the top hard chalk horizon agrees with a another in-house study. The 4D difference map shows strong correlation with areas of expected change in the effective stress field, due to production and water injection at the fractured reservoir, figure 8.25.

For the Valhall field, where the reflection events are gently dipping, the local 1D assumption is a good approximation. Pre-stack 3D CMP gathers can be used directly for AVOA surface fitting. However, when the structures are complex, traces in the 3D CMP gather no longer come from the same subsurface reflector, invalidating the underlying 1D assumption for AVOA analysis. Pre-stack depth migration is needed in this context to provide a local reflectivity as a function of local angle

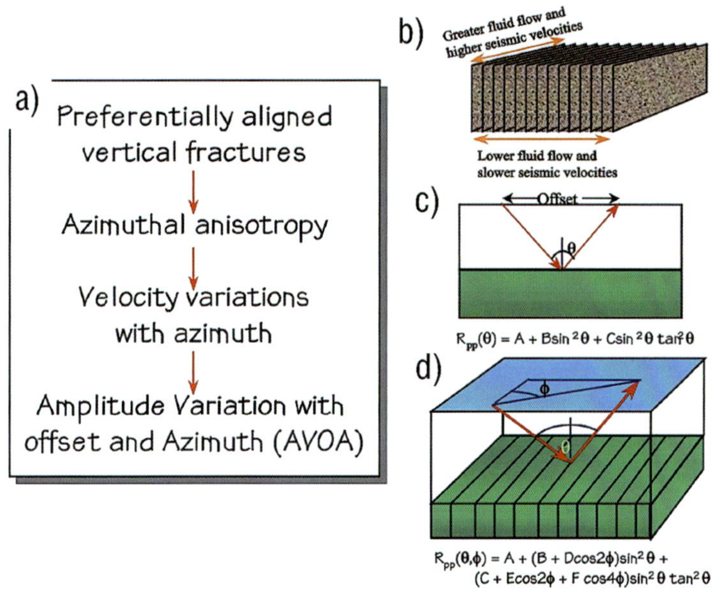

Figure 8.23: (a-b) The influence of aligned fractures on the elasticity and reflectivity of a medium. (c) Standard offset-dependent reflectivity and AVO equation. (d) Extension of the AVO concept and equation to AVOA, form Hall et al. (2002).

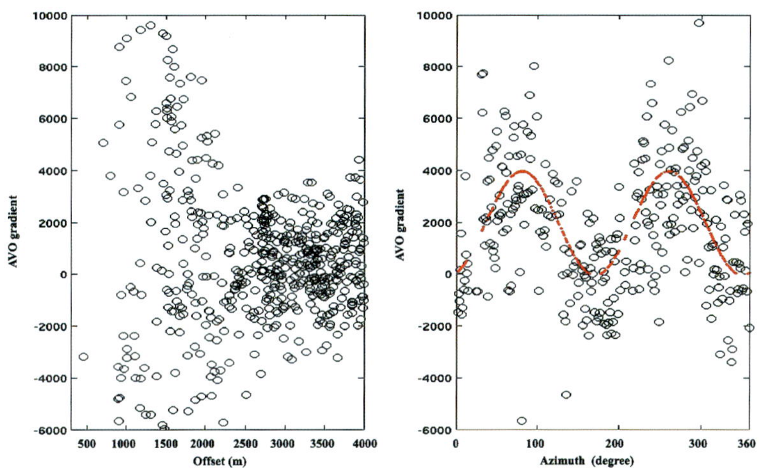

Figure 8.24: a) AVO gradient for the top reservoir horizon. When sorted by offset, as in the case of conventional AVO, the gradient appears to be "cloudy". b) When sorted by azimuth, a distinctive trend shows up. The trend is fitted in the least-squares sense by an elliptical function, an indication of the presence of anisotropy. The best-fit gradient is shown in dots and the data gradient in circles. From Xia et al. (2006).

of incidence and azimuth that can feed into AVOA analysis. Overburden effects can be treated by the migration as well.

The key to interpreting AVOA data is to incorporate the potential stress redistribution that might be associated with depletion or injection. Unless there is a very dominant regional stress trend, local production effects will easily manifest themselves in the AVOA, in the same manner in the shallow S-wave splitting effects, as we can see in figure 8.25. There appears to be a weak correlation to fault/ fractures but a strong correlation to the edge of the produced area and we might conclude that the AVOA observation can be used to define the drained area (or the 4D effects) from a single survey! The analysis was run on subsequent surveys, with consistent results and changes, Xia et al. (2006).

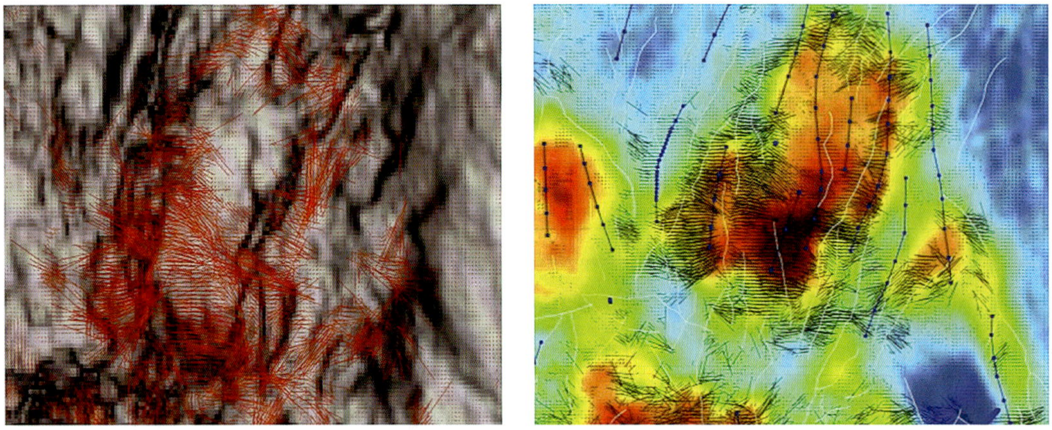

Figure 8.25: Left: Results of AVOA analysis of Valhall LoFS data, red sticks, align to some degree with the implied fracture density and orientation at the top hard chalk horizon, as seen in the black to white coherency map, and agreed well with a previous study. Right: the 4D difference map shows strong correlation with changes in the production induced time-lapse. The blue-green-yellow-red, small-large, map reflects changes in acoustic impedance in the reservoir for the 10 years prior to the LoFS survey. Figure from internal BP report, used with permission. (AVOA analysis courtesy of O.J. Askim).

8.6 Shear-wave time-lapse at the reservoir

The presence of high-quality and significant shear-wave splitting effects in the overburden in the 3D/4C dataset from Valhall has been discussed earlier. Isolating time-lapse effects at reservoir level has turned out to be more challenging. Maultzsch et al. (2006) describe a study in which a significant C-wave time-lapse anomaly was found, which coincided with the reservoir production, figure 8.26, however correlation with the P-wave time-lapse signal was not conclusive.

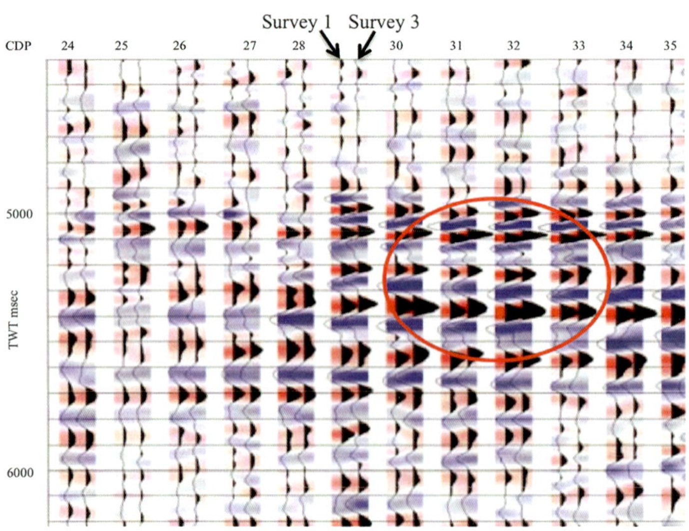

Figure 8.26: Comparison of traces from two surveys in the area where an anomaly was observed. An amplitude increase (30%) and a time-shift is apparent. From Maultzsch et al. (2006), used with permission. From Maultzsch et al. (2006).

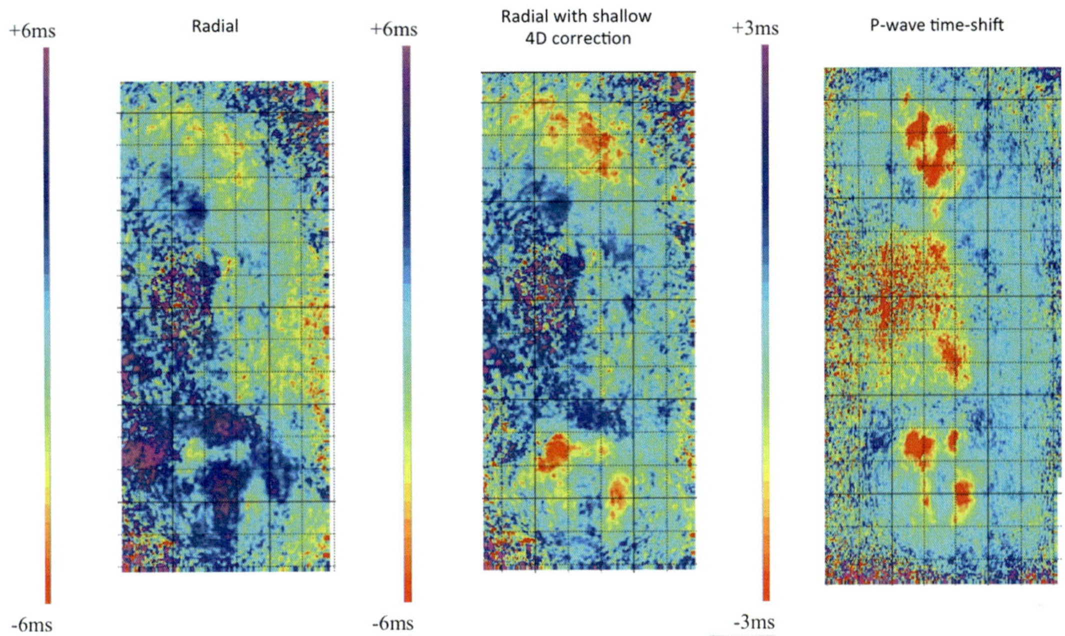

Figure 8.27 Left: top reservoir converted wave 4D time-shifts after depth migration: Center: same, but after compensation for shallow 4D effects. Right: top reservoir PP 4D time-shifts. From Zwartjes et al. (2008).

Zwartjes et al. (2008) studied shallow and deep time-lapse effects on the Valhall LoFS. His findings were consistent with what is shown in figures 8.15 and 8.17. The strong shallow shear-wave time-lapse effects mask the deeper effects, but after applying a receiver consistent 4D correction, they managed to present a top reservoir converted 4D time-shifts map that was consistent with the P-wave signal, figure 8.27. Zwartjes also used an updated velocity model comparable with the one for the analysis in Maultzsch (2006).

8.7 Time-lapse changes in critical reflection angle

Landrø et al. (2004) introduced a method for a more precise estimation of velocity changes in stiff rocks. Long offset time-lapse seismic monitoring has the potential to monitor the velocity change in stiff-rock reservoirs, when the conventional time-lapse seismic methods have difficulty. This method has also been tested at Valhall, Zadeh et. al. (2010). The method used maximum amplitude offset, instead of critical offset, in the interfering area of reflected and head wave to monitor P-wave velocity. The dataset used for the test was selected from an area of fairly uniform reservoir properties and depletion history, although azimuthal effects might not be ignored. The largest effects were found in the where the stacked data showed an increase in velocity in the reservoir, and a decrease in the velocity overburden, which agreed with the modeled response. Figure 8.28a illustrates the basic principle of critical offset monitoring, and figure 8, 28b shows examples of CMP gathers.

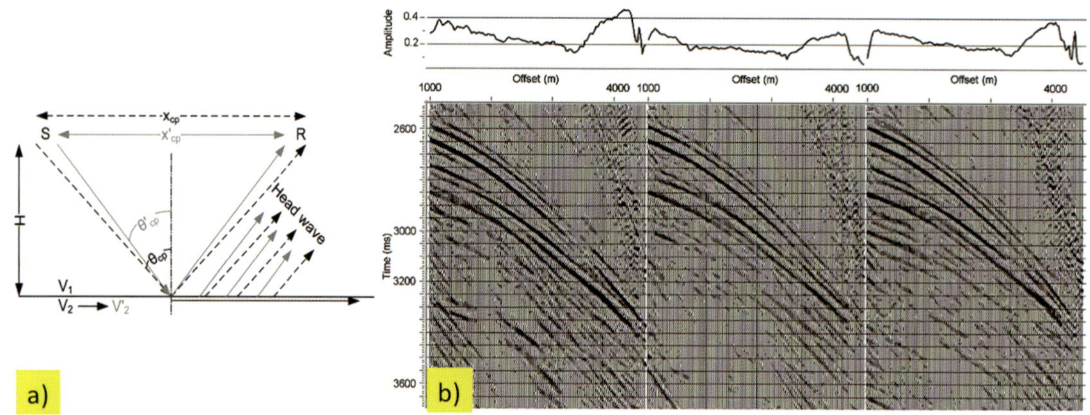

Figure 8.28: a) Basic principal for critical offset monitoring. A ray generated at source (S) hits the interface at depth H, from the source and receiver (R) location, is called critical when the incident angle is equal to velocity ratio at the interface (V1/V2). The corresponding offset is called critical offset (Xcp). If the velocity in the second layer changes (from V2 to V´p2) the critical offset is changed (from xcp x´cp). Head waves start from the critical incident ray point. b) Example of CMP gathers (from left to right: LoFS-1, LoFS-6 and LoFS-8) and amplitude analysis result (top curves). Top reservoir is located around 2650 ms at 1000 m offset. From Zadeh et al. (2010).

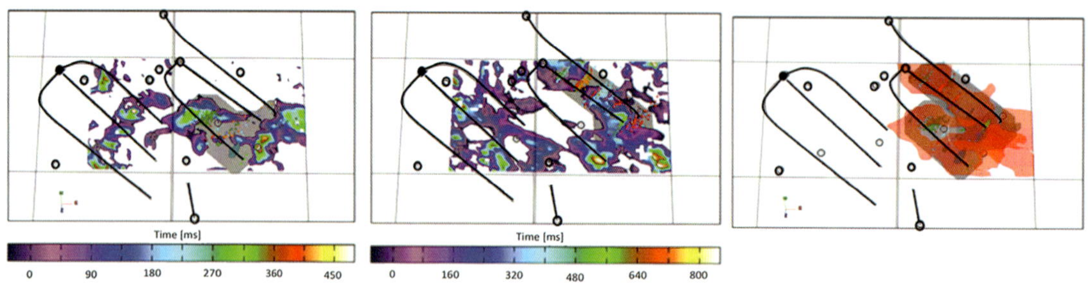

Figure 8.29: Shear-wave splitting representation of increased azimuthal anisotropy between monitors 2 and 1 (left) monitors 3 and 1 (middle), and their relative correlation (right, red) to the microseismic geobodies representing traditional passive monitoring of well stimulations. From Atkinson and Davis (2011).

8.8 Monitoring hydrofracturing using shear-wave splitting and passive monitoring

4D seismic monitoring started onshore. And in many cases the 4D seismic concept could be better suited on land. Source and receiver positioning may be better controlled onshore than using marine towed streamers and sources.

The reservoir characterization project at the Colorado School of Mines has been addressing the use of onshore 4C/4D application for more than a decade. Atkinson and Davis (2011) show that due to significant variations of the local stress regime during fracture stimulation, they could monitor the effects using stress-sensitive seismic attributes from time-lapse surveys. Time-lapse shear-wave splitting was used to define the extent of a hydraulically induced fracture network. The results were consistent with the observations from the passive seismic data recorded during fracturing, figure 8.29.

Shear-wave splitting is a challenging attribute to work with, but in this case the very low permeability did not allow the pressure increase to move into a larger part of the reservoir. Modeling showed that this would be needed to get conventional post stack P-wave amplitude and time-shift analysis to work. An alternative method to test could have been AVOA.

Atkinson and Davis concluded that the shear-wave splitting was a viable method but improved lateral sampling and more rigorous pre-processing is needed to improve the signal noise of the data.

8.9 Conclusion

In this chapter we have seen that passive seismic monitoring technology has the potential to become an important technology in the future. Data recorded in the wells turns out to be of especially high quality, and allows for detailed analysis. A technique that takes into account the anisotropic effects of the production induced changes appears to give meaningful insight and has the potential for also being used as a one of the surveys reflecting the production induced changes. Converted waves continue to be challenging, however, some very encouraging results have been produced lately.

9 The Business Value

"There are known knowns; there are things we know we know. We also know there are known unknowns; that is to say we know there are some things we do not know. But there are also unknown unknowns – the ones we do not know we do not know.
- Former United States Secretary of Defense Donald Rumsfeld.

In this chapter we will discuss some of the elements that should be considered in a business case for a seismic surveillance program. The business drivers will vary from field to field, and the elements required might vary in format from organization to organization. The key is to provide the information that impacts reservoir delivery and economical outcome. Some generic guidelines might be found by reviewing the historical development of existing fields, but the field specific case needs to be based on a workflow that links the need for data to reservoir delivery.

A key part of the business case is to qualify that the technology will work for the field and for the relevant production scenarios. The surveillance objectives might be changed during the field's life, the business case is expected to address if this has any impact on the selection of systems. Seismic surveillance is one element of the field's surveillance and links to other technologies might be crucial for reducing non-uniqueness in the observations.

Determination of the value expected from a 4D project can be broken up into three basic elements. The first is to define activities or value elements that will benefit from the 4D program. The second is to estimate the likely impact of 4D on these activities. The third element is to use the activity list and the impact of 4D on these activities as input into a value of information model or workflow, to determine the value of the 4D. This can be done several ways, from very simple to more sophisticated ones.

9.1 Introduction

The distinction between 'risk' and 'uncertainty' was introduced by the economist Frank Knight (1921). Risk refers to situations where the probability of outcomes can be established based on established theories or experiences with reliable and complete data. Uncertainty refers to situations in which the relevant data might be unavailable, associated with inaccuracies or non-uniqueness. The purpose of a surveillance scheme is to reduce these uncertainties and hence, make better decisions.

During the last decades, seismic surveillance of reservoir performance has come of age. The technology has been proved to add value, supporting reservoir management, increasing recovery and in optimizing new well locations. Figure 9.1 shows one example of a seismic surveillance value chain. In this case, the value is in improved subsurface understanding, a better reservoir model to guide the decisions and finally, value through drilling a better well than we would have done without this information.

The decision to implement a seismic surveillance scheme is a business-driven decision and needs to be backed by relevant support material or a business case. There is no single formula for building a successful business case; the elements might vary from field to field, flavored by the companies' experience and policies. Several of the major and national oil companies have a clear corporate strategy on the use of 4D seismic. The main reservoirs has normally been screened for its feasibility and there are even

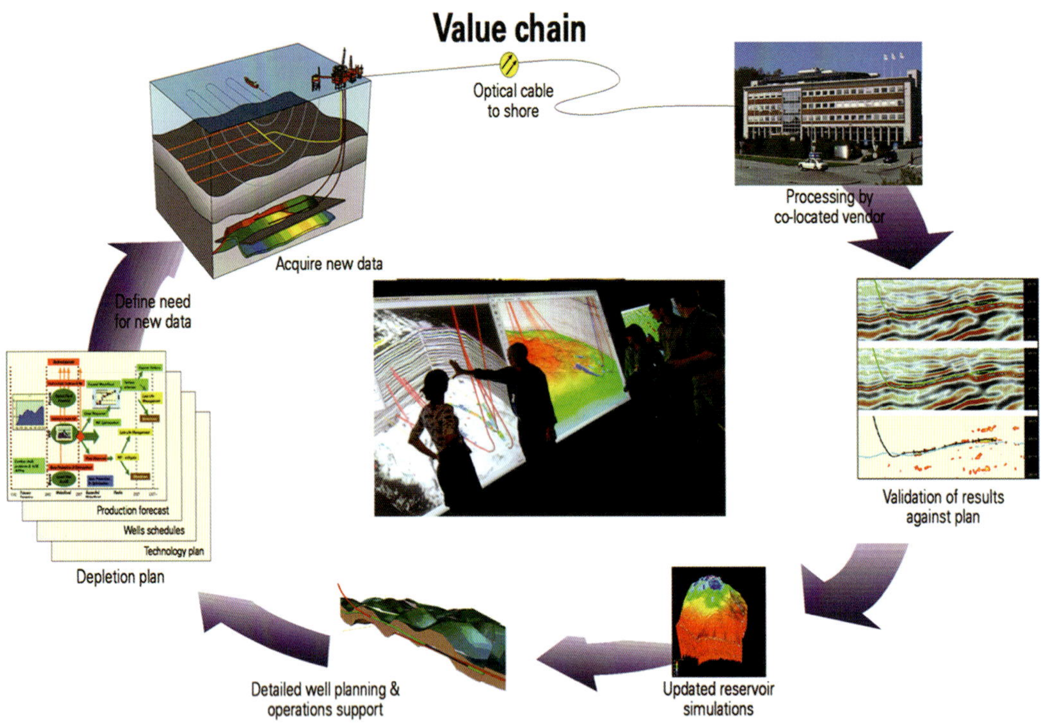

Figure 9.1: The seismic surveillance value chain.

dedicated support teams to support the planning and implementation of a surveillance program. Even so, the detailed understanding of how a surveillance program may help in maximizing the return of a field development program needs to be shared within the multidisciplinary reservoir management and delivery teams. To capture the full value of the technology, a common understanding of the technology and its potential across the multi-disciplinary teams working together to deliver the reservoir value is required. The team's ability to embrace the technology and commitment to make it work is one of the elements. Engaging the team in making a business case is an effective way of doing this.

The risks and opportunities of a field are covered in the field development plan. The key objective of doing surveillance is to assist in developing the field in a cost-effective and safe manner, while optimizing resources. Surveillance is supposed to help in quantifying and mitigating downsides and to help quantifying and capturing upsides. Figure 9.1 is illustrating the "value chain" of seismic surveillance. The need for information is articulated in the surveillance plan, which sets the stage for acquiring new data. The data is analyzed and a 4D response is extracted. The actual observations are matched with expected observations, as expressed by the reservoir model. If there are large mismatches, i.e., significant more compartmentalization is observed than expected, a major revision of the depletion plan might be needed, as more wells will need to be drilled to deliver the volumes of resources.

9.2 Why should we do seismic surveillance?

"The reservoir is our primary asset – surveillance is the maintenance strategy for enabling delivery", Grose (2007). The type of surveillance will necessarily have to address the actual field's risks, uncertainties and opportunities. These elements are typically addressed by a development organization though structured workflows and conversations that help to articulate and quantify each element. A significant part is to include learning from analogue fields, which means the need to categorize it according a relevant framework and we need to develop a common appreciation of the potential drivers.

9.2.1 Growth – maximizing the recovery

Capturing the upsides or growing the field, is typically a key objective for the seismic surveillance scheme. The report on petroleum resources on the Norwegian continental shelf (NPD 2005), lists the growth factors for the Norwegian oil fields, figure 9.2. On average, the fields have almost doubled in size, compared to the estimated reserves when the initial plan for development was submitted. According to this NPD report, the reported estimated resources may often be reduced in the early stages of the field's production, figure 9.3, but as more data is acquired, the estimate of the reserves normally increases, provided projects for improved recovery can be implemented! From 1995 to 2005 the overall increase in gross oil reserves in the NCS has been 1600 million Sm(3). "Half of this volume stems from projects for improved recovery.

How will seismic surveillance play a role in improved recovery? Figure 9.4 shows the predicted remaining reserves for NCS fields following expected cessation of production. The fields will be shut in when it is no longer economic to produce. There might be a number of factors influencing this. One may be that the reservoir model predicts that is no longer economic to drill new wells. 4D seismic may be used to identify baffles and as it is directly linked to the changes in the reservoir, may reduce the uncertainties in the reservoir model. The net impact is that new drilling targets are identified that meet the economic hurdles, leading to an increase in recovery from the field.

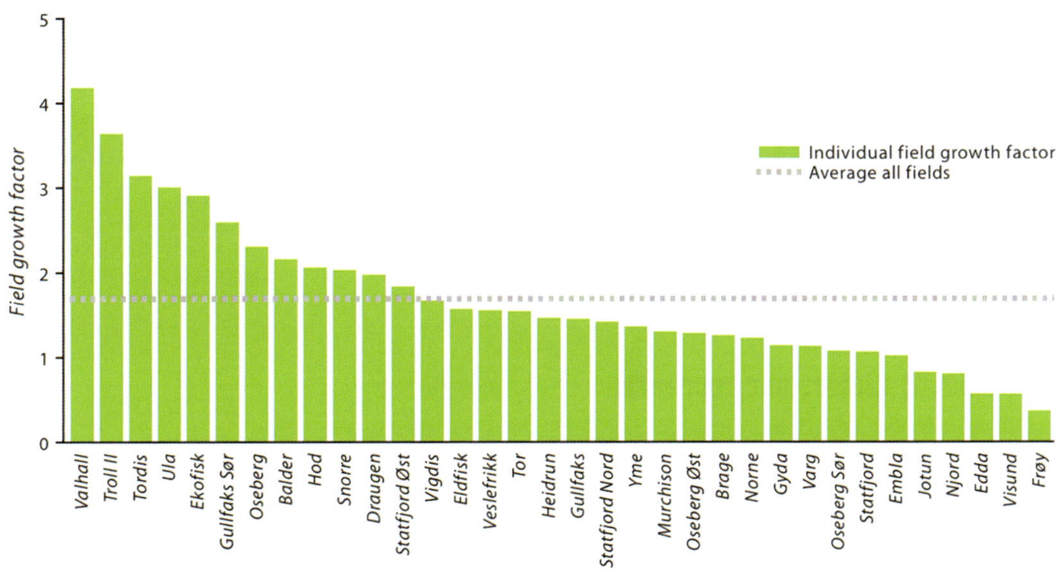

Figure 9.2: Field reserves growth factors. The growth factor is the ratio between the reserves estimates with production data to the volume in the original plan of development; NPD (2005), used by permission.

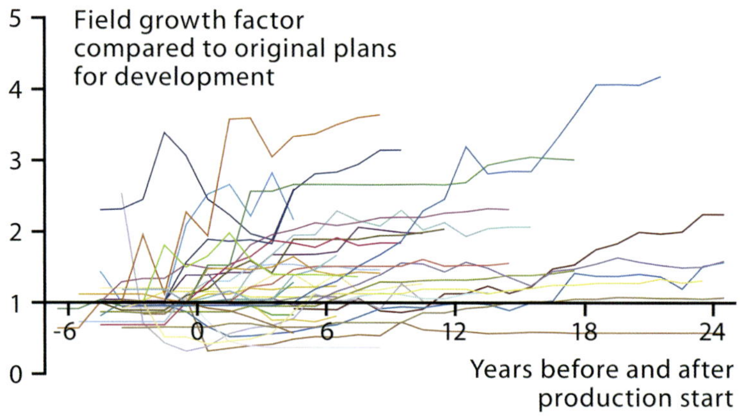

Figure 9.3: Trend in the growth in reserves of oil per field compared with their original plans for development and operation ; NPD (2005), used by permission.

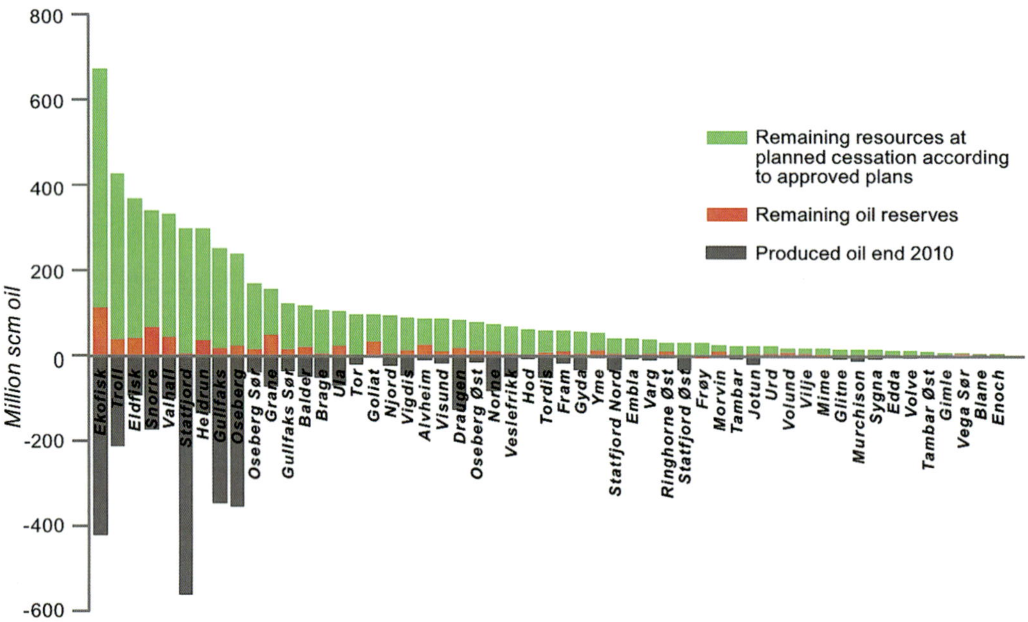

Figure 9.4: The estimated remaining oil after the field has been shut in, versus how much will have been produced; NPD (2005), used by permission.

9.2.2 Complexity factors and recovery

IOR refers to any activity that is used to increase the recovery of oil from a reservoir, economically, Stosur (2003). It includes the use of Enhanced Oil Recovery Methods, which are used to describe methods used to mobilize and produce residual oil that may not be produced by water flood alone. IOR methods include more wells, more advanced wells and well control, more water and gas injection, improved reservoir characterization, better use of data, extended lifetime and use of EOR methods. Miscible gas, Water Alternating Gas (WAG), low salinity flooding, surfactants and CO_2 injections are some relevant EOR technologies.

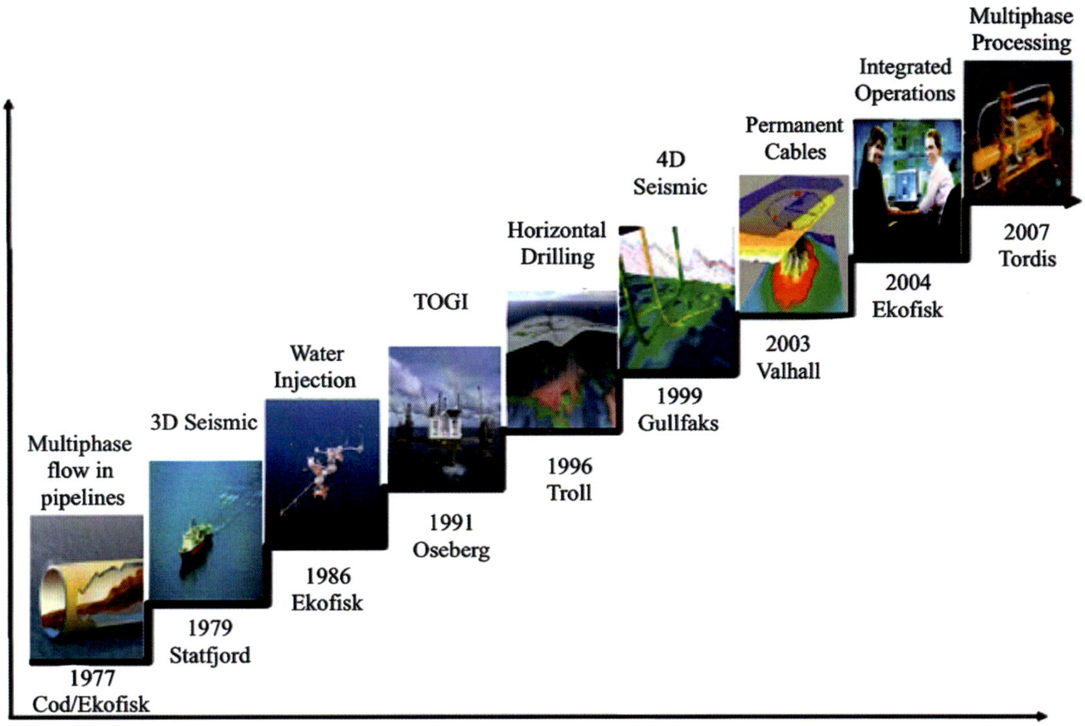

Figure 9.5: Example of important milestones implemented to improve recovery in Norwegian Oil fields; MPA Meld. St. (2010-2011); Source: Expert committee for improved recovery.

Figure 9.5 summarizes some of the technologies recognized by the Norwegian Ministry of Petroleum and Energy that have contributed to more recovery over time. 4D seismic and permanent monitoring is recognized to contribute to improving recovery through its contribution through better reservoir definition.

Is the value of seismic surveillance linked to the complexity of the fields? There is clearly a correlation between field complexity and recovery factors. This is articulated in various ways. Droomgole and Speers (1997) found on UKCS fields that for low complexity fields, the reserve estimate is under-estimated by 20%, medium complex fields are under-estimated by 10%, and complex fields are over-estimated by 15%. Droomgole and Spears based their assessment on static (geological) reservoir model parameters and according to Wickens and Kelly (SPE 2010), their nine parameters for defining a complexity score are subject to a high degree of subjectivity and access to specialists. Wickens and Kelly suggest a statistical approach based on four parameters: oil viscosity, vertical reservoir heterogeneity, STOOIP areal density, and structural compartmentalization. They show that there is a strong correlation between the complexity score and recovery factors.

Table 9.1, shows a complexity index system developed by the NPD to express the complexity of the reservoir and to articulate the challenges in achieving higher recovery factors. The complexity attributes are given a value based on objective limits and subjective assessments.

Smalley et al. (2007) presented BP's reservoir technical limits (RTL (™)), which is used as a conceptual framework with supporting software that is designed to help the asset team in assessing the field's maximum recovery and identifying the specific activities needed to deliver it. The process includes comparing recovery factors for the field in question to carefully selected analogue fields. In the RTL approach, the oil recovery factor is the product of four efficiency factors, figure 9.6. This is the

		Complexity Scores				
Complexity Attribute	**Description**	**Low complexity**				**High complexity**
		1	**2**	**3**	**4**	**5**
Average permeability	Describes the pore volume weighted average permeability in the main flow direction of the defined reservoir. mD	> 10.000	1000-10000	100-1000	10-100	< 10
Permeability contrast	Describes the permeability contrast between geological layers/facies types, and is geological layers/facies types, calculated as $\log_{10}[K_{max}/K_{min}]$	<1	1-2	2-3	3-4	>4
Structural complexity	Describes how fluid flow between wells is affected by fault density, fault throw, fault transmissibility, ...	The fault properties does not restrict fluid flow.				The fault properties restrict fluid flow significantly.
Lateral stratigraphic continuity	Describes the stratigraphic continuity of the flow units in the main flow direction within the defined reservoir	High degree of continuity				Highly discontinuous. Difficult to predict/ describe injector/ producer connecting flow units.
STOOIP density	STOOIP/area (mill. Sm3/km2)	>4.5	2 - 4.5	1 - 2	0.5 - 1	< 0.5
Coning tendency	Describes the coning problems associated with a gas cap or aquifer support. Large complexity only in cases where the oil band is thin.	No coning tendency.		Some coning problems from gas cap or aquifer		Thin oil zone and production severely restricted by gas or water coning problems.

Table 9.1 RCI complexity attributes, their description and complexity scores, NPD 2005. (Courtesy of NPD/ J. Bygdevoll)

pore-scale displacement efficiency (EPs), drainage efficiency (Ed); sweep efficiency (Es) and cut-off efficiency (Ec). Eps is the maximum recovery factor that we could get if we were able to drain the rock to the same extent as we do in the lab. Ed defines how, if a part of the reservoir is in pressure connection with a well, it is counted as connected, figure 9.6. E_s indicates to what degree the oil is being moved to the producers from the part of the reservoir the well is connected to. This is one of the most difficult parameters to quantify, as it is linked to all the parameters influencing the hydraulic and fluid state conditions in the reservoir, but also the one parameter that seismic surveillance can contribute the most to. And finally Ec, which is the production reduction due to field shut in. The outcome from an RTL is an opportunity set for improving the recovery from the field. Surveillance is important for verifying the chosen technology and seismic surveillance is one of the candidates but needs to be considered in each case. See Smalley et al. for more details.

The need for surveillance is linked to the complexity of our system. However, some of the celebrated 4D seismic successes, like Gullfaks and Draugen, are in the mid- to lower-end of the complexity spectrum. (Gullfaks (PUD 43%) had by 2010 produced 58% of its reserves, with an ambition to reach 70%, Draugen

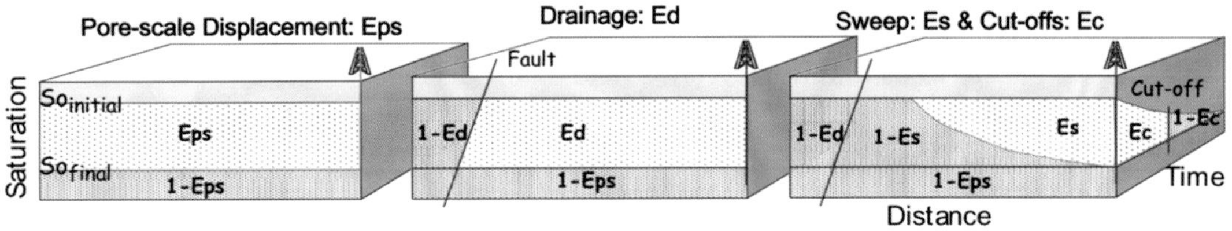

Figure 9.6: Illustration of the efficiency factors, Pore Scale Displacement (Eps), Drainage (Ed), Sweep (Es) and Cut-offs (Ec) that are used to understand the recovery factor and how it can be increased. Smalley et al. (2007), used by permission.

"PUD 23) today 50 - ambition 75". Complexity indexes may therefore not necessarily be a good indicator for whether we need surveillance or not, but may be useful to inform us on the frequency and quality of surveillance. A higher complexity index will in general call for more frequent surveys requiring high-end repeatability, especially if the reservoir is poorly defined with respect to size and communication. Candidates for Permanent Seismic monitoring today have typically been in the ranges of mid- to complex fields. The various elements going into the complexity index assessment will provide a useful framework for assessing seismic surveillance needs.

Identifying the complexity index will not improve the recovery in itself. Complex fields are prone to deliver below predictions and this adds challenges to a business case.

9.3 Value, cost and savings

Given that a sound understanding of the field's opportunities and uncertainties are defined and prioritized, the next step will be to define the surveillance and data gathering efforts that need to be in place to understand and mitigate the risk and to define and capture the elements. The specific value-adding actions or elements that seismic surveillance can contribute will vary from field to field. There is no recipe for success other than a careful and dedicated analysis of each individual field, but analogues and general experience can be used to help facilitate the field-specific work.

9.3.1 Defining the value categories

A starting point for establishing a value statement for a specific field might be to create a list of activities or elements where we think seismic surveillance may add value. This will need to be based on the existing depletion plan, which hopefully already includes a good description of the uncertainties, risks and opportunities. The outcome of this analysis will most likely result in a revision of the depletion plan, as we might realize that there are assumptions in the existing plan that are based on information being present or more detailed work might result in new insights into opportunities.
We might end up with lists of value-added activities like this:

- Identify and realize upside reserves from improved reservoir description.
- Better basis for strategic decisions such as optimization and expansion of the water flood (e.g., drilling targets and water management).
- Reduced drilling cost due to improved prognosis and improved well life statistics.
- Improve recovery by high grading low productivity areas.
- Improve well production rate by better positioning of new wells.
- Reduce risk of drilling wells in low productivity areas.

- Support well optimization.
- Reduce risk of lost time during well interventions.
- Prioritize well intervention work.
- Reduce number of new wells to fulfill effective field drainage plan.
- In water flooding, improve sweep efficiency by close monitoring of water movement.
- Reduce the need for sidetracks in the overburden.
- Etc.

At this stage we should not worry too much about these element overlapping, as the next step will categorize the information and demonstrate how we can make a better decision with information. Examples of categories could be:

Reduce Cost:
- Reduce number of uneconomic wells
- Reduce number of uncompleted wells
- Less time drilling overburden
- Longer economic well life

Improve Value:
- Better reservoir characterization
- Production acceleration
- Better wells (optimal locations, length, well life)
- Improved sweep (less undrained bypassed oil)
- Rate control
- Qualify EOR technology

These categories need to be aligned with the concurrent work and terminology used for the field. We can do that by i.e. elaborating its importance and criticality towards delivery of the depletion plan. Information from the field or analogs in the area might help in quantifying each category. The impact of seismic surveillance needs then to be quantified through a feasibility study as discussed in section 3.7, and if possible complemented by general use of 4D for the specific issues in a similar setting.

Grose (2007) uses base (production) management as an example in the case of a water-flood. The key categories may then be areal and vertical sweeps, compartmentalization, production optimization, well and system integrity management. The next step is to identify the data that can influence our understanding and delivery within these categories.

The more clarity we can gain from what the data can tell us, the more easily we can develop a framework for how often we should acquire the data. Seismic surveillance schemes need to be reconciled and qualified against the well observations and production data is really useful for production optimization.

To make a seismic surveillance scheme we need to interact closely with the wider team and help in assessing whether current development plans are sound or need revision. The whole process is a set of inner and outer loops of closing the plans using plan, do, measure, learn.

The depletion plan will normally include future drilling schedules, forecasts about production, production management activities and planning or progressing the opportunities or upside. Surveillance plans, plans for technical studies and engaging dedicated technology plan to be part of the future depletion strategy and should be included in net development plans. While this provides the framework, the implementation of a specific study is a business decision and we need be able to articulate the value to ensure that the right priorities are given.

9.3.2 Value of information

A value of information exercise is normally expected to be done to support a surveillance program. This is definitely useful for better understanding the impact on specific decisions from different but similar technologies, to understand the size of the prize, how quality of data impact value, etc. It is important to keep the line of sight to the field's overall delivery. An unexpected failure of a well could have complications for the whole development strategy, by driving uncertainties on whether the field will deliver. The statistics from NPD indicate that such a situation often occurs in the early stages after the field has been put on production, before "enough" information is available. It is beyond our scope to discuss VOI in this context. We will rather limit our discussion to some practical ideas, which are useful to understand the concept.

There are numerous technical papers suggesting innovative use of seismic methods, including case studies to demonstrate their technical relevance. Unfortunately, information about the economic impact is rarely presented. From published material, Gray (2011) concludes that the probability of drilling a successful well with 3D is 48–80%, while without 3D the success rate is 14–35%. The statistics for horizontal wells at Valhall, Halvorsen (1997), are that they stayed in the reservoir zone for 95% of the well length using 3D data *versus* 60% in zone without 3D, or where the image quality was too bad. The study was done in 1995 to justify use of OBC to image below the gas cloud, and compares wells drilled in poor seismic areas, but with "good" well control *versus* wells drilled in the flank based on high-quality seismic and less well control, figure 9.7 The incremental value of staying 95% in pay *versus* 60% in pay was estimated to be NOK 40 million, using the reservoir simulator to test the two scenarios and using an oil price of $12/b.

What 4D can offer beyond 3D will be field specific. If we use the Valhall case, the thickness in the high seismic quality area in the flank of the field is often below 10 m. Being able to steer the well into areas off thickness greater than 10 meters has significant value, impacting the flow capacity toward the wells, a well that is significant thinner than 10 meters may not be economic. As the 4D seismic works for these types of thickness in the actual acoustic setting, given there are pressure changes present, 4D is vital to reduce risk. In addition to this, we also have examples where we can see the 4D response and are able to define top and base reservoirs in an area where the 3D data is too noisy, see chapter 5.

Another example will be a producer drilled into the water swept area. As a mitigation this could be re-scheduled into an injector well or sidetracked into a new location, in reality it will be field-specific information that controls the options. Often the cost of recovering from the unfortunate situation will be as high as the cost of drilling the well. In the North Sea this may be on the order of $60m.

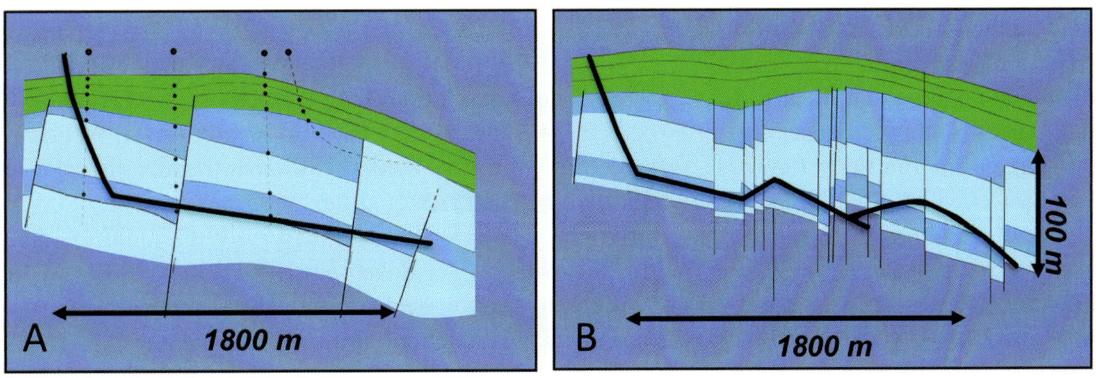

Figure 9.7: Planned wells based on well control A, versus actual B. Drilling the well included 3 sidetracks. Halvorsen (1999).

The final decision to drill a well is based on the reservoir model. For a field with some production history, we may find out that the wells have performed 75% of the time. That is every 4th well would not have been drilled if we had more information, as we would have been better off drilling another well. So if we had better information we could have saved the cost of one out of every four wells we are drilling.

The value of information method (VOI) provides a structured framework for making decisions about data gathering based on formal decision analysis methods. The standard approach is to evaluate and rank the project according to economic criteria like net present value (NPV) or internal rate of return (IRR).

The value of information (VOI) analysis was introduced to the oil and gas industry by Grayson in 1960. The basic concept is known in the subsurface community and elements of the methodology are used regularly to help in the decision-making process. The value of information is defined as, Coppersmith et al. (2006):

$$VOI = AssetValue_{WithInformation} - AssetValue_{WithOutInformation} \tag{9.1}$$

There have not been many papers presenting the Expected Monetary Value (EMV) of a seismic survey before it is acquired, but Waggoner (2000), Houck (2004) and Wills (2004) are relevant examples. Let us consider a 3D case for estimating the value of information. Berteussen (SEG 2006) used Bayesian theory to demonstrate how the Expected Monetary Value (EMV) of seismic data can be calculated. The EMV is the sum of the product of probability P and the Net Present value of a specific outcome, and can be formulated as:

$$EMV = E\{NPV\} = \sum_{i=1}^{N} NPV_i * P(NPV_i) \tag{9.2}$$

For a simple example we may consider that two outcomes of drilling a well at a cost of 15 m$ are:

a) the probability of a dry well is 0.55
or
b) the probability of an oil well worth 100 m$ is 0.45.

The EMV is then 0.45 *100+0.55*(–15) = 36.7 m$ and from this point of view, a reasonable thing to do is to drill the well, Figure 9.8A.

Consider that we now 'buy' additional information about the prospect, perhaps seismic data that may show a ***perfect*** direct hydrocarbon indicator (which implies that in 100% of the cases the presence of a DHI indicates there are hydrocarbons and in 100% of the cases where a lack of DHI indicates no hydrocarbons, there is none). The value of this perfect information will then be 0.55*15=8.25 MM$, basically saving us the risk of a dry well. The new information will not change odds that the prospect contains oil worth 100 MM$.

IF we on the other hand accept that the DHI is not necessarily perfect, but gives us the right answer in 90% of cases and a false positive in 10% of cases. This results in a significantly lower, $3.6 million versus $8.3 million for the perfect case, figure 9.8B. See Berteussen (2006) for further details. Of course, we can easily replace DHI with 4D information, to be used as a qualifier for an unswept area.

Ballin et al. (2005) give a detailed description of the value assessment study they did to evaluate options for future seismic acquisition programs to support the development of a number of gas fields in Trinidad. The value assessment process and the use of the decision tree allowed them to translate a complex problem into manageable smaller elements. The outcome of this study was to focus on localized seismic OBC acquisitions for static descriptions, rather than a permanent installation, which was the other option, as the value of the dynamic information for gas field was found to be too low.

Folstad (2011) presented how they used the value of information analysis to support the installation of the Ekofisk permanent seismic installation in 2011. They used the long-term forecast for the field as the

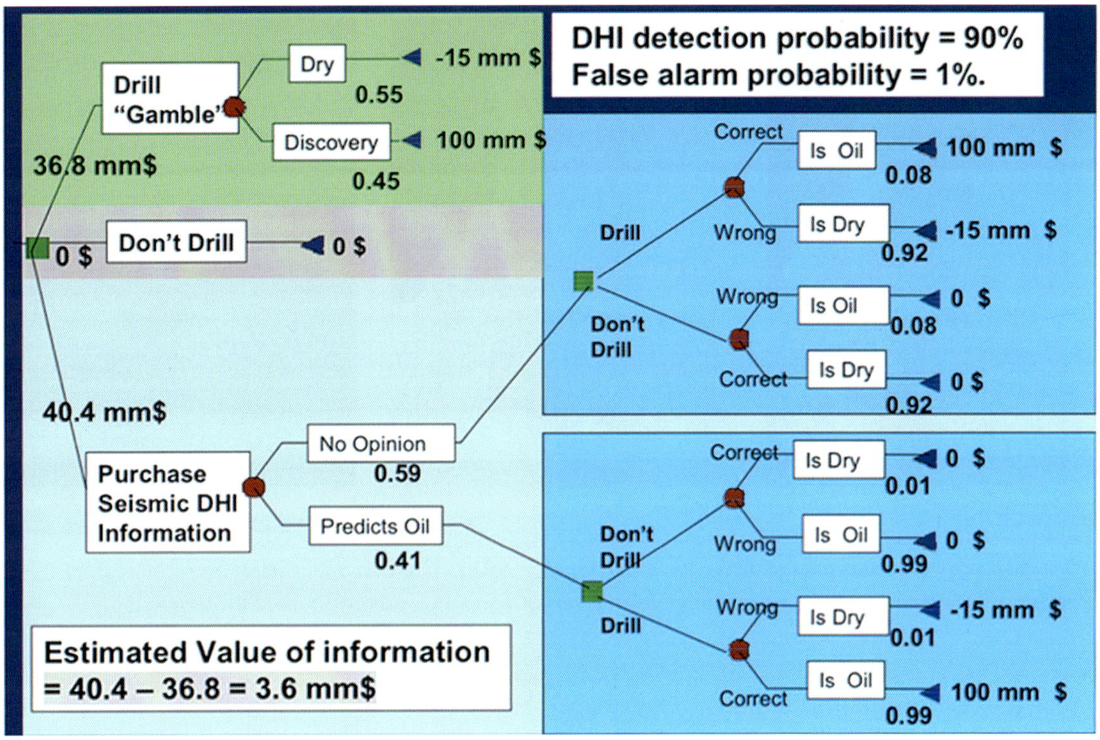

Figure 9.8 A& B: Estimated value of information. Modified from Berteussen (2006).

most likely case without seismic monitoring. The case including seismic monitoring generated additional revenue by the increased production from new wells, as the risk of drilling into partially swept reservoirs was reduced. In addition, they included savings in the modeling case owing to avoiding re-drills of wells that would be placed in a fully swept zone without 4D seismic available. The team also included an assessment of the predictive capabilities of 4D seismic under different noise levels. Noise added to the perfect seismic response from the reservoir simulator to simulate the NRMS level for data acquired by the streamer system and using a permanent recording system. The reference NRMS levels were from streamer surveys across the field, while published Valhall data was used as reference NRMS level for a permanent system, see figure 9.9. Decision trees were constructed and used with the relevant statistics to test alternative 4D monitoring options. Over a period of 15 years, the permanent monitoring system was the most attractive solution and a full field permanent monitoring system was installed in 2010.

Houck (2007) also looked into the value of improving repeatability. He concluded that improving repeatability always increases the value of the seismic data, however, for a given field the actual value of the seismic data also depends on the uncertainty in the reservoir simulator prediction. When the reservoir simulator predictions are reliable, the value of the seismic prediction is less and it is highest when the reservoir simulator prediction is low. One way of finding this out would be to acquire seismic data; the other would be to review similar reservoir complexity indicators, which might give some upfront idea of the risks and uncertainty.

9.3.3 Value examples

An impressive amount of papers have been presented on the technical successes of 3D and 4D seismic applications. The numbers of papers that quantify the economic impact are significantly less, but there are a significant number of papers that at least touch on the economic impact of 4D seismic projects.

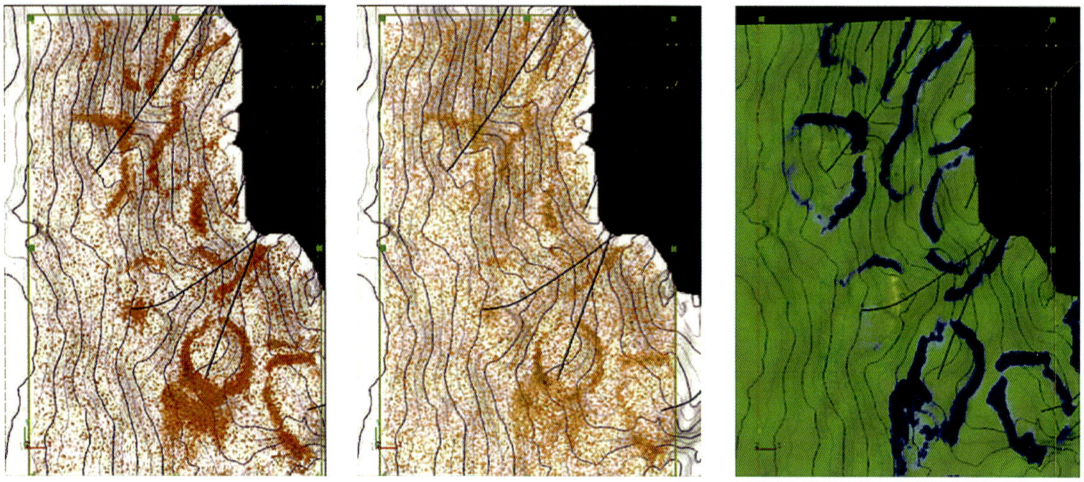

Figure 9.9: a) Modeled 4D seismic response from simulation model in c) with 7% NRMS noise. b) Modeled 4D seismic response from simulation model in c) with 15% NRMS noise. c) Water saturation change from Ekofisk simulation model over a 12-month injection period. From Folstad (2011), used by permission.

Some of this information is "common knowledge" based on "industry experts'" common views but without the objective technical work available to back it being disclosed. One of these informal "statements" says that if 4D seismic works for your field and you have a plan for active drilling or well intervention work, you should expect to increase recovery by 3–5%. While this could be a useful teaser, the likelihood for this to be true for the actual reservoir needs to be worked out from realistic scenarios.

9.4 The role of surveillance in future development plans

The field's development plan is our guide to making the reservoir deliver. The plan should provide a holistic approach to developing and operating the reservoir in a safe and cost-effective manner, using appropriate technical solutions. A fully integrated surveillance plan is an important part of maintaining the development plan. Grose (2007) suggested building the surveillance plan around the key activity sets, base management (production), delivery of new wells and resource progressions. For each of these the risks, the uncertainties and opportunities need to be identified and divided into more recognizable categories following the key activities, which may change through field life.

9.4.1 Seismic surveillance through field life

The main levers for capitalizing on the additional insights from 4D seismic may vary through the various stages of the field's development. High-quality 3D seismic surveys are essential in the early stages for appraisal and development of a field. The trend appears to be towards high-density and wide azimuth data acquisition (OBCs) for more efficient and cost-effective resource progressions, Davies et al. 20111. The main role of 4D in the early life of the field is generally recognized to be to ensure delivery, i.e., Foster (2007). Based on figure 9.2, only four of the fields on the NCS delivered less than predicted in the initial development plan (PUD), consequently, one may argue that the focus should be on value creation.

While drilling the wells, there is a need to process the drill cuttings. In marine environments this is typically shipped to shore, or injected in a dedicated drill cutting re-injection well. The re-injection process is closely monitored using downhole sensors, to establish trends. Sophisticated modeling might be

Field	Surveillance	$Value	Technical Enabler	Capture by	Comments
BP fields North Sea Until 2003	4D Streamer	**24 MStb/d 95 MMstb**	Fluid/Pressure AVO	Infill Wells	Bowman et al. (2005)
Gannet	4D streamer	Trebling oil production	History matching / Improved model	Well Conversion/ 2 infill well	Staples et al. 2005
Gullfaks	4D streamer	1billion	Inversion	15 Infill Wells	Cost of 4D 60MMUSD Sandø et al (2009)
Norne	4D streamer	240MM	Repeatability Integration Pressure Fluid Fronts	Infill Wells	Ousdal et al. (2009)
Oseberg	4D streamer	Extending Field Life	Inversion Vp/Vs Pressure	Monitor contacts	Sandø et al (2009)
Troll West	4D streamer	~100 mUSD(year	Thin Oil	Infill Wells	Sandø et al. (2009)
Valhall	4D Streamer PRM	Aspiration: improved recovery (2-3%) savings (1-2 wells)	Frequent data compaction water management	Infill wells Base Management Reservoir Managment	Barkved (2003 a)

Table 9.2: Example of 4D Seismic Value from the literature

used to predict and estimate future injection potential. Micro-seismic or dedicated 4D monitoring might provide useful complementary data for this.

During the mid-life of the field, enough information has been gathered to refine the reservoir model. This may lead to enough confidence to drill locations that initially were considered less economic. 4D seismic becomes important to help high-grade new drilling targets, and to help avoid swept areas if the field has water in certain sections or a strong aquifer. Cost-effective progression of added reserves becomes the focus of the 4D. In addition, a complementary surveillance scheme may be implemented to support the optimization of well performance.

In the late life of the field, low potential wells might turn out to make the economic limits. That is; there may be available capacity in the transport system, such that cost per unit becomes less, etc. For these wells to be drilled, the uncertainty needs to be low. There might be other issues related to uneven depletion and risks of pressure compartmentalization in the later life of the field. In this case, 4D seismic becomes an important support for infill well drilling and for flood optimization. At that stage it is also important to be aware of potential production induced effects in the overburden. Due to stress redistribution it may be more difficult to drill the wells, Kristiansen (2007).

In addition, there are safety or environmental reasons for the conduction of a seismic surveillance program, there might be specific legislation driving the need, i.e., for monitoring drill cutting re-injection programs.

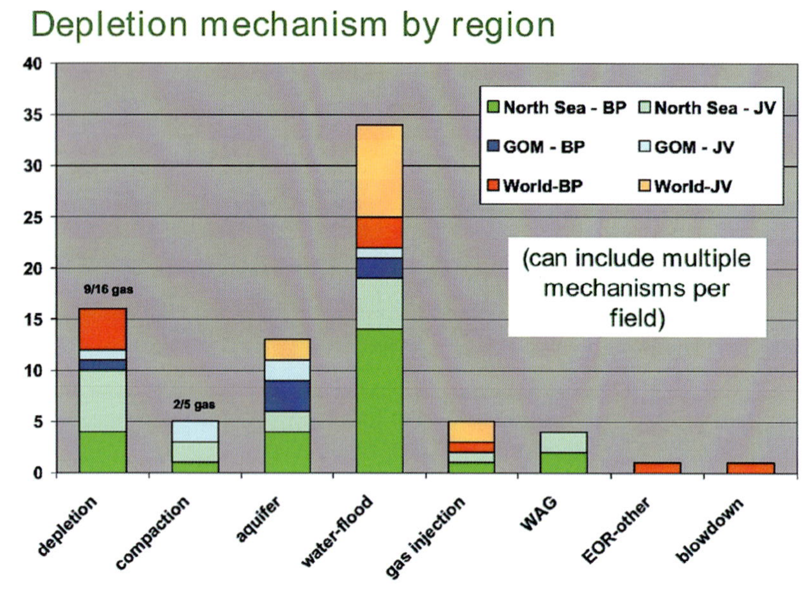

Figure 9.10: Breakdown of fields surveyed by depletion mechanism – by region Foster (2007).

According to Foster (2007), BP's 4-D seismic monitoring application is clearly dominated by water-flood for oil reservoirs, figure 9.10, and depletion for gas reservoirs. For oil reservoirs, monitoring of water movement in aquifer drives and water-flooding accounts for about 70%, 60% of monitoring is for secondary water-flood and gas injection, about 33% of monitoring for primary depletion, and 8% of monitoring is for tertiary or EOR recovery in WAG, steam flood and late-life schemes. As there might be more than one objective for a survey, these numbers do not add to 100 %, the intention is to indicate the relative importance. The main point is that most 4D seismic until 2005 was conducted to support water flood and driven by the possibility to understand how the water flood is performing, to understand possible impacts by faults and preferential pathways. The objective of focusing on primary depletion and aquifers was driven by the need to de-risk or high-grade infill locations.

Foster (2007) also compiled statistics on what stage in field life the reservoir was at when the 4D seismic program was implemented, figure 9.11. The majority of 4D seismic surveys were acquired in the late plateau / early decline of the field. At this stage, there might be additional capacity in the system to handle new oil and extending the time on the plateau is always of interest. We might argue that the best wells have been picked and more details were needed to find the remaining economic wells. The statistics are from BP but published cases from other operators indicate that this focus is probably representative for the industry. If we look more closely at the data, they reflect a more diverse story. The timing is probably also driven by the technical developments, most North Sea fields happen to be at this stage when the technology started to deliver high-quality data. Newer developments, mostly outside the North Sea, benefited from the North Sea experience and have apparently embarked on using 4D seismic at an earlier stage of development.

According to NPD's data, NPD report 2005, the "size of the reservoir" appears to fluctuate most during the first 6–12 years of field development. Although there is no study, it is reasonable to suggest that active use of 4D seismic would have accelerated learning, helped in refining the depletion strategy and most likely, managed to recover more of the field by getting the "tactics" right.

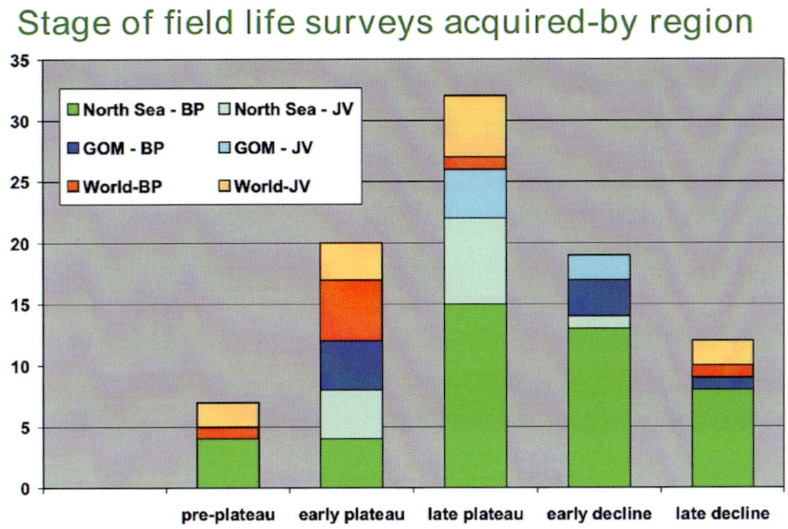

Figure 9.11: Distribution of 4D surveys acquired by stage of field-life by region Foster (2007).

9.5 Selection of seismic surveillance system

When planning a 4D seismic project, all relevant options need to be considered. At the point of proposing a project, provided we have done our homework, we should now have all the selection criteria in place. We should know what we will use the data for, what sensitivity is needed, when we need the surveys, what frequency of surveying is needed, turnaround of data, need and opportunity for passive use, logistical challenges and HSE factors.

Tables 9.3, 9.4 and 9.5 below are an illustration of how to reconcile the information from some of these elements. The tables show a summary of the ranking of a subset of the technical, logistic and HSE elements for alternative 4D seismic systems for one specific field. The red color indicates that this can not be done or delivered; yellow suggests that it is possible but challenging and green indicates that this is the best or good enough solution. Some of these issues require significant efforts to be properly assessed for the specific fields. Other fields might need additional elements.

9.5.1 Technical elements

Technical Comparison	Surface Tow	OBC	Node	Permanent
Converted / Shear Wave Recording	No	Yes	Yes	Yes
Repeat Positioning - Source	Reasonable, no close approach to structures	Good	Good	Good
Repeat Positioning - Receivers	Not usually attempted Q-fins can reduce error	+/- 10 m. Could do better with slow progress	+/- 5m	Exact
Coupling Repeatability	Good.Coupling to water!	Moderate. Poor where seabed is rough?	Good.Planted by ROV	Good. No change unless soil compaction.
Quality of Sensor Coupling	Good. Coupling to water!	Depends on seabed	Good	Good
Vector Fidelity	N/A	Dependent on cable design	Should be best	Good
Noise Levels	~5 microbars tow noise pluss swell noise	~5 microbars water flow and ambient noise	~1.5 microbars ambient noise	~1.5 microbars ambient noise
Passive Monitoring Capability	No	Receiver array not large enough (?)	Possible, but very inefficient	Yes
Source Sampling Density / Uniformity	25 metres in-line variable cross-line	75 metres in-line 75 m cross-line	50 x 50m carpet shooting	50 x 50m carpet shooting
Receiver Sampling Density / Uniformity	12.5 metres in-line variable cross-line	25 metres in-line 360m cross-line	300m x 300m regular grid	50m in-line 300m cross-line
Illumination range	Narrow Azimuth	Medium azimuth	Full 360 degrees	Full 360 degrees
Typical Fold (p-wave) Offset 0-4000m	80	300	135	800
R&D	N/A	Repeat OBC	Repeat Node	Shear-wave sources / other sources

Table 9.3. Showing most of the Technical elements considered.

9.5.2 Logistic elements

Logistical Comparison	Surface Tow	OBC	Node	Permanent
Survey Frequency	Seasonal	Seasonal	Seasonal	Continuous
Fishing Interaction	Workable	Problematic	High Risk . Not fishing friendly unless buried	Buried
Ensuring Availability	Retain one/two vessels? Long term contract?	Changing market	Long term contract for nodes and source boat	Long term contract for source boat.
Time Pressure	Vessel can't stop.	Daily costs Daily costs for vessels	Battery limitations	None
Weather Dependence	Normal towed streamer limitations	Limited by source and ability to move cable	Weather affects source	Weather affects source
Acquisition Rate (small field survey)	11 sq.kms/day 2 source/8 streamers	1.5 sq.kms/day with 2 OBC cables	2 sq.kms/day	9 sq.kms/day
Stations recorded per shot	2560	480	200	2000
Data Monitoring	Real Time	Real Time	QC samples	Real Time
Data QC	Shot quality check	Shot quality check	Hardly any	Real-time comparison with previous survey
Data Delivery to processing centre	Onboard processing or 2 weeks after last shot	After Nav computation 2 weeks after last shot	Nodes recovered, time checks, transcription	Direct stream to onshore processing
Repair / maintenance consequences	10% of survey time	Frequent cable repair often >10% survey	Bad nodes replaced no delay unless reshoot	>10% cable failure
Interaction with other Activities / Facilities	Close approaches. Undershooting	Cable lay and recovery near seabed structures	Nodes lay and recovery near seabed structures	One-off installation. Flexible shooting plan.

Table 9.4: Showing some of the logistic issues considered.

9.5.3 HSE issues

HSE Comparison								
	Surface Tow		OBC		Node		Permanent	
Man-hours exposure per 200 sq.km survey	36000		235000		72000		10000	
Major Safety Risk during surveys	Towing near platforms streamer and gun work		Cable lay/ recovery gun work		Node lay/ recovery gun work		Gun work	
Environmental Impact	Short term no seabed impact		Temporary cable lay on seabed		ROV installation of geophones on seabed		Cables trenched in continuous shooting	
Fisheries issues	short term obstruction and airgun 'scare' effect		longer obstruction and airgun 'scare' effect		short term obstruction and airgun 'scare' effect		airgun 'scare' effect should fade with time.	
Fuel /CO2 emissions per survey	High - one vessel working hard		High - two vessels for a long time		Medium - two vessels		Low - one small vessel	

Table 9.5. Showing some of the HSE issues considered

9.6 Conclusion

In this chapter we have looked at the story reflected in some of the statistical information available from the NCS and other sources. The majority of fields have had a significant growth in oil reserves since their start up. In most cases the growth happens after an initial reduction in reserves estimated immediately after the start up, probably reflecting uncertainties in reservoir understanding at this stage. The growth is partly due to obtaining more insights from the field during production. Seismic surveillance, 3D and 4D seismic are recognized as important contributors to the improved recovery from reservoirs at the NCS.

There is a strong correlation between recovery factors and complexity indexes for a reservoir. Seismic surveillance has had a role in developing fields with a medium to high complexity index. The outcome from reservoir uncertainty studies brings clarity to which reservoir properties impact the end results the most. Tools aimed at articulating and defining field performance attributes, like the complexity scores and processes aimed at defining the opportunities, may provide important benchmarks for sense checking. Reservoir Technical limits are important input for assessing and defining the need for surveillance.

General observations and analog fields are important in setting the stage for the issues to be considered when formalizing the field's surveillance strategy. The surveillance strategies are an integral part of the field's development plan and the outcome from the surveillance is a key source of information for improving development.

Not all reservoirs are candidates for seismic surveillance, a dedicated feasibility study possibly combined with pilots or experiences from analog fields are critical in setting the expectations. It is important to establish detectability levels and to choose systems/solutions accordingly.

Determination of the value expected from a 4D project can be broken up into three basic elements; define the value elements, estimate the likely impact of 4D on each element and finally, determine the value. An example was that 4D would reduce drilling costs by helping to properly land wells, thus reducing the number of sidetracks. Experience was used to estimate the reduction in the likelihood of a sidetrack for an average well. In this example the activity is drilling sidetracks and the impact of 4D on this activity is the reduction in the likelihood of a sidetrack. The third element to determine value is to then use the activity list and the impact of 4D on these activities as input into a value of information model or workflow, to determine the value of the 4D. This can be done in several ways, from very simple to more sophisticated.

The value estimation process may be as valuable in aligning understanding across the disciplines as in pointing out the criticality of the surveillance in estimating the resulting monetary value. The cost elements are typically easier to estimate. A rigorous analysis of the feasibility of alternative methods is also needed to guide the selection of which technology to use. Technical as well as logistic and HSE impact are important to understand.

The purpose of implementing a seismic surveillance scheme for a field is to help develop the field in a safe and cost-effective manner. The selection of seismic surveillance will need to be closely linked to the actual needs of the field.

10 Summary and Reflections

"The whole problem with the world is that fools and fanatics are always so certain of themselves, but wiser people so full of doubts."
- Bertrand Russell (1872 - 1970)

During the last decade, seismic surveillance of reservoir performance has come of age. The technology has been proved to add value, in supporting reservoir management, increasing recovery and optimizing new well locations.

The value creation from 4D seismic at the Norwegian Continental Shelf (NCS) alone has been estimated at more than NOK 22 billion over the last 10 years. These figures reflect the industry's experience and have set expectations of adding value using seismic surveillance technologies. Today, most major oil companies will include a rigorous assessment of seismic surveillance technology options to be used to improve value of their reservoir and to protect investments in most new developments.

Offshore, marine towed streamer survey has been by far the most utilized technique, but new technology and the business environment may favor alternative options. Recently, we have seen highly repeatable images from ocean bottom system (OBS), using nodes or cables. At Valhall the permanent seismic array has been in operation since 2003. An average of 2 surveys a year have been acquired, and the project has been a technical and commercial success.

In this chapter we summarize the Valhall experience as discussed in the examples and case studies presented in the previous chapters.

We will also use a few selected topics from the most recent publications, to reflect on where we might be heading with seismic surveillance in the future.

10.1 Introduction

The value creation from 4D seismic at the Norwegian Continental Shelf (NCS) is impressive. Estimated investments of 4-5 billion NOK have generated more than NOK 22 billion over the last 10 years, Meld. St. 28 (2011).

During the last 15 years, oil companies operating in the North Sea have played a key role in bringing 4D seismic applications based on marine tow streamer of age. To day almost all of the mid-size and large oil fields in the North Sea have 4D seismic coverage. Seismic surveillance plans for protecting the investment and to assist in capturing upsides have become a compulsory element in any new or major re-development program.

Marine tow streamer acquisition dominates by far the market as the most used options for 4D seismic acquisition, and is the primary source of the value creation we have seen in the North Sea. In the cases when it is preferred to put the sensors at the sea bottom we have seen that highly repeatable 4D images can be made based on using Nodes and Seabed cables, chapter 4.

The seismic surveillance toolbox contains a wide range of technologies and methods, which should be considered for use on reservoir where the seismic surveillance technology applies. Not all reservoirs are candidates for extensive seismic surveillance. The production induced changes needs to be of a nature that is detectable by the seismic signal, see chapter 3 and 4. In addition we need a business

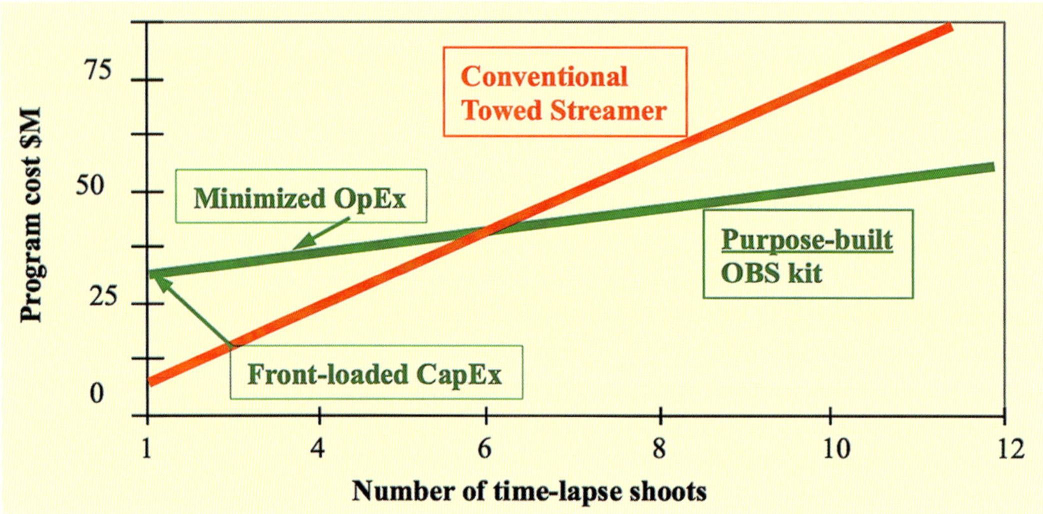

Figure 10.1: Notional Program 4D scenarios based on 2002 numbers.

rationale underpinning the surveillance program. As we discussed in chapter 9, this may be linked to the complexity of the field but also the various stages of the field life have different needs.

The price of oil, cost of drilling and other operational costs are influencing the value elements of a 4D seismic business case, and are outside our control. The objective of the team assessing possible seismic surveillance technologies for a field is to come up with the most cost effective and robust methods. The Marine tow streamer 4D seismic is a well-established technology, with a high number of reference cases from around the globe. Examples of use of retrievable OBC seismic system in time-lapse projects are increasing. Permanent seismic systems are expensive to install but the cost of a survey is low, see figure 10.1. The method has clearly an edge when frequent surveys are needed, and there is also a potential for using passive methods. Capturing the value from seismic surveillance goes beyond the methodology of acquiring and processing the data, and involves the teams who are integrating the findings into business decisions. We will not discuss this explicitly, but just make the point that the frequency of new data may impact the business processes, and impact the seismic technology awareness and skillsets in the organization.

Use of the 4D seismic methods started with an onshore permanent seismic array. The first marine permanently installed larger hydrophone seabed array was the Foinaven installation, while the first, field wide multi-component system was probably the Valhall LoFS array. Although an even larger array has been installed across the Ekofisk field recently, and future installations are being planned, the use of permanent systems in the industry is limited.

The Valhall LoFS project is a technical and commercial success. Future extension of the system is being planned. The high number of survey acquired across the field has open for new an innovative ways of linking the production /injection volumes to the 4D seismic responses, which are routinely being used. New ideas based on balancing the sampling in calendar time and space may open up new options. The use of passive recordings appears to pay of, as a tool for passive imaging of shallow subsidence effects and help in the history matching of geomechanical modeling.

In the following we will summarize our experience from the Valhall Life of Field Seismic projects. The intention is to share insights and experience that could be inspiring or of help for anyone involved in deciding on using seismic surveillance techniques.

10.2 The Valhall LoFS project - summary

In January 2000 BP rolled out the Life of Field Seismic, LOFS, concept to seismic contractors and manufacturing companies. The intention was to qualify a seismic recoding system for permanent installation at the seabed. Competitive bidding was used to identify a viable commercial option. A solution that included equipment from Geospace Engineering Resources International (GERI) was selected. At the time when the equipment was readily qualified for use in the field, a BP global screening exercise identified the Valhall field in Norway as the candidate where the system would add most value, Barkved et al. (2003a) The project, which included the recording system, installation and the cost on acquiring six surveys was approved by the Valhall partnership, at that time BP, Hess, Shell and Total in late 2002.

From 2003 to 2011, 14 seismic surveys have been acquired across the field. The installation of the equipment and the acquisition of the first six surveys were delivered within 2% of the budget of 45 mUSD.

The operation of this system is done from shore, only a few visits have been made offshore to maintain and upgrade the system since 2003. In this period we have also had two maintenance campaigns making repairs to the seafloor system.

The Valhall Life of field seismic program benefitted from BP's North Sea commitment to use 4D implementation as part of a strategic investment in their North Sea developments and assets, Marsh et al. (2003). Up to then there has been a rapid industry take-up of the first wave of 4D technology delivered through marine towed streamer acquisition. This had been driven from the North Sea, with the Gulf of Mexico a fast follower, Bowman et al (2005).

In 2003 there was a strong anticipation that additional 4D value will be obtained from permanent installation over what could be achieved from marine towed streamer acquisition due to the following reasons:

- Having access to multiple 4D images recorded over short intervals (e.g. 3 months)
- Achieving lower noise levels by installing equipment on the sea-bed
- Achieving lower noise levels because of the higher fold and wide azimuth acquisition which are possible with data acquired from an Ocean Bottom System (OBS)
- Getting better geometrical repeatability

The use of a permanent installation would also provide a robust scheme for surveillance as the justifications and the main investment can be done at a time when there are still sufficient resources left in the field to support it. Made for value addition or protection over the lifetime of the field, the installation is a core part of an integrated field surveillance plan. Once the cables are installed, the cost of the acquiring individual surveys would be low, and there would be fewer challenges in developing a business case for acquiring data very late in the field life. To support any life of field investment, we need a convincing business case, including a review of remaining resources and planned activities on the field. Figures will vary greatly from case to case, but it is believed that a convincing business case could be made for such installations if the field has >200MMstb of remaining reserves, and >$200M of remaining drilling expenditure. These figures would of course vary greatly depending on the value that the system would be expected to deliver. We expected the value to increase with more experience operating the system.

10.2.1 Business context models

When starting on the Valhall project, the business context models for a life of field installation were immature. In the tow streamer and retrievable OBS market there were well established business models for operating and acquiring data, and specialized service companies offered cost effective solutions based on a large and mature market. As a result the Valhall Partnership had to engineer a customized seismic acquisition model.

The Valhall field's standby vessel was upgraded to carry a purpose made air gun array. During acquisition the system is controlled from this vessel, and approved data is sent onshore via an optical network for optimal processing via BP's offices in Stavanger. Between surveys, the receiver array remains active and may be used to detect any micro-seismic activity. In this model, the partnership owned and was responsible for operating the system including the seismic source, while the owner of the supply vessel operated the source on behalf of the partnership. After the sixth survey, the model was slightly modified, so that the source is owned and operated by the Valhall license. After survey 11, the model was adjusted again, and now a seismic service company both owns and operates the containerized source. The processing has been delivered by PGS (in addition to BP's internal fast track products) for all 14 surveys. The initial model we used offered great flexibility, but as the activity in the field picked up, the stand-by vessel was tied up in stand by mode and alternatives had to be implemented. Now flexibility is maintained by utilizing a supply vessel on long-term charter.

The market has developed and we believe that in the near future there may be 3-4 companies offering a complete package including equipment, installation, and acquisition and processing services. This should allow for more options in the market place.

10.2.2 Technical elements

With the sensors on the seabed the seismic vessel will no longer need to pull the seismic cable and this means more flexibility to operate in poor weather and close to platform facilities, and less chances of interfering with other activities in the field. It also means, as we have discussed earlier, much improved positional repeatability.

The Valhall system is trenched into the seafloor, primarily due to fisheries, but we believe this also offers more consistent coupling between sensors and the sediments. While the cost of the equipment could be calculated with great precision, there have been larger uncertainties associated with the cost of installation. This is linked to the fact that there are a limited number of cases available as references, the fact that there is few if any construction vessels with custom built equipment for the installation of permanent cable systems, and uncertainties associated with seabed properties for trenching. At Valhall, the installation was delivered at budget, due to a focused project team and favorable seabed properties.

It may be now be seen as a curiosity, but the "Life of Field Seismic concept", initially included the use of tapes for data transfer, and dedicated operators offshore to operate the recording system. We "banned" tape and rigorously implemented a strategy around "no tapes in this project". Fortunately, we did have an optical fiber connection between the platform and the office, and we have managed to operate through 9 years with 5 visits offshore for upgrading and maintenance of the recording system.

Today the bandwidth of the communication links has increased, and we have capacity for continuous recording from shore. The "tape" operator becomes the "recording system supervisor" who ensures that all the data are being recorded with the right (navigation and environmental) attribute included. The "supervisor" used to be placed in the Stavanger drilling center, which clearly had its preferences in staying on top of the operation. However, experience proved that this service would be delivered more efficiently from the system manufacturers offices in Houston.

10.2.3 Interpretation and data management

It was clear from day one that we had to face a data management and documentation challenge. The first three surveys were interpreted and analyzed using a combination of interactive and scripted tools. Based on these findings, we defined a set of automated workflows utilizing BP internal software. Rigorous naming standards and standardized attribute sets were defined, which were available to the users the day after we got a new dataset. As more data become available, the workflows have been revised, and the data loading and analysis sequence has re-run and the database would be updated within days. Models were also treated in the same manner, ensuring that the same analyzing tools were utilized on the real as well as the synthetic data.

Modeling is essential in setting the expectation and for checking the validity of the interpretation. Both the geological model and the flow simulator model are compared with the seismic results.

The project included the intention of learning, while at the same time it had been approved on strict business criteria and needed to deliver key business objectives. This has been managed by keeping a strict line between critical deliveries and R&D elements. We have had the pleasure of working tightly with a number of universities and R&D institutions during this project. Unfortunately, discussing all of these projects become beyond the scope of this book.

10.2.4 Integration

4D seismic data has become a routine tool for the Valhall development team, who use them extensively in the day-to-day work, in understanding depletion mechanism, mapping the drained and undrained areas and tracking where the water has moved. The key to the success is the ability to interact across the disciplines and to engage the wider team in the discussion. Frequent acquisition maintains the focus on a readiness to use the data. In some cases, it's possible to frame a field issue into a question that could be answered by just a brief look at the 4D images.

The skills of integrating the 4D seismic observation into the "business decision" are not necessary generic, but need to be developed and maintained in the team. If the frequency of the new data gets too sparse, it might be difficult to maintain the skills in the team. While the ad-hoc interaction during 4D seismic interpretation, often takes place between the reservoir engineer and the geophysicist, it is critical to engage the geologist, geomechanics engineer, petrohysicist and production engineers in the final analysis. In reality, there is a need for strong cross-disciplinary skill for anyone involved in analyzing the data from seismic surveillance.

Seismic surveillance is not the only option for guidance on how the reservoir performs. It is possible that lack of seismic surveillance may be mitigated by observations in the wells and vice versa. A key for successful field monitoring is to be able to combine the various surveillance options and benefit from added synergies. This will require that the team have the right combined competencies, skills and experience.

10.2.5 Changes outside the reservoir

The Valhall LoFS program was initiated at the same time that the 4D seismic practitioners, re-invented what the engineers with geomechanical insights have known for ages, changes in pressure or temperature in the reservoir would eventually impact the rock outside the reservoir. For the vast majority of reservoir the induced changes were expected to be small, and could be viewed as second or third order effects. As the resolution of the 4D signal improves, this may no longer be the case. The awareness around the possible complications, and the potential for utilizing the geomechanical changes in the time-lapse analysis will help us addressing these issues in a proper manner.

Some of the early reports of 4D seismic response outside the reservoir came for the Valhall and the Ekofisk fields. While the geomechanical experts expected the effects, the magnitude of the effects was a big surprise. In the literature there are probably 20-25 fields that have reported significant 4D effects linked to stress-redistribution outside the reservoir, see table 10.1. At Valhall we have concluded that compaction down to 5 cm at the reservoir level may be detected, and will appear as a 0.3 msec time-shift between subsequent surveys. This type of observation can help quantifying possible impacts on pore-pressure outside the reservoir due to geomechanical effects, prior to re-drilling into a produced reservoir.

10.3 Future directions

10.3.1 Methods

The major oil companies with the help of the seismic service industry have driven the development of today's seismic surveillance technologies. Not only are there significant investments involved in acquiring the data, but also the logistic of managing the data has not been in favor of the independent research institutes or universities. The practitioners in the reservoir management and new well delivery team, has embraced any information that can be of help in predicting ahead of the bit in a producing assets. This has lead to implementation of pragmatic approaches, and engineering solutions that have been gradually improved as the teams have got the experience.

Fortunately the university has been picking up on processing and analysis of the seismic surveillance data over the last 10 years, as computing resources has increased and handling of the data has became more streamlined. A cross-section of recent technical papers on the subject from the academia may give some insights to what the future may hold. We will look at four elements. The first example is linked to processing and analysis, the second to high end interpolation for improving interpretation, the third to sampling and the frequency of acquisition, and finally to the use of passive imaging.

The full waveform inversion results from the Valhall LOFS system, represented an improved basis for imaging and an enabler for seeing the 4D seismic response below the gas cloud, but also the results provided an additional attribute for interpretations, Barkved et al. (2010). Few cases have been presented on time-lapse analysis based on full waveform analysis. We are sure that this will change in the near future, however full waveform inversion is a non-linear process, and matching end results might not be the solution. We have noticed that tomographic inversion of refracted waves (chapter 8.3) can be used to isolate changes in velocities in the order of +/- 5 m/s. By matching subsequent full waveform inversion applied to the each vintage data, the sensitivity appears to be an order of magnitude lower (+/- 50 m/sec) at best. The solution might be a strategy that the residual effects are inverted for, as has also been suggested by Zheng et al. (2011), see figure 10.2.

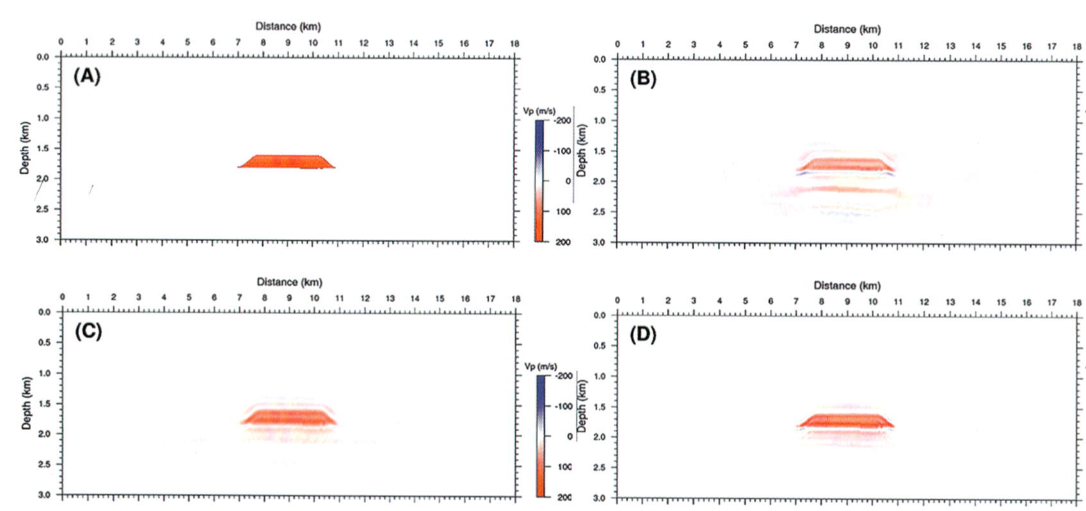

Figure 10.2: Change in P-wave velocity: (A) true change; (B) using the same starting velocity models; (C) using starting velocity models by smoothing the true repeat model; (D) residual data inversion with updated starting velocity models from inversion of baseline data, from Zheng et al. (2011).

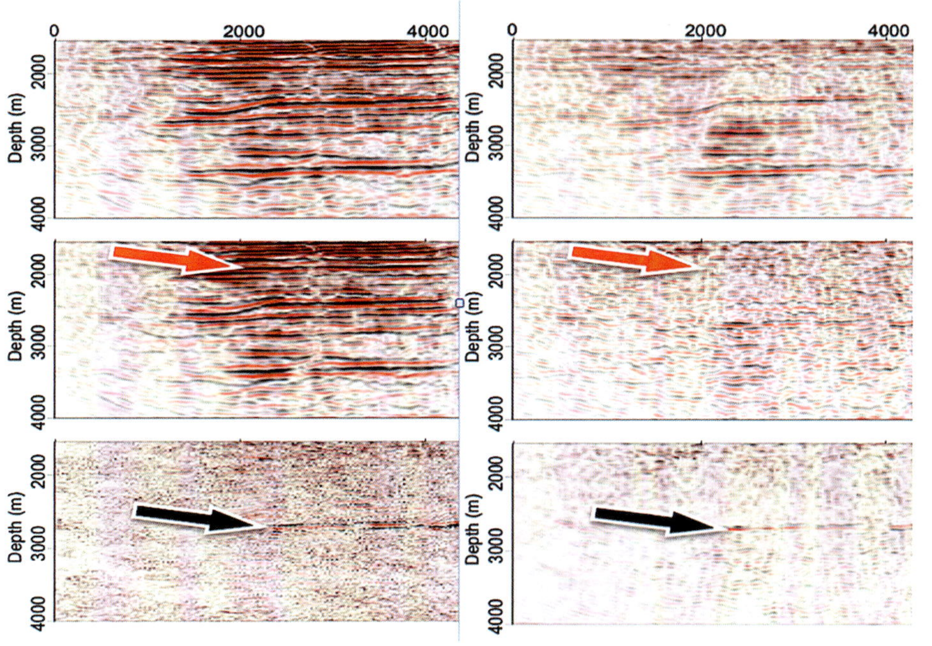

Figure 10.3: Left Time-lapse images from the recorded data sets after different processing steps, left panels are Time-lapse images with incomplete monitor data after different processing steps, right panels are same for complete monitor data. The top panel shows the image obtained after multiple attenuation and amplitude balancing. The middle panel shows the image after residual moveout correction, pre-stack and post-stack warping. The bottom panel shows the results after wave-equation inversion. The black arrow shows the reservoir position. From Ayeni and Biondi (SEG 2011).

Conventional Marine Towed Streamer seismic acquisition has a limitation if there is a significant footprint of the production facilities. Undershoots, which are used to image below production facilities, are often challenging and/or expensive to implement. Gboyega and Biondi propose a linearized wave equation- inversion formulation for time-lapse data sets. Their method accounts for differences in acquisition geometries and illumination patterns, and band-limited wave- propagation effects. They claim that this not only provides better estimates of production- related changes than migration but also demonstrates how obstruction artifacts may be attenuated by wave-equation inversion. If this turns out to be a robust technique, the undershoot limitations might be less of an issue in the future, figure 10.3.

The concept of temporal and spatial sampling for avoiding aliasing is well understood. However we don't have as similar tool for defining the right sampling or time period between successive 4D seismic acquisitions for a reservoir. Sparse sampling from 4D is an important issue, as we could potentially drive the cost down significantly, or may there be more value to gained by spanning the acquisition of a single survey over a longer time period?

Harris et al., (2007) proposed a strategy, called True4D, for quasi-continuous reservoir monitoring with sparse data. They introduced two concepts for processing sparse but continuously recorded datasets: (1) dynamic imaging and inversion, (2) data evolution, and (3) model evolution. Arogunmati and Harris (2011) look at the practicality of an approach to data evolution i.e., inversion or imaging using data estimated from sparse recordings. The True4D approach is implemented by acquiring as little as 5% of the conventional 3-D survey data volume at small time intervals. Unrecorded data at each time interval are then estimated using recorded data at all intervals to produce an image with good spatial resolution of the subsurface. They

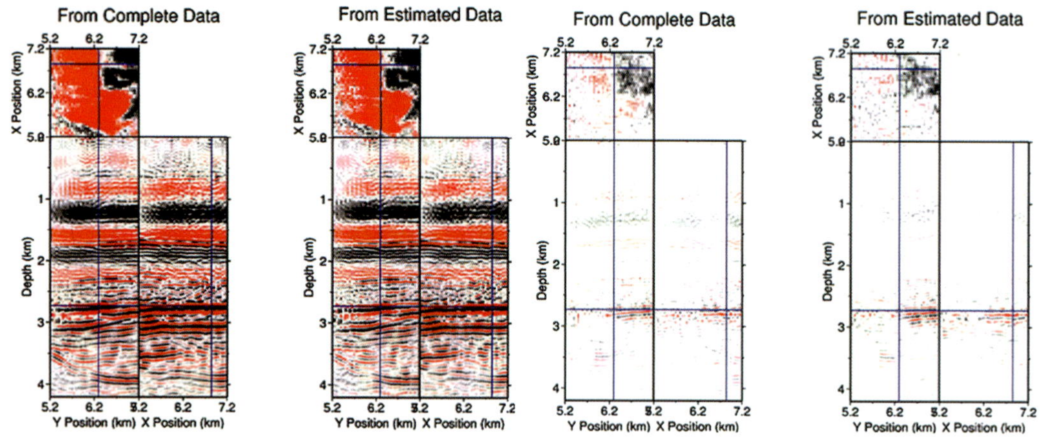

Figure 10.4: Migrated image from complete Survey 11 field data (left first) and the migrated image from data estimated using 33% of the field data (second). Third: Time-lapse difference image obtained from the migrated image from complete Survey 11 field data and the migrated image from data estimated using 33% of the field data (right), from Arogunmati and Harris (2011).

show that the True4D concept may be effective for reservoir monitoring where calendar-time temporal resolution of the reservoir changes is taken to be more important than spatial resolution, figure 10.4.

As shown in chapter 8, there have been some impressive developments in use of passive recordings, and new results are coming out continuously, i.e., de Ridder (2011), and Mordret et al. (2011). They show that by calculating cross correlation between sensor pairs combinations, using low frequencies, as suggested by Dellinger and Yi (2009), they manage to retrieve the Scholte waves from using passive noise, by constructing a virtual source from all the receivers, see 8.3.3. By tomographic inversion, the wave velocity may be reconstructed from following the various pathways between the receivers. Figure 10.5 (left) shows Scholte-wave virtual sources that are generated based on 29 hours of data using seismic interferometry. Very-low frequency is used (0.4–1.75 Hz), and Scholte-wave group arrival times are inverted for. The top 200 m near surface at Valhall contains buried channels about 100-200 m wide that have been imaged with active seismic. An image obtained contains anomalies that match these channels. When continuous recordings are made in real-time, tomography images of the shallow subsurface can be formed or updated on a daily basis, forming a very low cost near-surface monitoring system using seismic noise.

As we have discussed earlier, the significant subsidence at Valhall result in a change in stress state at the seafloor. The new state will be more compressive in the center of the bowl, while the edge of the bowl will result in weak extensions, Kristiansen (1998). Stress state changes will impact polarized wave modes, and we may deduce the effect fitting the velocity to a parabolic function of azimuth, as first demonstrated by Olofson et al. (2003). The new observation is that these phenomena also can be retrieved from passive recordings, figure 10.6. The resulting "shear wave splitting" maps can then be used to history match the geomechanical model, as a complement to bathymetry data.

For the fields like Valhall, where continuous monitoring of the shallow overburden is a main application, this may have some influence on choice of system. However these type of analysis might add to an improve overburden description which might help in utilizing shear waves for active reservoir surveillance, as demonstrated by Zwartjes et al. (2008).

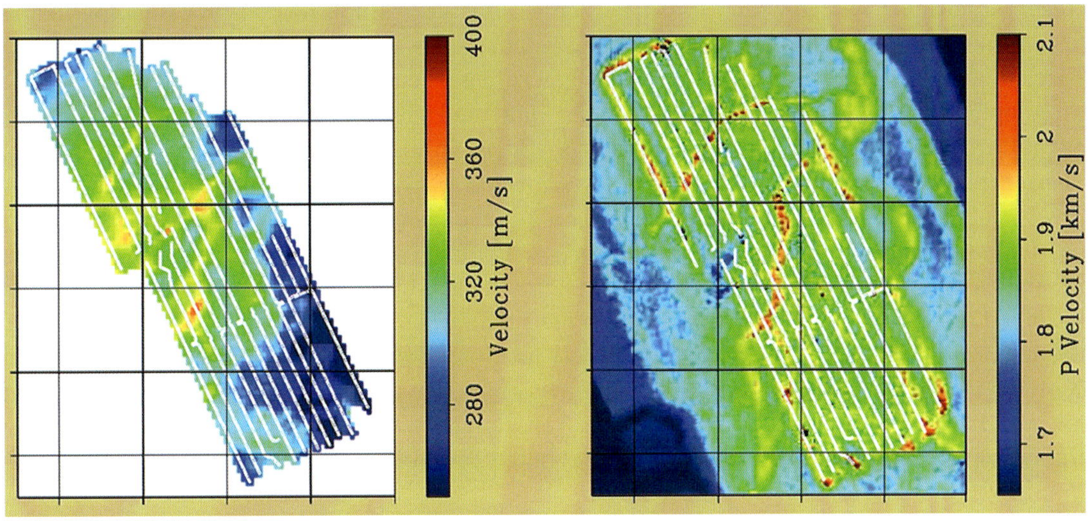

Figure 10.5: Left picture shows Scholte wave velocities from low frequency passive data, right picture shows P-wave velocity 170 m below seabed, from Full waveform inversion. From de Ridder (2011) and Sirgue et al. (2008).

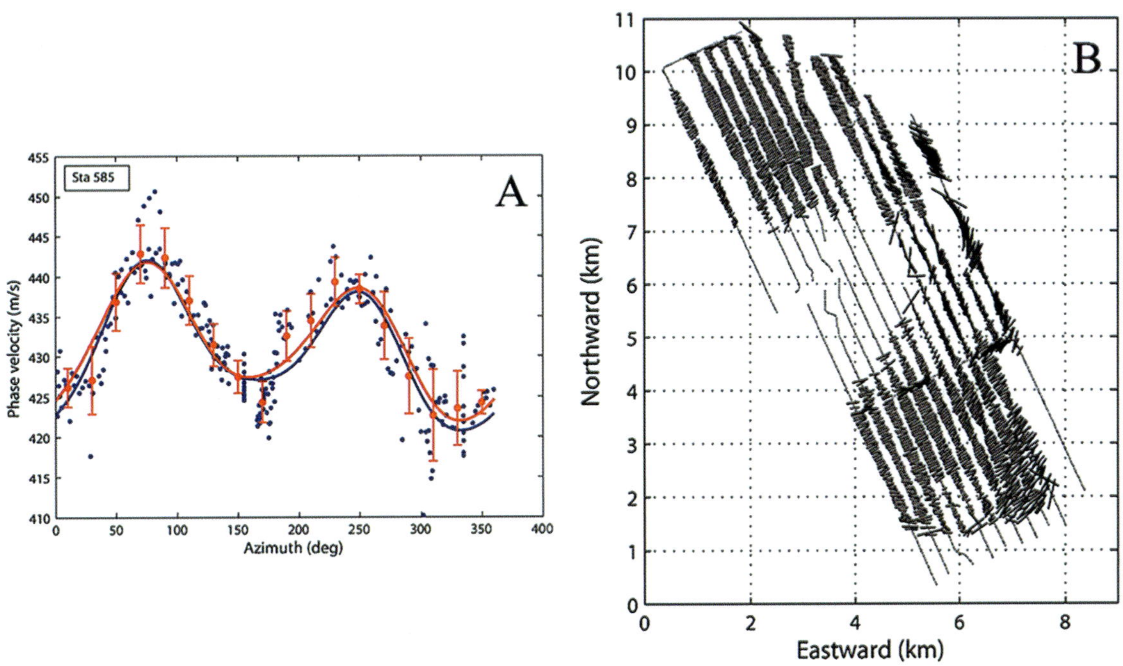

Figure 10.6: A) Azimuthal distribution of the phase velocity at 0.8 s for one of the 2500 stations. The blue dots are the phase velocity measurement, the red dots with error bars are the phase velocity averaged over 20°. B) Map of the azimuthal anisotropy. Mordret et al. (2011).

10.3.2 Systems

The recent innovation we have seen in the marine seismic acquisition market on extending the bandwidth, and improved imaging using wide azimuth techniques will probably influence the choice of future technology for some reservoir. This may suggest that placing the receiver on the seabed may be a preferred option, as the applications favors wide azimuth and very long offset data. Current 4D seismic techniques are biased towards imaging the matrix effects. In many cases, the impact of fractures and high permeable layers might be inferred from these images. However, alternate methods like AVOA and Shear-wave splitting might be able to detect the nature of the fracture system more directly, and therefore be more valuable in utilizing or blocking the effects of the fractures, like a an early water-breakthrough. These applications will clearly benefit from wide azimuth data, which again is probably most cost effectively solved by placing the recording system on the seafloor.

The majority of available seismic surveys across offshore reservoirs are streamer survey; this clearly influence on the choice of monitoring survey. For new and re-developments, wide azimuth ocean bottom seismic (OBS) surveys is getting more common. Consequently, it will be natural to expect that seismic time-lapse surveys based on OBS systems will be more common in the future.

10.3.3 Integrating seismic surveillance data into business decisions

Cross discipline integration is key in turning the seismic surveillance data into value. The development of 4D seismic methods has benefitted strongly by improvement in acquisition and processing technology. Taking the 4D seismic images to the next level includes the involvement of additional disciplines, finding the optimal way of working effectively together could be more challenging. Quantitative analysis, i.e. through history matching of the reservoir model, becoming more common, but we should keep in mind that more qualitative analysis still has merits, and probably accounts for most of the value adding from time-lapse seismic studies.

In the Valhall subsurface team, 4D images and maps, and how they compare to the prediction from the reservoir flow model, have become an important source for context for almost any discussions related to reservoir performance. The frequent acquisition implies that the 4D images are continuously updated and well within the expiration date for the dynamic information revealed, which adds to the confidence to use the data on specific well issues, e.g. where is the most likely entry point of water break-trough in a horizontal well.

It is important to stress that the trust in the seismic time-lapse data also is built up over time. An important aspect is that the frequency allows to follow how detailed changes developed, and the observations is then judged against what is reasonable based on geological understating and fluid changes. It goes without saying, but the driving interest for the seismic information is based on the need to understand the details of reservoir for optimizing water sweep and location of new infill producing wells.

10.4 Conclusions

During the last decades, seismic surveillance has become an important technology helping to develop oil and gas reservoirs. The Value creation, as being documented on the Norwegian Continental Shelf is impressive.

Today, most major oil company will include extensive studies in how to make best use of seismic surveillance for development of new field, and major re-development plans.

Offshore, marine tow streamer continuous to dominate the 4D seismic marked. The use of OBS systems is emerging, and impressive results from using cables and nodes have been presented.

At Valhall the permanent seismic installation (LoFS) has been in operation since 2003. The project has been a technical and commercial success and there are plan to extend the system to cover larger areas.

Through the last ten years, there has been a growing awareness of production induced stress state changes outside the reservoir. New tools have been developed to make use of the 4D signal associated with these changes.

We have apparently by no means reached what the technology may offer through a variety of acquisition systems - permanent versus marine tow versus retrievable ocean bottom cables or nodes. There is and will continue to be great value in delivering more resources in a safe and cost effective manner through the acquisition, processing, analysis and integration of 4D seismic data.

Field	Area	Geology	References	Operator
Ekofisk	North Sea	Chalk	Guilbot and Smith (2002)	ConocoPhillips
Valhall	North Sea	Chalk	Hall et al. (2002) Barkved et al. (2003)	BP
Shearwater	North Sea	HTHP	Hatchell et. al. (2003)	Shell
Offshore Sarewak	SE Asia	Carbonates	Hatchell and Bourne (2005)	Shell
Heron	North Sea	HTHP	Hatchell and Bourne (2005)	Shell
Europa	GoM	Deepwater turbidite	Tura et al. (2005)	Shell
Mars	GoM	Deepwater turbidite	Tura et al. (2005)	Shell
Genesis	GoM	Deepwater turbidite	Hudson et al. (2005) Rickett et al. (2006)	Shell
South Arne	North Sea	Chalk	Schiøtt and King (2006)	Hess
Skua	North Sea	HTHP	Staples et al. 2006	Shell
Egret	North Sea	HTHP	Staples et al. 2006	Shell
Elgin	North Sea	HTHP	Hawkins et al. (2007)	Total
Franklin	North Sea	HTHP	Hawkins et al. (2007)	Total
	Reefs	Reefs	Barker et al. (2008)	Shell
Kirstin	North Sea	HTHP	Hansen et al. (2009)	Statoil
In Salah		CO^2 sequestration	Mathieson et al. (2010)	BP
Groningen	The Netherland	Permian Sandstone	Geertsma (1973).	Shell / NAM
Holstein	GOM	Stacked Pliocene turbidities sand	Ebaid et al. (2012)	BP

Table 10.1: Examples of published cases from reservoir/fields there are reported significant production/injection induced 4D effects linked to stress-redistribution outside the reservoir

References

Ali, N. and Alcock, T. (1994). Valhall Field, Norway – The first ten years, North Sea oil and gas reservoirs – III. NTH: 1994, 25- 40.

Albertin, U., Sava, P.C., Etgen, J., Maharramov, M., (2006), "Adjoint wave-equation velocity analysis", Proceedings of the 76th Annual International Meeting, Society of Exploration Geophysicists.

Albrechtsen, T., Andersen, S.J., Dons, T., Engstrøm, F., Jørgensen, O. and Sørensen, F.W. (2001). Halfdan: Developing non- structurally trapped oil in North Sea chalk. *SPE Annual Technical Conference 2001*, New Orleans, Louisiana, 30 September – 3 October, SPE 71322.

Al-Jenaibi, M., Soroka, W.L., Al-Jeelani, A.B., Hafez, H.H., Kleiss, E. and Melville, P., Adco (2006). 4D seismic monitors fluid changes over time in a carbonate reservoir. *Abu Dhabi International Petroleum Exhibition and Conference*, Abu Dhabi, 5-8 November 2006, UAE, SPE 100479-MS.

Andersen, M.A. (1995). Petroleum research in North Sea chalk, *Monograph*, 1995, Rogaland Research.

Arogunmati, A. and Harris, K.M. (2010). A data-estimation based approach for quasi-continuous reservoir monitoring using sparse surface seismic data. *EAGE* 2010.

Arogunmati, A. and Harris, K.M. (2011). Reservoir monitoring with true 4D surface seismic data. *SEG Annual Meeting 2011*, San Antonio, Texas, September 18 - 23.

Artman, B. (2007). Passive seismic imaging. *Ph.D. Thesis, Stanford Exploration Project report*, Stanford University, Stanford, California, SEP-128.

Askim, O.J. (2003). Seismic forward modeling in a chalk reservoir with permanent monitoring. *65th EAGE Conference & Exhibition, Extended Abstracts*, Stavanger, 2003, A16.

Askim, O. J., Brandsberg-Dahl, S., Barkved, O.I. (2004). 4D seismic analysis using pre stack depth migration. *66th EAGE Conference & Exhibition*, 07 June 2004.

Askim, O.J., Barkved, O. I., Hornby, B., Williams, B. and Bostick, F.X. (2008).Time lapse VSP using permanent in-well seismic receivers. *NPF Biennial Geophysical Seminar*, Kr.Sand 2008.

Atkinson, J. and Davis, T. (2011). Multi-component time-lapse monitoring of two hydraulic fracture stimulations in the Pouce Coupe Field unconventional reservoir. *First Break* , **29**, 91-97.

Avseth, P., Mukerji, T. and Mavko, G. (2005). Quantitative seismic interpretation – Applying rock physics tools to reduce interpretation risk. *Cambridge University Press*.

Avseth, P.A., van Wijngaarden, J., Johansen, T.A. and Mavko, G. (2006). Seismic fluid prediction in heterogeneous reservoirs. *68th EAGE Conference & Exhibition*, 12 June 2006.

Ayeni, G. and Biondi, B.(2011). "Wave-equation inversion of time-lapse seismic data sets. *SEG Annual Meeting 2011*, San Antonio.

Aylor, W. K. (2005). Business performance and value of exploitation 3-D seismic. *Leading Edge*, July 1995, **14**(7),797-801.

Ballin, P. R., Ward, G.S., Whorlow, C.V. and Khan, T. (2005). Value of information for a 4d-seismic acquisition project. *SPE Latin American and Caribbean Petroleum Engineering Conference*, Rio de Janeiro, Brazil.

Barkved, O.I., Kunz, T. and Whitman, D. (1998). Automated extraction of faults and porous reservoir bodies - examples from the Valhall Field. *NPF*, Kr.sand ,1998.

Barkved, O.I., Mueller, M.C. and Thomsen, L. (1999). Vector interpretation of the Valhall 3D/4C OBS dataset. *EAGE Extended Abstracts*, 61.

Barkved, O.I., Gaucher, E., Hornby, B., Kristiansen, T. and Maisons, C.(2002). Analysis of seismic recordings during injection using in-well permanent sensors. *64th EAGE Annual Conference and Exhibition.*

Barkved, O. I., Bærheim, A., G.B., Howe, D.J., Kommedal, J.H. and Nicol, G. (2003a). Life of field seismic implementation – First at Valhall. *65th EAGE Conference and Exhibition, Extended Abstracts*, Stavanger.

Barkved, O. I., Buer, K., Halleland, K.B., Kjelstadli, R., Kleppan, T. and Kristiansen, T. (2003b). 4D seismic response of primary production and waste injection at the Valhall Field. *65th EAGE Conference and Exhibition*, Stavanger.

Barkved O., Heavey P., Kjelstadli R., Kleppan, T. and Kristiansen, T. G.(2003c). Valhall Field – Still on plateau after 20 years of production. *Offshore Europe Conference*, SPE 83957.

Barkved, O.I., Kommedal, J.H. and Thomsen, L.(2004). The role of multi-component seismic data in developing the Valhall Field, Norway. *66th Meeting of the EAGE*, Paris, 2004, E040.

Barkved, O. I . (2004b). Continuous seismic monitoring. *74th annual SEG meeting, Extended Abstracts*, 23, 2537

Barkved, O.I. and Kristiansen, T. (2005a). Seismic time-lapse effects and stress changes: Examples from a compacting reservoir. *The Leading Edge*, **24**, 1244.

Barkved, O., Kristiansen, T. and Fjær, E. (2005b). The 4D seismic response of a compacting reservoir—examples from the Valhall Field, Norway. *75th SEG annual meeting*, Exptended Abstracts, 24, 2508

Barkved, O.I., Kommedal, J.H., Kristiansen, T.G., Buer, K., Kjelstadli, R.M., Haller, N., Ackers, M., Sund, G. and Bakke, R.(2005c). Integrating continuous 4D seismic data into subsurface workflows. *67th EAGE Conference and Exhibition*, Madrid.

Barkved, O.I., Kommedal, J.H., Howe, D. and Bærheim, A. (2005). Seismic on Demand – a world first at Valhall. *International Petroleum Conference*, Cairo, 2005.

Barkved, O.I, Buer, K. and Kristiansen, T.G. (2005d). Valhall permanent seismic monitoring—reducing geological model uncertainties using 4-D seismic. *AAPG International Conference*, Paris, France, 2005.

Barkved, O.I., Kristiansen, T. G., van Gestel, J. and Pettersen, R.S.H. (2008). Observed velocity changes due to unloading in the shale cap-rock from amplitudes and time-shift analysis. *NPF Biennial Geophysical Seminar*, Kristiansand, Norway.

Barkved, O. I., van Gestel, J., Bergsvik, L.S., Stockden, I. and Kommedal, J. H. (2009). Seismic PLT - linking seismic time-lapse responses to production and injection data. *71st EAGE Conference and Exhibition, Extended Abstracts.*

Barkved,O. I., Heavey, P., Kommedal, J.H., van Gestel, J., Pettersen, R.S.H. and Kent, C. (2010). Business impact of full waveform inversion at Valhall. *SEG Extended Abstracts*, 29, 925

Barr, F. J. and Sanders, J. I. (1989). Attenuation of water-column reverberations using pressure and velocity detectors in a water-bottom cable. *59th Annual International Meeting, Expanded Abstracts*, Society of Exploration Geophysicist, 653–656.

Batzle, M. and Wang, Z. (1992). Seismic properties of pore fluids. *Geophysics*, **57**(11), 1396-1408.

Bauer, A., Lehr, C., Korndorffer, F., van der Linden, A., Dudley, J., Addis, T., Love, K. and Myers, M. (2008). Stress and pore pressure dependence of sound velocities in shales: Poroelastic effects in time-lapse seismic. *78th SEG Annual International Meeting, Expanded Abstracts*, 1630–1634.

Beaudoin, G. and Ross, A.A. (2007). Field design and operation of a novel deepwater, wide-azimuth node seismic survey. *The Leading Edge*, 2007, 494-503.

Beaudoin,G.J., Reasnor, M.D., Pfiste, M. and Openshaw, G. (2010). First wide-azimuth time-lapse seismic acquisition using ocean bottom seismic nodes at Atlantis Field. *72nd EAGE Exhibition and Conference*.

Ben-Menahem, A. (2005). A concise history of mainstream seismology; origins, legacy, and perspectives. *Bulletin of the Seismological Society of America*.

Berryman, J.G. (1980). Long wavelength propagation in composite elastic media. *Journal of the Acoustical Society of America*, **68**, 1809-1831.

Berryman, J. G. (1999). Tutorial: Origin of Gassmann's equations. *Geophysics*, **64**, 1627-1629.

Berteussen, K. A., Barkved, O.I., Jahren, L., Kommedal, J.H. and Rathore, J. (1993). Experience with sea bottom acoustic sensors on the Valhall Field. *55th EAGE Convention*, Stavanger.

Berteussen, K.A. (2006). Quantifying the value of 4D data before the survey. *SEG Annual Meeting*, New Orleans, Louisiana, 1-6 October 2006.

Biot, M.A. (1956).Theory of propagation of elastic waves in a fluid saturated porous solid. *Journal of the Acoustical Society of America*, *28*,168-191.

Bjørlykke, K., Høeg, K. and Mondol, N.H. (2010). Introduction to geomechanics- stress and strain in sedimentary basins. *Petroleum Geoscience: From Sedimentary Environments to Rock Physics*, Springer 2011.

Bourbié, T., Coussy, O. and Zinsner, B. (1987). Acoustics of Porous Media. Gulf Publishing Company.

Bowman, M.B.J., Marsh, M. J. and Whitcombe, D. N. (2004).What's 4D worth? – a BP Perspective. *SPE/EAGE workshop "What do petroleum engineers expect from timelapse seismic and do geophysicists answer the right questions?"*, Copenhagen, March 2004.

Brandsberg-Dahl, S., O'Brien, M.J., Whitmore, D., Etgen, J.T. and Murphy, G.E. (1999). Multicomponent modeling of the Valhall field. *EAGE extended abstract*s, **61**.

BRIE (1995)

Buer, K. and Trice, R. (2004). Data integration and numerical fracture models for water flood management -A case study of the Valhall Chalk Reservoir, North Sea.*Fractured Reservoirs: International conference,* 16-17 November 2004, Geological Society of London.

Bygdevoll J. (2010). Introduction to Reservoir Complexity Index (RCI). Lecture by the Norwegian Petroleum Directorate, 18 March, 2010.

Caley, A., Kendall, J-M., Jones, R., Barkved, O.I. and Folstad, P. G. (2001). Monitoring fractures in 4D using microseismic data. *63rd EAGE Conference and Technical Exhibition, Proceedings*, Amsterdam, Netherlands, 2001.

Calvert, R.W., Hatchell, P. and Jones, C. (2002). 4D Acquisition Geometry Requirements and QA. *EAGE, Extended Abstracts*, 2002.

Calvert, R. (2005). Insights and methods for 4D reservoir monitoring and characterization. *SEG/EAGE Distinguished Instructor Short Course #8 Lecture Notes*, Tulsa, 2005.

Calvert, R. (2005b). 4D Technology: where are we, and where are we going? *Geophysical Prospecting*, **53**, 161-171

Campbell, S., Lacombe, C., Brooymans, R., Hoeber, H. and White, S. (2011). Foinaven: 4D processing comes up trumps. *The Leading Edge*, **30**(9), 1034 -1040.

Campbell, S., Ricketts, T., Davies, D. M., Slater, C.P., Lilley, G.G., Brain, J., Stammeijer, J. and Evans, A.C. (2005). Improved 4D seismic repeatability — a West of Shetlands towed streamer acquisition case history. *75th annual SEG meeting, SEG Extended Abstracts*, **24**, 2394.

Carlson, R. L. and Gangi, A.F. (1985). Effect of cracks on the pressure dependence of P wave velocities in crystalline rocks. *Journal of Geophysical Research*, **90**, 8675–8684.

Castagna, J.P. and Swan, H.W. (1997). Principles of AVO crossplotting. *The Leading Edge*, **16**(4), 337-344.

Castagna, J. (1985). Relationships between compressional-wave and shear-wave velocities in elastic silicate rocks. *Geophysics*, **50**(4), 571-581.

Chambers, K., Kendall, J.-M. and Barkved, O. I. (2010). Investigation of induced microseismicity at Valhall using the Life of Field Seismic array. *The Leading Edge*, **29**(3), 290–295.

Clarke, R., Askim, O., Pursley, K. and Vu, P. (2005). 4D rapid turnaround for permanent 4C installation. *75th annual SEG meeting, Extended Abstracts* , **24**, 2452

Connolly, P.(1999). Elastic Impedance. *The Leading Edge*, **18**(4), 438-452.

Cooper, M.M., Evans, A.C., Lynch, D.J., Nevill, G. and Newley, T. (1999). The Foinaven field: managing reservoir development uncertainty prior to start-up. Fleet, A.J. and Boldy, S.A.R. (eds.), *5th Conference of the Petroleum Geology of Northwest Europe, Proceedings*, 1999a.

Cooper, M., Westwater, P., Thorogood, E., Kristiansen, P. and Christie, P. (1999b). Foinaven active reservoir management: Towed streamer and buried sea-bed detectors in deep water for 4D seismic. *69th SEG Annual Meeting, Extended Abstracts*, Houston, Texas, **18**, 1632.

Cooper, M., Bouska, J., Thorogood, E., O'Donovan, A., Kristiansen, P. and Christie, P. (1999c). Foinaven Active Reservoir Management: The Time Lapse Signal. *69th SEG Annual Meeting, Extended Abstracts*, Houston, Texas.

Coopersmith, E.M., Burkholder, M.K. and Schulze, J.H. (2006).Value of information lookbacks -Was the information you gathered really worth getting? *SPE Annual Technical Conference and Exhibition*, San Antonio, Texas, 24-27 September, SPE 101540.

Corzo, M., MacBeth, C. and Barkved, O.I. (in Press). Estimation of Pressure Changes in a Compacting Reservoir from Time Time-Lapse Seismic Data. *Geophysical Prospecting*.

Davies, D., Mannaerts, H., McGarrity, J., Ibram, M., McKenzie, C., Campbell, S., Alexander, G., Lozano, A., Kommedal, J., Barkved, O.I. and van Gestel, J.P., BP, (2011). High-density OBC-a step change in reservoir imaging-A BP North Sea View. *Offshore Europe*, 6-8 September 2011, Aberdeen, UK, 146144-MS.

de Meersman, K., Kendall, J-M. and van der Baan, M. (2007). Valhall microseismicity revisited: relocated sources, seismic multiplets and S-wave splitting. *69th EAGE Meeting*, Proceedings, A038, 2007.

Deflandre, J.P., Laurent, J. and Blondin, E. (1993). Use of permanent downhole geophones for microseismic surveying of gas storage reservoir. *55th EAGE Meeting*, Stavanger, Norway.

Dellinger, J. (2008). Low frequencies using conventional sensors: "sign-bit" recording revisited. *78th annual SEG meeting, Extended Abstracts*, 149–153.

Dellinger, J. and Yu, J. (2009). Low-frequency virtual point source interferometry using conventional sensors. *71st EAGE Conference*.

de Ridder, S. and Dellinger, J. (2011). Ambient seismic noise tomography at Valhall. *81st Annual Meeting SEG, Expanded Abstracts*, **30**, San Antonio, USA, 1597-1601.

Domenico, S.N. (1976). Effect of brine-gas mixture on velocity in an unconsolidated sand reservoir. *Geophysics*, **41**(5), 882-894.

Dong, Y. and Oliver, D.S. (2003). Quantitative use of 4D seismic data for reservoir description. *SPE Annual Technical Conference and Exhibition*, Denver, USA, 2003, SPE 84571.

Dons, T., Jørgensen, O. and Gommesen, L. (2007). Observations and quantitative analyses of waterflood patterns in a chalk reservoir using seismic data, Halfdan Field, Danish North Sea. SPE 108531.

Dromgoole, P. and Speers, R. (1997). Geoscore: a method for quantifying uncertainty in field reserve estimates. *Petroleum Geoscience*, **3**, 1-12.

Duffaut, K. and Landrø, M. (2007). Vp/Vs ratio versus differential stress and rock consolidation - A comparison between rock models and time-lapse AVO data. *Geophysics,* **72**(5), C81-C94.

Dubois et al., (1998). Well seismic monitoring by using permanent downhole geophones: Equipment, results and prospects. *EAGE Reservoir Monitoring Workshop*, Leipzig, 1998.

Dvorkin, J., Mavko, G. and Nur, A. (1991). The effect of cementation on the elastic properties of granular material. *Mechanics of Materials*, **12**, 207-217.

Dvorkin, J., Nur, A. and Yin, H. (1994). Effective properties of cemented granular materials. *Mechanics of Materials*, **18**, 351-366.

Dvorkin, J. and Nur, A. (1998). Time-Average Equation Revisited. *Geophysics*, **63**, 460- 464.

Dvorkin, J. and Nur, A. (1996). Elasticity of high-porosity sandstones, theory for two North Sea data sets. *Geophysics*, **61**, 559-564.

Dyer, B. C., Barkved, O.I., Jones, R.H. and Folstad, P.G. (1999). Microseismic monitoring of the Valhall reservoir. *EAGE 61st Meeting*.

Dyce, M., Whitcombe, D., McKenzie, C. and Hodgson, L. (2004). The Quantification of 4D Noise. *66th EAGE Conference Workshop: Can time lapse seismic be more quantitative?* 7 June 2004.

Ebrom, D., Krail, P., Ridyard, D. and Scott, L. (1998). 4-C/4-D at Teal South. *The Leading Edge,* **17**, 1450.

El Ouair, Y. and Strønen, L.K. (2006). Value creation from 4D seismic at the Gullfaks Field: Achievements and new challenges. *76th SEG annual meeting*.

Farmer, C. L., and Barkved, O. I., (1999), "Influence of syndepositional faulting on thickness variations in chalk reservoirs - Valhall and Hod Fields", Norway, Petroleum Geology of Northwest Europe, Proceedings of the 5th Conference (Barbican), The Geological Society of London.

Falcone, G., Gosselin, O., Flavien, M., Marrauld, J. and Zhakupov, M. (2004). Petroelastic modeling as key element of 4D history matching: A field example. *SPE Annual Technical Conference and Exhibition*, Houston, USA, 2004, SPE 90466.

Fjær, E. (2006a). Modeling the stress dependence of elastic wave velocities in soft rocks. *41st USRM Symposium*, paper ARMA/USRM 06-1070.

Fjær, E. and Kristiansen, T. G. (2009). An integrated geomechanics, rock physics and seismic model. *71st EAGE Conference and Exhibition*, Amsterdam, The Netherlands.

Fjær, E., Larsen, I. and Scheldt, T. (2005). What is the Poisson's ratio of a soft rock? *40th U.S. Symposium on Rock Mechanics (USRMS)*, Anchorage, USA, 25 – 29 June, 2005.

Fjær, E., Holt, R.M., Horsrud, P., Raaen, A.M. and Risnes, R. (2008). Petroleum related rock mechanics. 2nd Edition, Elsevier 2008.

Fjær, E. and Holt, R.M. (1994). Rock acoustics and rock mechanics: Their link in petroleum engineering. *The Leading Edge*,**13**(4), 255-258.

Fjær, E. (2009). Static and dynamic moduli of a weak sandstone. *Geophysics*, **74**, WA103-WA112.

Fjær, E. (2006). Geomechanics and geophysics for reservoir management. *Revue Européenne de Génie Civil*, **10**, 703-730.

Flatebø, R., Pettersen, R.S.H., Kvernstrøm, J. and Ulen, V. (2010). Improved well delivery using advanced technology and integrated processes in the planning phase. *72nd EAGE Conference and Exhibition*, 2010.

Folstad, P.G. (2010). Fiber optic sensor technology for permanent seismic monitoring at Ekofisk. *Biennial Geophysical Seminar Norwegian Petroleum Society, Expanded Abstracts*, 100-103.

Folstad, P.G. (2010). Monitoring of the Ekofisk field. *GeoExpro*, **3**.

Folstad, P.G., Haugvaldstad, H. and Jeangeot, G. (2011). Ekofisk PRM - The technical case for this brand new installation. *EAGE Workshop on Permanent Reservoir Monitoring (PRM) - Using Seismic Data*, 28 February, 2011.

Folstad, P.G.. (2011). Ekofisk - Justification for a permanent monitoring system. *EAGE Workshop on Permanent Reservoir Monitoring (PRM) - Using Seismic Data*, 28 February 2011.

Foster, D. (2007). The BP 4-D story: Experience over the last 10 years and current trends. *International Petroleum Technology Conference*, Dubai, UAE, 4-6 December, 2007.

Foster, D., Fowler, S., McGarrity, J., Riviere, M., Robinson, N., Seaborne, R. and Watson, P. (2008). Building on BP's large-scale OBC monitoring experience—The Clair and Chirag-Azeri projects. *The Leading Edge*, **27**(12), 1632-1637.

Fuck, R.F., Tsvankin, I. and Bakulin, A. (2011). Influence of background heterogeneity on traveltime shifts for compacting reservoirs. *Geophysical Prospecting*, **59**, 78-89.

Gardner, G.H.F., Gardner, L.W. and Gregory, A.R. (1974). Formation velocity and density - the diagnostic basics for stratigraphic traps. *Geophysics*, **39**, 770-780.

Gassmann, F. (1951). Elastic waves through a packing of spheres. *Geophysics* , **16**, 673-685.

Gassmann, F. (1951). Elasticity of porous media: Uber die elastizitat poroser medien. *Vierteljahrsschrift der Naturforschenden Gesselschaft*, **96**, 1-23.

Geertsma, J. (1973). A basic theory of subsidence due to reservoir compaction: The homogeneous case. *Royal Dutch Society of Geologists and Mining Engineers*, Transactions, **22**, 43-62.

Grayson, C.J., JR. (1960). Decisions under uncertainty: Drilling decisions by oil and gas operators. *Harvard Business School, Division of Research*, Boston, USA.

Greaves, R.J. and Fulp, T.J. (1987). Three-dimensional seismic monitoring of an enhanced oil recovery process. *Geophysics*, **52**(9), 1175-1187.

Greenberg, M. L., and J. P. Castagna, (1992), "Shear-wave velocity estimation in porous rocks: Theoretical formulation, preliminary verification and applications", Geophysical Prospecting, 40, no. 2, 195–209, 1992

Griffin, T., Best, K.D., Thingvoll, T., Stockden, I. and Tjetland, G. (2007). Waterflood performance in a depleted fractured Chalk reservoir. SPE 108687

Griffin, T., Kristiansen, T. G., Tjetland, G. and Buer, K. (2007). Reservoir management aspects of early waterflood response after 25 years of depletion in the Valhall Field. *International Technology Conference*, Dubai, 4-6 December, 2007, IPTC 11276

Grion, S., Keggin, J., Ronen, S. and Caldwell, J. (2000). Seismic repeatability benchmarks. *Offshore Technology Conference*, OTC 12099.

Grose, T. (2007). Surveillance – maintaining the field from cradle to grave. *Offshore Europe*, Aberdeen, UK, 4-7 September, 2007, SPE108498.

Guilbot, J. and Smith, B. (2002). 4D-constrained depth conversion for reservoir compaction estimation. Application to the Ekofisk field. *The Leading Edge*, **21**, 302-308.

Guderian, K., Kleemeyer, M., Kjelstad, A., Pettersson, S.E. and Rehling, J. (2003). Draugen Field - successful reservoir management using 4D seismic. *65th EAGE Meeting*.

Han, D. (1986). Effects of porosity and clay content on acoustic properties of sandstones and unconsolidated sediments. Ph.D dissertation, Stanford.

Hall, S.A., Kendall, J.M. and Barkved, O.I. (2002a). Fractured reservoir characterization using P-Wave AVOA analysis in 3D OBC data. *The Leading Edge*, **21**.

Hall, S., MacBeth, C., Barkved, O. I. and Wild, P. (2002). Time-lapse seismic monitoring of compaction and subsidence at Valhall through cross-matching and interpreted warping of 3D streamer and OBC data. *72nd SEG annual meeting, Expanded Abstracts*, **21**, 1696.

Hall, S.A., MacBeth, C. and Barkved, O. I. (2003). Interpreted warping for 4D monitoring of production-induced subsidence at Valhall. *65th EAGE conference, Extended Abstracts* ,Stavanger, Norway.

Halvorsen, H. (1997). USE of 4C OBS data on Valhall - The business case. *SEG Workshop*.

Harris, J. M., Zoback, M. D., Kovscek, A. R. and Orr, F. M. (2007). Geologic storage of CO2: Global climate and energy project technical report……

Harris, P. (2005). Prestack repeatability of time-lapse seismic data. *SEG, Expanded Abstracts*, 2005.

Hashin, Z., and S. Shtrikman, (1963) , "A variational approach to the elastic behavior of multiphase materials", Journal of the Mechanics and Physics of Solids, 11, no. 2, 127–140

Hatchell, P. and Bourne, S. (2005). Rocks under strain: Strain-induced time-lapse time shifts are observed for depleting reservoirs. *The Leading Edge*, December, 1222-1225.

Hatchell, P.J., Kawar, R.S. and Savitski, A. A. (2005). Integrated 4D seismic, geomechanics and reservoir simulation in the Valhall oil field. *67th EAGE International Meeting*, C012.

Hatchell, P.J., van den Beukel, A., Molenaar, M.M., Maron, K.P., Kenter, C.J., Stammeijer, J.G.F., van der Velde, J.J. and Sayers, C.M. (2003). Whole Earth 4D: Reservoir monitoring geomechanics. *73rd SEG Annual Meeting, Expanded Abstracts*, 1330-1333.

Helgerud, M. B., Tiwari, U., Woods, S., Homonko, P., Bucki, A., Laugier, B., Hicks, E., Hoeber, H. and Khan, J. (2011). Optimizing seismic repeatability at Ringhorne, Ringhorne East, Balder and Forseti with QC driven time-lapse processing. *SEG, Expanded Abstracts*, 2011.

Herwanger,J. and Koutsabeloulis, N. (2011). Seismic Geomechanics. How to build and calibrate geomechanical models using 3D and 4D seismic data. *EAGE, Education Tour Series*, 2011.

Herwanger, J. and Horne, S. (2005). Linking geomechanics and seismics: Stress effects on time- lapse multi-component seismic data. *67th EAGE International Meeting*, C011.

Herwanger, J., (2008), "R we there yet?": 70th Annual International Meeting, EAGE, Expanded Abstracts.

Herwanger, J. V., Schiøtt, C. R., Frederiksen, R., If, F., Vejbæk, O. V. , Wold, R., Hansen, H. J., Palmer, E. and Koutsabeloulis, N., (2010)," "Applying time-lapse seismic methods to reservoir management and field development planning at South Arne, Danish North Sea", in Petroleum Geology: From Mature Basins to New Frontiers—Proceedings of the 7th Petroleum Geology Conference v.7, 2010.

Hodgson, N. MacBeth, C. Duranti, L. Rickett, J. and Nihei, K. (2007). Inverting for reservoir pressure change using time-lapse time strain: Application to Genesis Field, Gulf of Mexico. *The Leading Edge*, **5**, 2007.

Hodgson, N. (2009). Inversion for reservoir pressure change using overburden strain measurements determined from 4D seismic. Doctoral Thesis, Heriot-Watt University.

Hoeber, H., Butt, S., Davies, D., Campbell, S.and Ricketts, R. (2005). Improved 4D seismic processing: Foinaven case study. *75th SEG annual meeting, Expanded Abstracts*, **24**, 2414.

Holt, R.M. and Fjær, E. (2003). Wave velocities in shales: a rock physics model. *65th EAGE Conference and Exhibition, Extended Abstracts*.

Holt, R.M., Bakk, A., Fjær, E. and Stenebråten, J.F. (2005b). Stress sensitivity of wave velocities in shale. *75th SEG Annual International Meeting, Expanded Abstract*, Houston, USA.

Holt, R.M., Nes, O.-M. and Fjær, E. (2005). In situ stress sensitivity of wave velocities in reservoir and overburden rocks. *The Leading Edge*, December 2005,1268-1274.

Holt, R. M., Fjær, E., (2009), "R- From laboratory data to 4D seismic interpretation", SEG Society of Exploration Geophysicists, Expanded Abstracts. Houston, USA: Society of Exploration Geophysicists

Hornby, B., Barkved,O.I., Knudsen, S., Bostick, F. and Williams, B. (2007). Reservoir monitoring using permanent in-well seismic. *International Technology Conference*, Dubai, 4-6 December, 2007.

Houck, R.T. (2004). Predicting the economic impact of acquisition artifacts and noise. *The Leading Edge*, **23**, 1024–1031.

Houck, R.T. (2007). Time-lapse seismic repeatability – How much is enough? *Touch Briefings,Exploration and Production – Oil & Gas Review*, 2007, Issue II.

Hornby, B., Barkved, O. I., Askim, O.J., Bostick III, F.X. and Williams, B.A. (2007). Permanent fiber-optic borehole seismic installation and imaging at Valhall. *69th EAGE Conference and Exhibition*, 11 June, 2007.

Hornby, B., Bostick, III, F., Williams, B., Lewis, K. and Garossino, P. (2005). Field test of a permanent in-well fiber-optic seismic system. *Geophysics*, **70**(11).

Huang, Y. and MacBeth, C. (2011). Direct correlation of 4D seismic and well activity for a clarified dynamic reservoir interpretation. *Geophysical Prospecting*, in press.

Huang, Y. and MacBeth, C., Barkved, O.I. and van Gestel, J. (2011). Enhancing dynamic interpretation at the Valhall Field by correlating well activity to 4D seismic signatures. *First Break*, **29**(3), 37-44.

Jack, I.(1998). Time lapse seismic in reservoir management. *The SEG/EAGE Distinguished Instructor Short Course*, **2**, Lecture Notes, 240, Tulsa,1998.

Kjelstadli, R.M., Lane, H.S., Johnson, D.G., Barkved, O.I., Buer, K. and Kristiansen, T.G. (2005). Quantitative history match of 4D seismic response and production data in the Valhall Field. *Offshore Europe*, Aberdeen, 2005, SPE 96317.

Jørgensen, O. (2002). Using flow induced stresses for steering of injection fractures. *SPE/ISRM Rock Mechanics Conference*, Texas, USA, 20-23 October, 2002, SPE/ISRM 78220.

Kommedal, J.H., Barkved, O.I. and Thomsen, L. (2007). Acquisition of 4 component OBS data-a case study from the Valhall Field . *EAGE 59th Conference, Expanded Abstracts*, I, Paper B, 1997.

Kommedal, J., Barkved, O.I. and Howe, D. (2004). Initial experience operating a permanent 4C seabed array for reservoir monitoring at Valhall. *74th SEG Annual International Meeting, Expanded Abstracts*, 2004.

Knight, F. (1921). Risk, Uncertainty and Profit. Augustus M. Kelley, Bookseller, Reprinted 1964, New York.

Kommedal, J.H., Semb, P.H. and Manning, E. (2006). A case of SI attenuation in 4D seismic data recorded with a permanently installed array. *76th SEG annual meeting, Expanded Abstracts*, **25**.

Kommedal, J.H. Barkved, O.I., Van Gestel, J. and Pettersen, R.S.H. (2007). Processing strategies for multiple repeat 4D seismic. *77th SEG Annual Meeting, Expanded Abstracts*, San Antonio, Texas.

Koster, K., Gabriels, P., Hartung, M., Verbeek, J., Deinum, G. and Staples, R. (2000). Time-lapse seismic surveys in the North Sea and their business impact. *The Leading Edge*, **19**(3), 286-293.

Kragh, E. and Christie, P. (2002). Seismic repeatability, normalized rms, and predictability. *The Leading Edge*, July, 640–647.

Kristiansen, P., Christie, P., Bouska, J., O'Donovan, A., Westwater, P. and Thorogood, E. (2000). Foinaven 4D: Processing and analysis of two designer 4Ds. *70th SEG annual meeting, Expanded Abstracts*, **19**.

Kristiansen, P. (2000). Foinaven 4-D towed streamer and buried sea-bed detectors in deep water for 4-D seismic. *62nd EAGE Conference, Expanded Abstracts*, Glasgow, UK.

Kristiansen, T.G. (1998). Geomechanical characterization of the overburden above the compacting chalk reservoir at Valhall. *Eurock conference*, Trondheim, Norway, 1998, SPE/ISRM 47348.

Kristiansen, T.G., Barkved, O.I. and Pattillo, P. D. (2000). Use of passive seismic monitoring in well and casing design in the compacting and subsiding Valhall Field, North Sea. *SPE European Petroleum Conference*, Paris, France, 24–25 October, 2000, SPE 65134.

Kristiansen, T.G. (2004). Drilling wellbore stability in the compacting and subsiding Valhall Field. IADC/SPE 87221, 2004.

Kristiansen, T.G., Barkved, O.I., Buer, K. and Bakke, R. (2005). Production induced deformations outside the reservoir and their impact on 4D seismic. *International Petroleum Technology Conference*, Doha, Qatar, 21–23 November, IPTC 10818.

Kristiansen, T.G. and Flatebø, R.E.(2009). 60 days ahead of schedule -reducing drilling risk at Valhall using computational geomechanics. *SPE/IADC Drilling Conference and Exhibition*, Amsterdam, The Netherlands, 17–19 March, SPE/IADC 119509.

Kristiansen, T. G. (2007). Drilling wellbore stability in the compacting and subsiding Valhall Field: A case study. *SPE Drilling and Completion*, **22**(4), December, 2007.

Kristiansen, T.G. (2009). Computational mechanics based workflow and toolkit to manage deformable reservoir developments. *Offshore Europe*, Aberdeen, UK, 8-11 September, 2009, SPE 124131-MS.

Kristiansen, T.G. and Plische, B.(2010). History matched full field geomechanics model of the Valhall Field including water weakening and re-pressurisation. *SPE EUROPEC*, 2010, Barcelona, Spain.

Lane, H.S., Kjelstadli, R.M., Barkved, O.I., Johnson, D.T., Askim, O.J. and van Gestel, J. (2006). Constraining reservoir uncertainty with frequent 4D seismic data at Valhall Field. OTC18222, Houston, Texas.

Landes, M., Shapiro, N.M., Singh, S. and Johnston, R. (2009). Studying shallow seafloor structure based on correlations of contineous seismic records. *79th SEG Annual Meeting*.

Landrø, M. (1999). Repeatability issues of 3-D VSP data. *Geophysics*, **64**(6),1673-1679.

Landrø, M., Solheim, O.A., Hilde, E., Ekren, B.O. and Strønen, L.K. (1999). The Gullfaks 4D seismic study. *Petroleum Geoscience*, **5**, 213-226.

Landrø, M. and Janssen, R. (2002). Estimating compaction and velocity changes from time- lapse near and far offset stacks. *64th EAGE Meeting*.

Landrø, M. and Kvam, O. (2002). Pore pressure estimation what can we learn from 4D? *CSEG Recorder*, September, 2002.

Landrø, M. and Stammeijer, J. (2004). Quantitative estimation of compaction and velocity changes using 4D impedance and traveltime changes. *Geophysics*, **69**, 949-957.

Landrø, M., Nguyen, A.K. and Mehdizadeh, H. (2004). Time lapse refraction seismic – a tool for monitoring carbonate fields? *74th SEG Annual International Meeting*, 2295-2298.

Landrø, M. and Amundsen, L. (2011). Using water layer normal modes to detect shallow gas and CO2 leakage. EAGE Workshop on Permanent Reservoir Monitoring (PRM) - Using Seismic Data, 28 February, 2011.

Leonard, R.C. and Munns, J. W. (1987). Valhall Field. In Geology of Norwegian oil and gas fields, Graham and Trotman, 153-163.

Lumley, D., Meadows, M., Cole, S. and Adams, D. (2003). Estimation of reservoir pressure and saturations by crossplot inversion of 4D seismic attributes. *73rd SEG Annual International Meeting, Expanded abstract,* Dallas, 2003.

MacBeth, C., Stephen, K.D. and McInally, A. (2005). The 4D seismic signature of oil–water contact movement due to natural production in a stacked turbidite reservoir. *Geophysical Prospecting*, **53**(2),183-203.

MacBeth, C., Floricich, M. and Soldo, J. (2006). Going quantitative with 4D seismic analysis. *Geophysical Prospecting*, **54**, 303-317.

MacBeth,C., Hajnasser, Y., Stephen, K. and Gardiner, A. (2011). The effect of meso-seismic shale beds on the reservoir's stress sensitivity to seismic waves. *Geophysical Prospecting*, **59**, 90-110.

Mann, K.O. and Lane, H.R. (2005). Graphic correlation. *Volume 53 of SEPM Short Course*, (Society for Sedimentary Geology, 1995. ISBN 1565760239, 9781565760233.

Marsh, J.M., Whitcombe, D.N., Raikes, S.A., Parr, R.S. and Nash, T. (2003). BP's increasing systematic use of time-lapse seismic technology. *Petroleum Geoscience*, **9**, 7-13.

Maritvold, R. (1990). Frigg field reservoir management. *North Sea Oil and Gas Reservoirs II*, 155–163.

Maultzsch, S., Mancini, F. and Payne, S. (2006). Observations of time-lapse converted-wave anomalies on Valhall LoFS Data. *68th EAGE International Meeting.*

Mavko, G., Mukerji, T. and Dvorkin, J. (1998). The Rock Physics Handbook. *Cambridge University Press*, Cambridge, United Kingdom, 51,1998.

Maxwell, S.C., Bossu, R.,Young, R.P. and Dangerfield, J. (1998). Processing of induced microseismicity recorded in the Ekofisk reservoir. *SEG, Expanded Abstracts*,1998.

Maxwell, S. (2001). Microseismic hydraulic fracture imaging: The path toward optimizing shale gas production. *The Leading Edge*, **3**(3), 340-346.

Maxwell, P. W. (1999). A new seismic sensor—at last! *The Leading Edge*, **18**, 1182.

Mehdi Zadeh, H., Landrø, M. and Barkved, O.I. (2011). Long-offset time-lapse seismic: Tested on the Valhall LoFS data. *Geophysics,***76**(2), O1-O13.

Meunier, J., Huguet, F. and Meynier, P. (2001). Reservoir monitoring using permanent sources and vertical receiver antennae: The Céré-la-Ronde case study. *The Leading Edge*, **20**, 622.

Meunier, J. and Huguet, F. (1998). Céré-la-Ronde: A laboratory for time-lapse seismic monitoring in the Paris Basin. *The Leading Edge*, **17**(10),1388-1394.

Meld. St. 28 (2010 - 2011). An industry for the future – Norway's petroleum activities. White paper on Norways Petroleum Activity, Ministry of Petroleum and Energy, Norway, 2011.

Mezghani, M., Fornel, A, Langlais, V. and Lucet, N. (2004). History matching and quantitative use of 4D seismic data for an improved reservoir characterization. *Annual Technical Conference and Exhibition*, Houston, 2004, SPE 90420.

Milne, J. (1885). Earthquakes and other Movements. 1886 – 434 pages

Bowman, M.B. J., Marsh, J.M. and Whitcombe, D.N. (2004). What's 4D worth? – a BP Perspective. Paper presented at SPE/EAGE workshop "What do petroleum engineers expect from timelapse seismic and do geophysicists answer the right questions?", Copenhagen, March, 2004.

Milne, J. (1895). The movements of horizontal pendulums. *Nature*, **53**, 180.

Mordret, A., Matthieu Landès, M., Shapiro, N.M., Singh, S.C., Philippe Roux, P. and Barkved, O.I. (2011). Seismic surface-wave tomography from cross-correlations of ambient noise at The Valhall oil field. *AGU Fall Meeting*, 5 – 9 December, 2011, San Francisco, USA.

Murphy, W.F. (1982). Effects of microstructure and pore fluids on acoustic properties of granular sedimentary materials. Ph.D. thesis, Stanford University.

Nakstad, H. and Langhammer, J. (2011). Successful North Sea fibre optic PRM installation. EAGE Workshop on Permanent Reservoir Monitoring (PRM) - Using Seismic Data", EAGE Workshop, 28 February, 2011, Trondheim Norway.

Muyzert, E., Kommedal, J. H. and Olofsson, B. (2002). Near-surface S-velocities, statics, and anisotropy estimated from Scholte Waves. *EAGE, Extended Abstracts.*

NPD, (2005), The petroleum resources on the Norwegian continental shelf. Published by the Norwegian Petroleum Directorate.

Nur, A. (1984). Monitoring of thermal EOR fronts by seismic methods. (co-authored with C.A. Tosaya), Society of Petroleum Engineers, 1984)

Nur,A. (1987). Seismic rock properties for reservoir description and monitoring. *Seismic Tomography*, 1987.

Nur, A. (1989). Four-dimensional seismology and (true) direct detection of hydrocarbons: The petrophysical basis. *The Leading Edge*, **8**, 30-36.

Olofsson, B., Probert, T., Kommedal, J.H. and Barkved, O.I. (2003). Azimuthal anisotropy from the Valhall 4C 3D survey. *The Leading Edge*, December, 2003, 1228-1235.

Osdal, B., Husby, O., Aronsen, H.A., Chen, N. and Alsos, T. (2006). Mapping the fluid front and pressure buildup using 4D data on Norne Field. *The Leading Edge*, **25**(9),1134 –1141.

Olsen, T.R., Buer, K. and Øxnevad, I.E.I, (2005), "Impact of Heterogeneity on Primary and Secondary Depletion in the Valhall Field, Norwegian North Sea". NGF Abstracts and Proceedings, No 4, 2005.

Pannett, S., Slager, S., Stone, G. and Dekker, S. (2004). Constraining a complex gas-water dynamic model using 4D seismic. *Annual Technical Conference and Exhibition*, Houston, 2004, SPE 89793.

Páramo, P., van Gestel, J., Manning, T., Howe, D. and Clarke, R. (2007). Valhall life of field seismic: quantitative analysis of 4D attributes on decimated datasets. *77th SEG Annual Meeting, Expanded Abstracts*, San Antonio, Texas.

Pettersen, R.S.H., Barkved, O. I. and Haller, N. (2006). Time-lapse seismic inversion of data from a compacting chalk reservoir. *76th SEG Annual Meeting, Expanded Abstracts*, **25**.

Pettersen, R.S.H., Flatebø, R., Kvernstrøm, J. and Ulen, V. (2010). Improved well delivery using advanced technology and integrated processes in the planning phase. *72nd EAGE Conference and Exhibition.*

Pettersen, Ø. and Kristiansen, T.G. (2009). Improved compaction modeling in reservoir simulation and coupled rock mechanics--Flow simulation, with examples from the Valhall Field. *SPE Reservoir Evaluation and Engineering*, **12**(2).

Powley, K.D., Peng, C.P. and Ali, N. (1992). Simulation of a North Sea Chalk reservoir with dynamic rock and natural fracture properties: Case study of the Valhall Field. SPE 24914.

Pullin, N., Matthews, L. and Hirsche, K. (1987). Techniques applied to obtain very high resolution three-dimensional seismic imaging at an Athabasca tar sands thermal pilot. *The Leading Edge*, **6**(12), 10-15.

Raymer, L. L., Hunt, E. R., and Gardner, J. S., (1980) ," An improved sonic transit time-to-porosity transform", SPWLA 21 Ann. Logging Symp. 1-12.

von Rebeur-Paschwitz, E. (1989). The earthquake of Tokio 18 April 1889. *Nature*, **40**, 294-295.

Rickett, J. and Lumley, D. (2001). Cross-equalization data processing for time- lapse seismic monitoring: A case study from the Gulf of Mexico. *Geophysics*, **66**(4), 1015–1025.

Rickett, T. and Barkved, O.I. (2011). Clair permanent reservoir monitoring - A pilot that shows potential. EAGE Workshop on Permanent Reservoir Monitoring (PRM) - Using Seismic Data", 28 February 2011, Trondheim Norway.

Ricketts, T.A., Dyce, M. and Whitcombe, D. N. (2008). 4D chi significance - Enhanced AVO imaging of 4D fluid and pressure changes. *70th EAGE Conference and Exhibition*, 2008.

Rigsby, T.B., Cafarelli, W.J. and O'Neill, D. J. (1987). Bottom cable exploration in the Gulf of Mexico: A new approach. *SEG, Expanded Abstracts*, 1987.

Rod, M.H. and Jørgensen, O. (2005). Injection fracturing in a densely spaced line drive waterflood -The Halfdan example. *14th Europec Biennial Conference*, Madrid, Spain, 13-16 June, 2005, SPE 94049.

Ronen, S., van Waard, R. and Keggin, J. (1999). Repeatability of sea bed multi-component data. *69th SEG Annual International Meeting, Expanded Abstracts*, 1330-1333.

Rosland, B.O., Tree, E.L., Barkved, O.I. and Kristiansen, P. (1999). Acquisition of 3D/4C OBS data at Valhall. *EAGE Expanded Abstracts*, 61.

Røste, T., Landrø, M. and Hatchell, P. (2007). Monitoring overburden layer changes and fault movements from time-lapse seismic data. *69th EAGE Conference and Exhibition*.

Røste, T., Stovas, A. and Landrø, M. (2005). Estimation of layer thickness and velocity changes using 4D prestack seismic data. *67th EAGE Meeting*.

Rutledge, J.T., Phillips, W.S. and Mayerhofer, M.J. (2004). Faulting induced by forced fluid injection and fluid flow forced by faulting: An interpretation of hydraulic-fracture micrsoeismicity. *Bulletin of the Seismological Society of America*, Carthage Cotton Valley Gas Field, Texas, **94**(5), 1817–1830.

Rutledge, J.T. and Phillips, W.S. (2003). Hydraulic stimulation of natural fractures as revealed by induced microearthquakes, Carthage Cotton Valley gas field, east Texas. *Geophysics*, **68**(2), 441-452.

Sandø, I.A., Munkvold, O. and Elde, R. (2009). Two decades of 4D geophysical developments – experiences, value creation and future trends. Eleventh International Congress of the Brazilian Geophysical Society.

Sandø, I.A., Munkvold, O. and Elde, R. (2009). 4D Geophysical Data. *GeoExpro*, 6(8), 2011.

Sayers, C. (2010). Geophysics under stress: Geomechanical applications of seismic borehole acoustic waves - the SEG/EAGE distinguished instructor short course #13 lecture notes. *SEG*, Tulsa, 2010.

Saunders, M., How., D. and Kommedal, J. (2004). Source signature repeatability and source array directivity — Measurements from the Valhall 4D survey. *74th SEG annual meeting, Expanded Abstracts* **23**, 37.

Schissele, E., Forgues, E., Echappe, J., Meunier, J., de Pellegars, O. and Hubans, C. (2009). Seismic repeatability: Is there a limit? *71st EAGE Conference and Exhibition*, Amsterdam, The Netherlands.

Segall P. (1992). Induced stresses due to fluid extraction from axisymmetric reservoirs. *Pure and Applied Physics*, **139**, 536–560.

Shuey, R.T. (1985). A simplification of the Zoeppritz equations. *Geophysics*, **50**, 609–614.

Sirgue, L., Barkved, O.I., van Gestel, J., Askim,O.J. and Kommedal, J. H. (2009). 3D waveform inversion on Valhall wide Azimuth OBC. *71st EAGE Conference and Exhibition, Expanded Abstracts*, Amsterdam, The Netherlands.

Sleefe, G.E., Warpinski, N.R. and Engler, … (1993). Observations of broad-band micro-seisms during reservoir stimulation. *SEG, Expanded Abstracts*, 1993.

Smit, F., Ligtendag, M., Wills, P. and Calvert, R. (2006). Toward affordable permanent seismic reservoir monitoring using the sparse OBC concept. *The Leading Edge*, **25**.

Stammeijer, J. and Landrø, M. (2003). Quantitative estimation of compaction and velocity changes using 4d impedance and travel time changes. *65th EAGE Conference and Exhibition, Expanded abstracts*, Stavanger, 2003, A10.

Staples, R., Hague, P., Weisenborn, T., Ashton, P. and Michalek, B. (2005). 4D seismic for oil-rim monitoring. *Geophysical Prospecting*, **53**, 243-251.

Stopin, A., Hatchell, P.J., Corcoran, C., Beal, E., Gutierrez, C. and Soto, G. (2011). First OBS to OBS time lapse results in the Mars basin. *SEG Annual Meeting*, 18 – 23 September, 2011, San Antonio, Texas.

Strachan, M., Kinn, S., Eide, Ø., Mason, C. and Chen, D.C.-K. (2005). Improving drilling performance in the Chalk Reservoir of the Valhall Field. *SPE Annual Technical Conference and Exhibition*, 9-12 October 2005, Dallas, Texas.

Tadema, H. J. (1959). Mechanism of oil production by underground combustion. Proc. 5th World Petr. Congress, sec.II, 279-287.

Teanby, N., Kendall, J-M., Jones, R.H. and Barkved, O.I. (2004). Stress-induced temporal variations in seismic anisotropy observed in microseismic data. *Geophysical Journal International*, **156**, 459–466.

Thomsen, L. (1988). Reflection seismology over azimuthally anisotropic media. *Geophysics* , **53**(3), 304–313.

Thomsen, L. (1986). Weak elastic anisotropy. *Geophysics*, **51**(10), 1954– 1966.

Thomsen, L.,(1996), Poisson was not a Rock Physicist, Either!, Geophysics: The Leading Edge of Exploration, 15(7), 852-855, 1996.

Thomsen, L., Barkved, O.I., Haggard, W., Kommedal, J. and Rosland, B. (1997). Converted wave imaging of Valhall Reservoir. *EAGE Expanded Abstracts*, **59**.

Tjentland, G., Kristiansen, T.G. and Buer, K. (2007). Reservoir management aspects of early waterflood response after 25 years of depletion in the Valhall Field. *International Petroleum Technology Conference*, 4-6 December, 2007, Dubai, UAE.

Tura, A. and Lumley, D.E. (1999). Estimating pressure and saturation changes time-lapse AVO data. *SEG, Expanded Abstracts,* 1999.

van Gestel, J., Barkved, O.I. and Kommedal, J.H. (2007). Valhall life of field seismic automated workflow. *77th SEG Annual Meeting, Expanded Abstracts*, San Antonio, Texas.

van Gestel, J., Clarke, R. and Barkved, O.I. (2008). Rapid turn-around for processing and analysis of frequent time-lapse surveys. *70th EAGE Conference and Exhibition*, Workshop Cycle Time Reduction, Rome, Italy.

van Gestel, J., Kommedal, J.H., Barkved, O.I., Mundal, I., Bakke, R. and Best, K.D. (2008a). Continuous seismic surveillance of the Valhall Field. *The Leading Edge*, 260-265.

van Gestel, J., Best, K.D., Barkved, O.I. and Kommedal, J.H. (2008b). Integrating frequent time-lapse data into the reservoir simulation modelling of the Valhall field. *70th EAGE Conference and Exhibition*, *Expanded Abstracts*, Rome, Italy.

van Gestel, J., Best, K.D., Barkved, O.I. and Kommedal, J.H. (2009). Life-of-field seismic system adds value to reservoir simulation of Valhall field. *Offshore*, 69 (2).

van Gestel, J., Tjetland, G., Parikh, M.J., Kommedal, J.H. and Barkved O.I. (2010). Business impact of frequent time-lapse observations around Valhall water injection well. *72nd EAGE Conference and Exhibition, Extended Abstracts*.

van Gestel, J., Best, K.D., Barkved, O.I. and Kommedal, J.H. (2011). Integration of the life of field seismic data with the reservoir model at the Valhall Field. *Geophysical Prospecting*, **59**(4), 673-681.

Vasco, D.W., Datta-Gupta, A. and He, Z. (2003). Reconciling time- lapse seismic and production data using streamline models: The Bay Marchand Field, Gulf of Mexico. *Annual Technical Conference and Exhibition,* Denver, USA, 2003, SPE 84568.

Verm, R. and Hilterman, F. (1995). Lithology color-coded seismic sections: The calibration of AVO crossplotting to rock properties. *The Leading Edge*, **14**, 847-853.

Waggoner, J. (1998). Lessons learned from 4D projects. SPE 49144.

Waggoner, J.R., Cominelli, A., Seymour, R.H. and Stradiotti, A. (2003). Improved reservoir modelling with time lapse seismic data in a Gulf of Mexico gas condensate reservoir. *Petroleum Geoscience*, **9**, 61-72.

Walker, G., Allan, P., Trythall, R., Parr, R., Marsh, M., Kjelstadli, R., Barkved, O.I., Johnson, D. and Lane, S. (2006). Three case studies of progress in quantitative seismic-engineering integration. *The Leading Edge*, **25**, 1161.

Wang, Z., and Nur, A., (1992), "Seismic and acoustic velocities in reservoir rocks" , vol. 2: Soc. Expl. Geophys. Geophysics Reprint Series 10.

Wang, Z. (2001). Fundamentals of seismic rock physics. *Geophysics*, **66**, 398–412.

Watts, G., Barkved, O.I. and Dickens, J. (2006). Seismic surveillance in the field-of-the-future. *SPE Intelligent Energy Conference and Exhibition*, Amsterdam, The Netherlands, 2006, SPE 99827.

Watts, G., Jizba, D., Gawith, D. and Gutteridge, P. (1996). Reservoir monitoring of the magnus field through 4D seismic analysis. *Petroleum Geoscience*, 2, 361-372.

Whitcombe, D., Marsh, M., Clifford, P., Dyce, M., McKenzie, C., Campbell, S., Hill, A., Parr, R., Pearse, C., Ricketts, T., Slater, C. and Barkved, O.I. (2004). The systematic application of 4D in BP's North West Europe operations – 5 years on. *74th SEG annual meeting*.

Whitcombe, D. N. and Fletcher, J.G. (2001). The AIGI crossplot as an aid to AVO analysis and calibration. *7th SEG annual meeting*, San Antonio, USA, 9-14 September, 2001.

Whitcombe, D., Connolly, P.A., Reagan, R.L. and Redshaw, T. C. (2002). Extended elastic impedance for fluid and lithology prediction. *Geophysics*, **67**(1), 63-67.

Weatherby, B.B. (1940). The history and development of seismic prospecting. *Geophysics* , **5**(3),215-230.

Wickens, L.M. and Jonge, G. (2006). Increasing confidence in production forecasting through risk-based integrated asset modeling. *JPT 58*,**12**, 61-62.

Williams, G.J.J., Mansfield, M., MacDonald, D.G. and Bush, M.D. (2004). Top-down reservoir modeling. *SPE Annual Technical Conference and Exhibition*, Houston, USA, 26-29 September, 2004, SPE-89974.

Wills, P.B. and Hector A. (2004). Information is costly, but how valuable is it? *Annual Technical Conference and Exhibition*, SPE.

Wills P.B., Hatchell P J. and Bourne, S.J. (2008). Time-lapse measurements of shallow horizontal wave velocity over a compacting field. *70th EAGE International Meeting.*

Woeber, A.F., Katz, S., & Ahrens, T.J. ,(1963), " Elasticity of selected rocks and minerals", Geophysics, 28, 658–663.

Wride, V.C. (1995). Structural features and structural styles from the five countries area of the North Sea Central Graben. *First Break*, **13**(10), 395-407.

Wyllie, M.R.J., Gregory, A.R. and Gardner, L.W. (1956). Elastic wave velocities in heterogeneous and porous media. *Geophysics*, **21**, 41-70.

Wyllie, M.R.J., Gregory, A.R. and Gardner, G.H.F. (1958). An experimental investigation of factors affecting elastic wave velocities in porous media. *Geophysics.*

Zadeh, H.M., Landrø, M. and Barkved, O.I. (2011). Long-offset time-lapse seismic: Tested on the Valhall LoFS data. *Geophysics*, **76**(2), O1–O13.

Zhang, J. J. and Bentley, L.R. (2000). Change of elastic moduli of dry sandstone with effective pressure. *SEG 2000, Expanded Abstracts.*

Zheng, Y., Barton, P. and Singh, S. (2011). Strategies for elastic full waveform inversion of time-lapse Ocean Bottom Cable (OBC) seismic data. *Institut de Physique du Globe de Paris*, SEG 2011.

Zoeppritz, K. (1919). über reflexion und durchgang seismischer Wellen durch Unstetigkerlsfläschen. über Erdbebenwellen VII B, Nachtrichten der Königlichen Gesellschaft der Wissenschaften zu Göttingen, *Journal of Mathematical Physics*, K1: 87-84.

Xia, G., Thomsen, L. and Barkved, O.I. (2007). Fracture detection from seismic P-wave azimuthal AVO analysis – application to Valhall LoFS data. Proc. Intl. Symp. In-situ Rock Stress, *International Society for Rock Mechanics*, Trondheim, 2006, Also: *Journal of Seismic Exploration*, 12IWSA Proceedings, 2007.

Zwartjes, P.M., Wills, P.B., De Maag, J.W. and Hatchell, P.J. (2008). Shallow and deep time-lapse effects on Valhall LoFS converted wave data. *70th EAGE Annual Conference*, G040.

NPD fact-pages, http://factpages.npd.no

Index